Topics in
Current Physics

22

Topics in Current Physics Founded by Helmut K. V. Lotsch

1 **Beam-Foil Spectroscopy**
Editor: S. Bashkin

2 **Modern Three-Hadron Physics**
Editor: A. W. Thomas

3 **Dynamics of Solids and Liquids by Neutron Scattering**
Editors: S. W. Lovesey and T. Springer

4 **Electron Spectroscopy for Surface Analysis**
Editor: H. Ibach

5 **Structure and Collisions of Ions and Atoms**
Editor: I. A. Sellin

6 **Neutron Diffraction**
Editor: H. Dachs

7 **Monte Carlo Methods** in Statistical Physics
Editor: K. Binder

8 **Ocean Acoustics**
Editor: J. A. DeSanto

9 **Inverse Source Problems** in Optics
Editor: H. P. Baltes

10 **Synchrotron Radiation** Techniques and Applications
Editor: C. Kunz

11 **Raman Spectroscopy** of Gases and Liquids
Editor: A. Weber

12 **Positrons in Solids**
Editor: P. Hautojärvi

13 **Computer Processing of Electron Microscope Images**
Editor: P. W. Hawkes

14 **Excitons** Editor: K. Cho

15 Physics of **Superionic Conductors**
Editor: M. B. Salamon

16 **Aerosol Microphysics I** Particle Interactions
Editor: W. H. Marlow

17 **Solitons**
Editors: R. Bullough, P. Caudrey

18 **Magnetic Electron Lens Properties**
Editor: P. W. Hawkes

19 Theory of **Chemisorption**
Editor: J. R. Smith

20 **Inverse Scattering Problems** in Optics
Editor: H. P. Baltes

21 **Coherent Nonlinear Optics** Recent Advances
Editors: M. S. Feld and V. S. Letokhov

22 **Electromagnetic Theory of Gratings**
Editor: R. Petit

23 **Structural Phase Transitions I**
Editors: K. A. Müller and H. Thomas

24 **Amorphous Solids** Low-Temperature Properties
Editor: W. A. Phillips

25 **Mössbauer Spectroscopy II** The Exotic Side of the Effect
Editor: U. Gonser

26 **Crystal Cohesion and Conformational Energies**
Editor: R. M. Metzger

Electromagnetic Theory of Gratings

Edited by R. Petit

With Contributions by
L. C. Botten M. Cadilhac G. H. Derrick
D. Maystre R. C. McPhedran M. Nevière
R. Petit P. Vincent

With 182 Figures

Springer-Verlag Berlin Heidelberg New York 1980

Professor Roger Petit

Laboratoire d'Optique Electromagnétique,
Faculté des Sciences et Techniques de Saint-Jérome,
Université d'Aix-Marseille III, Rue Henri-Poincaré,
F-13397 Marseille Cédex 13, France

ISBN 3-540-10193-4 Springer-Verlag Berlin Heidelberg New York
ISBN 0-387-10193-4 Springer-Verlag New York Heidelberg Berlin

Library of Congress Cataloging in Publication Data. Main entry under title: Electromagnetic theory of gratings.(Topics in current physics ; 22). Bibliography: p. Includes index. 1. Diffraction gratings. 2. Electromagnetic theory. I. Petit, Roger, 1931 –. II. Botten, L. C. III. Series.
QC 417.E 43 535'.4 80-18246

Offset printing and bookbinding: Konrad Triltsch, Graphischer Betrieb, Würzburg
2153/3130-543210

Preface

When I was a student, in the early fifties, the properties of gratings were generally explained according to the scalar theory of optics. The grating formula (which predicts the diffraction angles for a given angle of incidence) was established, experimentally verified, and intensively used as a source for textbook problems. Indeed those grating properties, we can call optical properties, were taught in a satisfactory manner and the students were able to clearly understand the diffraction and dispersion of light by gratings. On the other hand, little was said about the "energy properties", i.e., about the prediction of efficiencies. Of course, the existence of the blaze effect was pointed out, but very frequently nothing else was taught about the efficiency curves. At most a good student had to know that, for an echelette grating, the efficiency in a given order can approach unity insofar as the diffracted wave vector can be deduced from the incident one by a specular reflexion on the large facet. Actually this rule of thumb was generally sufficient to make good use of the optical gratings available about thirty years ago.

Thanks to the spectacular improvements in grating manufacture after the end of the second world war, it became possible to obtain very good gratings with more and more lines per mm. Nowadays, in gratings used in the visible region, a spacing smaller than half a micron is common. The spacing and the wavelength are therefore of the same order of magnitude and, in such conditions, one can suppose that, even for perfectly conducting gratings, the distribution of the diffracted energy among the spectral orders is probably dependent on the polarization. This idea, which will certainly seem trivial to microwave people, appeared relatively recently in optics texts. I myself found it for the first time in Concepts of Classical Optics in 1958. In this book, the authors (Madden and Strong) reported experimental results which clearly showed the influence of the polarization on the efficiencies. They observed that "when the gratings were studied using polarized light, with the electric vector parallel to and perpendicular to the grooves, a different efficiency curve was obtained for the two polarizations". Similar observations were made in the Soviet Union by Yakovlev and Gerasimov, in France by Hadni and his collaborators, and probably by many others all over the world. Thus, around 1959, a new theory of gratings was needed which had to take into account the vectorial character of light. Such a theory has been derived from Maxwell equations during the last fifteen years and we now refer to it as the electromagnetic theory of gratings. Numerous papers have been published on the subject but some of them describe notable efforts which, unfortunately are now best forgotten. Moreover,since not all the authors used the same mathematical techniques,the pioneering work, as a whole,is diffucult to read. The purpose of this book is to make the new theory comprehensible to nonspecialists and to give a general survey of its applications. To be as coherent as possible,I wanted the chapters to be written by authors who are in the habit of working together even if some of then belong to very distant laboratories. Very often we had to abandon the historical point of view for the sake of clarity; consequently, some papers which indeed played an important role in the development of the theory, have been only briefly mentioned or even not at all. We apologize in advance for these omissions and, more generally, we would like to emphasize that we do not consider our reference lists to be exhaustive; it must particularly be noted that practically no reference is made to the theoretical work done in the Soviet Union, because the contributors are unfamiliar with it, due to the language barrier.

We deal only with plane gratings because the electromagnetic theory of concave gratings has not yet been elaborated. The first chapter is a tutorial introduction which should be read by all readers; it can be looked upon either as a theoretical summary for experimenters and instructors or as an introduction to Chaps.3 and 4 in which integral methods and differential methods are studied in greater depth. Chapter 2 gives a survey of some mathematical topics in a rather concise manner. It can be skipped by readers mainly interested in experimental and practical aspects. I hope that it will be considered as a constructive step towards the resolution of fundamental questions which are still to be clarified. Chapter 5 is again a theoretical one, in which the author reviews phenomena apparently very different from each other, but which are all connected with the excitation of various surface waves. On the other hand, Chap.6 is entirely devoted to applications; it contains a great number of efficiency curves allowing us to answer most questions arising in spectroscopy; also given are a few proposals for less classical uses of gratings. Finally, Chap.7 is concerned with what we have called "crossed gratings". We have grouped under this name doubly periodic structures which can be deduced from an elementary cell by translations $n\underline{d} + m\underline{d}'$ (n and m are integers, $\underline{d}$ and $\underline{d}'$ given vectors). They include, for example, plates pierced with holes at regular intervals and corrugated surfaces resembling a waffle. They give rise to no really new fundamental problems but often involve the handling of cumbersome algebraic expressions and require voluminous and costly computations. That is probably why they did not attract the attention of researchers until recently; it was desirable to master the theory of singly periodic gratings and its numerical implementation before taking on doubly periodic surfaces. Today several computer programs do exist but there is still useful work to be done in this field; even with the biggest computers the curiosity of engineers cannot yet be fully satisfied. Progress is often linked with the use of a new numerical algorithm or with tricks of programming. One is therefore not surprised to see the authors sometimes using notations obviously derived from FORTRAN. Readers not particularly fond of numerical details may pass straight on to the sections containing the results. On the other hand, program-lovers will perhaps be delighted to find themselves in familiar territory.

Thanks are due to my fellow contributors for finding time and energy to write their chapters and for keeping their initial enthusiasm in spite of the editor's critical remarks. On behalf of the French authors, I would also like to thank all those who, in Marseille, helped us to prepare this book for publication. Mrs. M. Frizzi, our efficient secretary, has typed six chapters. The greater part of the illustrations were done by Mr. P. Tardy and Mr. L. Roussel, but two of my collaborators, J.P. Hugonin and P. Vincent, were also asked for some figures. Mr. S. Goodchild, "Lecteur" at the University of Provence, has been willing to try to make our so-called English intelligible, a variety of English far removed, no doubt, from the one he has been used to throughout his literary studies.

Finally I wish to express my sincere thanks to the publishers for helping me with so much patience in working out the final text of each chapter.

Marseille, August, 1980 *R. Petit*

Contents

List of Contributors

Botten, Lindsay C.

School of Mathematical Sciences, The New South Wales Institute of Technology, P.O. Box 123, Broadway, N.S.W., 2007, Australia

Cadilhac, Michel

Laboratoire d'Optique Electromagnétique, Faculté des Sciences et Techniques, Centre de St.-Jérome, F-13397 Marseille Cédex 13

Derrick, Graham H.

Department of Theoretical Physics, The University of Sydney, Sydney, N.S.W., 2006, Australia

Maystre, Daniel

Laboratoire d'Optique Electromagnétique, Faculté des Sciences et Techniques, Centre de St.-Jérome, F-13397 Marseille Cédex 13

McPhedran, Ross C.

Department of Theoretical Physics, The University of Sydney, Sydney, N.S.W., 2006, Australia

Nevière, Michel

Laboratoire d'Optique Electromagnétique, Faculté des Sciences et Techniques, Centre de St.-Jérome, F-13397 Marseille Cédex 13

Petit, Roger

Laboratoire d'Optique Electromagnétique, Faculté des Sciences et Techniques, Centre de St.-Jérome, F-13397 Marseille Cédex 13

Vincent, Patrick

Laboratoire d'Optique Electromagnétique, Faculté des Sciences et Techniques, Centre de St.-Jérome, F-13397 Marseille Cédex 13

1. A Tutorial Introduction

R. Petit

With 13 Figures

This chapter is especially devoted to beginners, that is to say to those who are perhaps experienced in optics but who are not specialists in microwave theory. We first derive the well-known grating formula from Maxwell equations and the associated boundary conditions. Then we give an idea of the different methods which can be used to predict theoretically the efficiencies of a grating, when the groove shape and the index of the material are known.

Quite often we make use of the theory of distributions which seems to be a mathematical tool well adapted to the grating problem. The reader who is not yet accustomed to this rather abstract theory will have to either study carefully Appendix A, or trust us for the demonstration of certain important results.

1.1 Preliminaries

1.1.1 General Notations

Throughout this volume we use a rectangular coordinate system Oxyz and we denote by $\hat{\underline{x}}$, $\hat{\underline{y}}$ and $\hat{\underline{z}}$ the unit vectors of Ox, Oy and Oz. A point M is located by the vector $\underline{r} = \underline{OM}$ whose components are x, y, z. We only study time-harmonic electromagnetic fields with pulsatance ω. Consequently, any vector function $\underline{a}(\underline{r},t)$ is systematically represented by its associated complex vector function $\underline{A}(\underline{r})$ assuming a time dependence in $\exp(-i\omega t)$

$$a(r,t) = \mathrm{Re}[A(r)\exp(-i\omega t)] \quad . \tag{1.1}$$

Of course, this rule holds also for scalar functions. For example, $\exp(i2\pi\lambda^{-1}\hat{\underline{u}}\cdot\underline{r})$ is the complex representation (we sometimes say the complex amplitude) of a scalar plane wave of unit amplitude, propagating in the direction of the unit vector $\hat{\underline{u}}$, with the wavelength λ.

1.1.2 Time-Harmonic Maxwell Equations

Let us recall these "equations" which are in fact relations between the complex field vectors $\underline{E}(\underline{r})$, $\underline{D}(\underline{r})$, $\underline{H}(\underline{r})$, $\underline{B}(\underline{r})$ and their partial derivatives. As is well known, we have in any continuous[1] medium

$$\text{curl } \underline{E} = i\omega\underline{B} \quad , \tag{1.2}$$
$$\text{curl } \underline{H} = \underline{j} - i\omega\underline{D} \quad , \tag{1.3}$$
$$\text{div } \underline{D} = \rho \quad , \tag{1.4}$$
$$\text{div } \underline{B} = 0 \quad , \tag{1.5}$$

where $\rho(\underline{r})$ and $\underline{j}(\underline{r})$ are functions representing the charge density and the current density.

The vector fields $\underline{D}(\underline{r})$ and $\underline{B}(\underline{r})$ can be eliminated if we introduce the permittivity $\varepsilon(\underline{r})$ and the permeability $\mu(\underline{r})$

$$\underline{D}(\underline{r}) = \varepsilon(\underline{r})\underline{E}(\underline{r}) \quad , \tag{1.6}$$
$$\underline{B}(\underline{r}) = \mu(\underline{r})\underline{H}(\underline{r}) \quad . \tag{1.7}$$

In view of the materials used to make gratings, it will always be supposed that μ is a constant equal to the permeability μ_0 of vacuum. The permittivity ε will be real for lossless dielectrics and complex for absorbent materials such as metals or lossy dielectrics. In a homogeneous and isotropic medium, ε does not depend on $\underline{r}$; it is a constant which characterizes the medium.

If, in some region R, we have neither charge nor current ($\rho=0, \underline{j}=0$) and if ε is constant all over the region, then, taking the curl of the two members of (1.2), we get

$$-\Delta\underline{E} + \text{grad}(\text{div } \underline{E}) = i\omega\mu_0 \text{ curl } \underline{H} \quad .$$

But from (1.4) div $\underline{E}$ is zero, and from (1.3) curl $\underline{H}$ can be expressed in terms of $\underline{E}$. Finally it appears that $\underline{E}$ verifies the Helmholtz equation

$$\Delta\underline{E} + k^2\underline{E} = 0 \quad , \tag{1.8}$$

with

$$k^2 = \varepsilon\mu_0\omega^2 \quad . \tag{1.9}$$

---

1 By continuous medium, we mean a medium in which ε and μ are continuous functions of $\underline{r}$.

Likewise, starting from (1.2,3,5), we can show that $\underline{H}$ verifies the same equation

$$\Delta\underline{H} + k^2\underline{H} = 0 \quad . \tag{1.10}$$

The number k can be used to characterize the material which fills the region R. In vacuum, we have

$$k_0^2 = \varepsilon_0\mu_0\omega^2 \tag{1.11}$$

and consequently

$$k = k_0(\varepsilon/\varepsilon_0)^{\frac{1}{2}} \quad .$$

But, by definition[2], $(\varepsilon/\varepsilon_0)^{\frac{1}{2}}$ is the optical index ν, so that

$$k = k_0\nu \quad . \tag{1.12}$$

It is easy to verify that the plane wave $\underline{E}_0 \exp(2\pi\lambda^{-1}\hat{\underline{u}}\cdot\underline{r})$ is a solution of the Helmholtz equation (1.8) provided $4\pi^2\lambda^{-2} = k^2$ and whatever the unit vector $\hat{\underline{u}}$.

In a dielectric, where k^2 is a real and positive number, we have therefore the important relation

$$k = 2\pi\lambda^{-1} \quad . \tag{1.13}$$

For a given pulsatance ω, we can define a wavelength λ_0 in vacuum and a wavelength λ in a dielectric of index ν. From (1.12,13) it turns out that

$$\lambda = \lambda_0\nu^{-1} \quad . \tag{1.14}$$

1.1.3 Boundary Conditions

This is the usual term for information about the jumps of the field components when crossing a surface S which is the boundary between two continuous media. To be more precise, let us refer to Fig.1.1: points P_1, P_2 and M are, respectively, located in medium 1, in medium 2, and on the surface S; $\hat{\underline{n}}_{21}$ is the normal unit vector pointing from medium 2 towards medium 1. Let us denoty by $\underline{E}_j$, $\underline{H}_j$, $\underline{D}_j$, $\underline{B}_j$ the limits of $\underline{E}(P_j)$, $\underline{H}(P_j)$, $\underline{D}(P_j)$, $\underline{B}(P_j)$ (j=1,2) when P_j tends to M. Then we have

$$\hat{\underline{n}}_{21} \wedge (\underline{E}_1 - \underline{E}_2) = 0 \quad , \tag{1.15}$$

2 This definition is not clear when ε is complex but we can forget this difficulty in this first chapter.

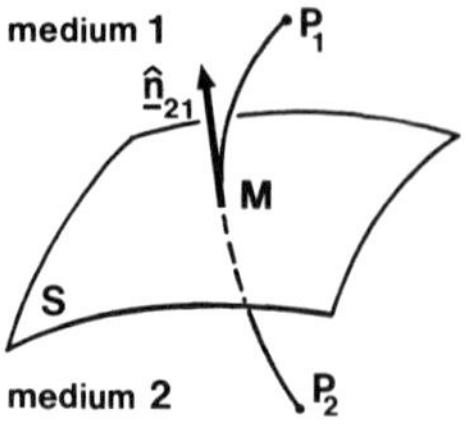

Fig. 1.1. The boundary conditions

$$\hat{\underline{n}}_{21} \wedge (\underline{H}_1-\underline{H}_2) = \underline{j}_S \quad , \tag{1.16}$$

$$\hat{\underline{n}}_{21} \cdot (\underline{D}_1-\underline{D}_2) = \rho_S \quad , \tag{1.17}$$

$$\hat{\underline{n}}_{21} \cdot (\underline{B}_1-\underline{B}_2) = 0 \tag{1.18}$$

where $\underline{j}_S$ and ρ_S are the surface current density and the surface charge density on S. Actually, in practice, $\underline{j}_S$ is not zero only if one of the two media, let us say medium 2, is perfectly conducting. In this case $\underline{E}(P_2)$, $\underline{H}(P_2)$ and consequently $\underline{H}_2$, are zero and (1.16) becomes

$$\hat{\underline{n}}_{21} \wedge \underline{H}_1 = \underline{j}_S \quad . \tag{1.19}$$

Very often ρ_S is also zero and then (1.17) expresses the conservation of the normal component of $\underline{D}$.

1.1.4 Electromagnetism and Distribution Theory

It is generally said that Maxwell equations (1.2-5) and boundary conditions are the fundamental laws of electromagnetic theory. There is in fact another, more elegant and more concise way, for stating these laws. We can *keep only Maxwell equations* and say that they are valid *in the sense of distributions*. Of course, this statement needs to be clearly understood. It means that any Maxwell equation which, up to now, was considered as an equality between two vector functions (equality valid in a certain region of space), must now be considered as an equality between two Schwartz distributions, i.e., two functionals (see Appendix A). In this way, we have to look upon $\underline{E}$, $\underline{D}$, $\underline{H}$, $\underline{B}$ as vector distributions associated with vectors functions which are continuous in the complement of S (Fig.1.1). The differential operators curl and divergence, when applied to these distributions, gives new distributions as explained in Appendix A.4. For example, if $\sigma_{\underline{D}}$ is the jump of $\underline{D}$ on S

$$\operatorname{div} \underline{D} = \{\operatorname{div} \underline{D}\} + \hat{\underline{n}} \cdot \sigma_{\underline{D}} \delta_S \quad . \tag{1.20}$$

As for ρ and $\underline{j}$ they are now distributions describing the sources using the following conventions: $\rho_S \delta_S$ and $\underline{j}_S \delta_S$ are used to represent, respectively, a surface charge and a surface current on a surface S, volume charge or current being still represented by functions $\rho(\underline{r})$ and $\underline{j}(\underline{r})$ (or rather by the associated distributions). We have to emphasize that the statement of the boundary conditions (1.15-18) is now useless because *these conditions are included in Maxwell equations*. For example, let us rewrite (1.2) taking into account a formula recalled in Appendix A. We obtain

$$\{\text{curl } \underline{E}\} + \hat{\underline{n}}_{21} \wedge (\underline{E}_1 - \underline{E}_2)\delta_S = i\omega\mu\underline{H} \quad ,$$

hence

$$\hat{\underline{n}}_{21} \wedge (\underline{E}_1 - \underline{E}_2)\delta_S = i\omega\mu\underline{H} - \{\text{curl } \underline{E}\} \quad . \tag{1.21}$$

But, as shown in specialized mathematics books, a distribution associated with a function [here the right-hand member of (1.21)] cannot be equal to a distribution of the type $\underline{A}\delta_S$ except if they are both zero. In other terms (1.21) implies

$$i\omega\mu\underline{H} - \{\text{curl } \underline{E}\} = 0 \tag{1.22}$$

and

$$\hat{\underline{n}}_{21} \wedge (\underline{E}_1 - \underline{E}_2) = 0 \quad .$$

We recognize the first boundary condition (1.15), and (1.22) is nothing other than the first Maxwell equation which, in the sense of functions, is only valid almost everywhere (everywhere except on S). Likewise, starting from (1.3-5) we can find the other boundary conditions (1.16-18). Thanks to distributions, we can therefore forget the boundary conditions if we know by heart the formulae given at the end of Appendix A.4.

1.1.5 Notations Used in the Description of a Grating

Since we do not intend to discuss the resolving power, a grating is always assumed to be infinitely wide and is represented by a cylindrical surface $y = f(x)$ whose generatrices are parallel to the z axis (Fig.1.2). Of course $f(x)$ is a periodic function and its period d is the grating period, also called "grating spacing". We often make use of the associated spacial pulsatance $K = 2\pi/d$. The graph of $f(x)$ is a line P which describes the grating profile. We call ν_1 the index of the homogeneous and isotropic material which lies above P in region 1 $[y > f(x)]$; ν_1 is supposed to be real and very often is equal to unity. On the other hand, the index ν_2 of region 2 $[y < f(x)]$ is either real (for transmission dielectric gratings) or complex (for conducting gratings or lossy dielectric gratings). We can also suppose that region 2 is filled with a perfectly conducting metal; in this case ν_2 is meaningless and we speak of a perfectly conducting grating.

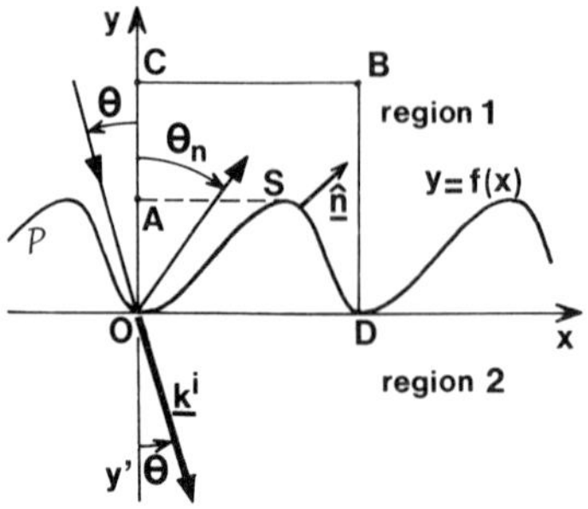

Fig. 1.2. Here θ and θ_n are defined in such a way that they are both positive. $\overline{OD}=d$, $\overline{OC}=c$, $\overline{OA}=a$, $c>a$

1.2 The Perfectly Conducting Grating

1.2.1 Generalities

In this section, we suppose that a perfectly conducting grating is illuminated under the incidence θ by a plane wave of unit amplitude propagating in region 1 (Fig.1.2). Except in Sect.1.2.13, the incident wave vector $\underline{k}^i = k_1\hat{\underline{u}}$ lies in the xOy plane, as usual in most practical applications; consequently the electromagnetic field is non-dependent upon z. The components of $\underline{k}^i$ upon Ox, Oy, Oz are, respectively, α, $-\beta$ and γ with

$$\alpha = k_1 \sin\theta \quad , \quad \beta = k_1 \cos\theta \quad , \quad \gamma = 0 \quad . \tag{1.23}$$

We assume that we are in one or the other of two fundamental cases of polarization, namely P polarization or S polarization. In P polarization the electric vector field is parallel to the grooves ($\underline{E}=E_z\hat{\underline{z}}$) and one speaks sometimes of $\underline{E}\|$ polarization. In S polarization, also called $\underline{H}\|$ polarization, it is the magnetic field which is parallel to the grooves ($\underline{H}=H_z\hat{\underline{z}}$). In order to treat the two cases simultaneously we can use a function u(x,y) [we write also u(P) [3]] which is defined as E_z in P polarization or H_z in S polarization. This function, which we call the field, is equal to zero in region 2 [$y<f(x)$] and, in view of (1.8,10), verifies the scalar Helmholtz equation in region 1

$$\Delta u + k_1^2 u = 0 \quad , \quad \text{if} \quad y > f(x) \quad . \tag{1.24}$$

Moreover u must satisfy a boundary condition which can be deduced from (1.15). Because $\underline{E}$ is null in region 2, we have

$$\hat{\underline{n}}_{21} \wedge \underline{E}_1 = 0 \quad . \tag{1.25}$$

---

3 Here P designates a point located in region 1 (do not confuse with P polarization) and in (1.26,29) M is a point on the grating surface.

In P polarization, $\underline{E} = u\hat{\underline{z}}$ and it is easy to see that (1.25) implies

$$\lim_{P \to M} [u(P)] = 0 \quad . \tag{1.26}$$

We recognize a Dirichlet boundary condition which we can write more simply

$$u[x, f(x)] = 0 \quad . \tag{1.27}$$

In S polarization, $\underline{H} = u\hat{\underline{z}}$ and (1.3) allows us to express $\underline{E}$ in terms of u in region 1

$$\underline{E} = \frac{i}{\omega\varepsilon_1} \operatorname{curl}(u\hat{\underline{z}}) = \frac{i}{\omega\varepsilon_1} \operatorname{grad} u \wedge \hat{\underline{z}} \quad . \tag{1.28}$$

Then, (1.25) becomes

$$\hat{\underline{n}}_{21} \wedge \lim_{p \to M} \operatorname{grad}[u(P)] \wedge \hat{\underline{z}} = 0 \quad ,$$

or, using the double vector product formula [4]

$$\hat{\underline{n}}_{21} \cdot \lim_{P \to M} \operatorname{grad}[u(P)] = 0 \quad . \tag{1.29}$$

We recognize a Neumann boundary condition which can be written using the normal derivative

$$\left.\frac{du}{dn}\right|_{y=f(x)}^{1} = 0 \quad . \tag{1.30}$$

The superscript 1 is used to recall that this normal derivative must be evaluated in region 1; the = sign that appears in (1.30) must not make us forget the limiting process clearly explained in (1.29).

1.2.2 The Diffracted Field

Let us define the diffracted field as the difference between the total field u and the incident field $u^i = \exp[i(\alpha x - \beta y)]$

$$u^d \overset{\text{def}}{=} u - u^i \quad . \tag{1.31}$$

It is worth noting that u^d, defined in this way, has no physical significance in region 2; on the other hand, it will be our unknown function in region 1. We know that u^i, which is a plane wave, verifies everywhere the Helmholtz equation. Consequently,

4 $\underline{a} \wedge (\underline{b} \wedge \underline{c}) = (\underline{a} \cdot \underline{c})\underline{b} - (\underline{b} \cdot \underline{c})\underline{a}$.

(a.b)c

starting from (1.24) and taking into account the linearity of the Δ operator, we can show that

$$\Delta u^d + k_1^2 u^d = 0 \quad , \qquad \text{for} \quad y > f(x) \quad . \tag{1.32}$$

From (1.27,30,31), it turns out that u^d verifies one or the other of the following boundary conditions:

$$\text{in P polarization:} \quad u^d[x,f(x)] = -u^i[x,f(x)] \quad , \tag{1.33}$$

$$\text{in S polarization:} \quad \left.\frac{du^d}{dn}\right|_{y=f(x)} = -\left.\frac{du^i}{dn}\right|_{y=f(x)} \quad . \tag{1.34}$$

• Moreover, to express mathematically very well-known experimental facts, we are led to add a radiation condition: we assume that, when y tends to infinity, u^d must be bounded and described as a superposition of outgoing plane waves (such a radiation condition does not hold of course for the total field because u^i is an incoming plane wave).

Finally, we are therefore faced with the following problem in mathematics: find a function which verifies the Helmholtz equation (1.32), a boundary condition on P [(1.33) or (1.34) depending on the polarization] and the outgoing wave condition (OWC). We will assume this problem, that we call the grating problem, to be a well-posed problem which has one, and only one, solution. To my knowledge, the existence of a solution — which seems obvious to physicists — has never been established by mathematicians. On the other hand, there exist some uniqueness theorems and the reader interested in theoretical questions can refer to Chap.2.

As a consequence of uniqueness, we are now going to show that $v(x,y) \stackrel{\text{def}}{=} u^d(x,y) \exp(-i\alpha x)$ is a periodic function of period d with respect to x, i.e., that we have

$$v(x+d,y) = v(x,y) \quad , \tag{1.35}$$

or equivalently

$$w(x,y) \stackrel{\text{def}}{=} u^d(x+d,y)\exp(-i\alpha d) = u^d(x,y) \quad . \tag{1.36}$$

Due to uniqueness, this equality will be established as soon as we have proved w(x,y) to be a solution of the grating problem. We therefore have to show that w(x,y) satisfies the Helmholtz equation, the boundary condition and the radiation condition. The first and last points are as good as obvious if we bear in mind that they hold for u^d which is itself a solution. As for the boundary condition, it is a consequence of the properties of u^d and of an obvious property of u^i, namely

$$u^i(x+d,y) = u^i(x,y)\exp(i\alpha d) \quad . \tag{1.37}$$

In P polarization, for example, we can write

$$
\begin{aligned}
w[x,f(x)] &= u^d[x+d,f(x)]\exp(-i\alpha d) \quad , \\
&= -u^i[x+d,f(x)]\exp(-i\alpha d) \quad , \\
&= -u^i[x,f(x)] \quad .
\end{aligned}
$$

In short, as a consequence of (1.37), $u^d(x,y)$ is the product of $\exp(i\alpha x)$ with a periodic function $v(x,y)$. We will say that u^d is a pseudo-periodic function which probably seems obvious to those familiar with the famous Floquet-Bloch theorem.

1.2.3 The Rayleigh Expansion and the Grating Formula

We can represent $v(x,y)$ by a Fourier series; then

$$
\begin{aligned}
u^d(x,y) &= \exp(i\alpha x)\sum_{n=-\infty}^{+\infty} v_n(y)\exp(inKx) \\
&= \sum_{n=-\infty}^{+\infty} v_n(y)\exp(i\alpha_n x) \qquad (1.38)
\end{aligned}
$$

with

$$\alpha_n = \alpha + nK = k_1 \sin\theta + nK \quad . \qquad (1.39)$$

Let a be the maximum value of $f(x)$. If $y > a$, $u^d(x,y)$ verifies the Helmholtz equation for any x. If we write this equation using (1.38), it results that

$$\sum_{n=-\infty}^{+\infty}\left[\frac{d^2v_n}{dy^2}+\left(k_1^2-\alpha_n^2\right)v_n\right]\exp(inKx) = 0 \quad .$$

The left-hand side can be looked upon as the Fourier series of the null function, which implies that, for any value of the integer n

$$\frac{d^2v_n}{dy^2} + \Omega_n v_n = 0 \qquad (1.40)$$

with

$$\Omega_n = k_1^2 - \alpha_n^2 \quad . \qquad (1.41)$$

U being the set of integers for which Ω_n is positive, let β_n defined by

$$\beta_n = \begin{vmatrix} \Omega_n^{\frac{1}{2}} & \text{if} \quad n \in U \\ (-\Omega_n)^{\frac{1}{2}} & \text{if} \quad n \notin U \end{vmatrix} \quad . \tag{1.42}$$

Then, the general solution of (1.40) is

$$v_n(y) = A_n \exp(-i\beta_n y) + B_n \exp(i\beta_n y) \quad . \tag{1.43}$$

The radiation condition leads to $A_n = 0$ because $A_n \exp(inKx - i\beta_n y)$ would represent an incoming wave for $n \in U$ and would not be bounded for $n \notin U$. Finally, from (1.38,43) we get an expansion of the diffracted field in terms of plane waves

$$u^d(x,y) = \sum_{n=-\infty}^{+\infty} B_n \exp(i\alpha_n x + i\beta_n y) = \sum_{n=-\infty}^{+\infty} B_n \psi_n(x,y) \quad . \tag{1.44}$$

Because this expansion seems to have been used for the first time by RAYLEIGH [1.1] it will hereafter be called the Rayleigh expansion. In fact, each term of this expansion represents a propagating plane wave only if $n \in U$; if $|n|$ is sufficiently large, $n \notin U$ and the associated term represents an evanescent wave $B_n \exp(-\beta_n y)\exp(i\alpha_n x)$ which propagates along the Ox axis and which is exponentially damped with respect to y. In classical optics, we are mainly interested in the propagating waves. Such a wave, called the diffracted wave in the nth order, is described by

$$\psi_n(x,y) = B_n \exp\left[i\alpha_n x + i\left(k_1^2 - \alpha_n^2\right)^{\frac{1}{2}} y\right] \quad . \tag{1.45}$$

But, because $|\alpha_n/k_1|$ is less than unity, we can put

$$\frac{\alpha_n}{k_1} = \sin\theta_n = \sin\theta + \frac{nK}{k_1} \quad , \quad -\frac{\pi}{2} < \theta_n < \frac{\pi}{2} \quad . \tag{1.46}$$

Then

$$\left(k_1^2 - \alpha_n^2\right)^{\frac{1}{2}} = k_1 \cos\theta_n \quad ,$$

and (1.45) becomes

$$\psi_n(x,y) = B_n \exp ik_1(x \sin\theta_n + y \cos\theta_n) \quad , \tag{1.47}$$

which proves that θ_n is the angle of diffraction (see Fig.1.2). As for (1.46), *it is nothing other than the famous grating formula* which is more frequently written in either of the two following forms:

$$\sin\theta_n = \sin\theta + n\,\frac{\lambda_1}{d} \quad , \tag{1.48}$$

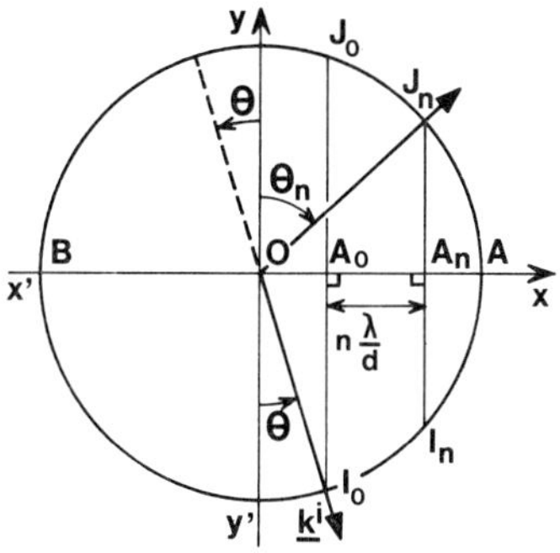

Fig. 1.3. The wavelength in region 1 is simply called λ. The incident wave vector k^i cuts the circle of unit radius in I_0. If $\overline{A_0A_n} = \bar{n}(\lambda/d)$, the nth diffracted plane wave is propagating along OJ_n

or

$$k_1 \sin\theta_n = k_1 \sin\theta + nK \quad . \tag{1.49}$$

Equation (1.48) explains the very useful geometrical construction shown in Fig.1.3. Equation (1.49) is very often used in integrated optics: one says that, in projection on the x axis, one passes from the incident wave vector to n^{th} diffracted wave vector by adding n times the grating space pulsatance K.

The only thing left to do now is the determination of the complex coefficients B_n which are closely related to efficiencies e_n. Let us consider (see Fig.1.2) the rectangle BCC'B' whose side CC', of unit length, is parallel to the z axis. Let ϕ^i and ϕ_n^d represent the flux, through this rectangle, of the Poynting vector associated with the incident wave and the nth diffracted wave. We can define e_n as ϕ_n^d/ϕ^i; then we find easily that

$$e_n = B_n \overline{B_n} \frac{\cos\theta_n}{\cos\theta} \quad . \tag{1.50}$$

Before speaking of the methods available for the computation of the e_n, we are going to establish two very general relations holding for all gratings. They are both consequences of a lemma that we must first prove.

1.2.4 An Important Lemma

If u and u' are two functions which satisfy the Helmholtz equation and one of the boundary conditions [(1.33) or (1.34)] we have

$$\int_0^d \left(u \frac{\partial u'}{\partial y} - u' \frac{\partial u}{\partial y}\right) dx = 0 \quad ,$$

for any fixed value of y greater than $a = \max f(x)$.

Here is the proof. Starting from the two Helmholtz equations

$$\Delta u + k_1^2 u = 0 \quad , \qquad y > f(x) \quad ,$$

$$\Delta u' + k_1^2 u' = 0 \quad , \qquad y > f(x) \quad ,$$

we get

$$u\Delta u' - u'\Delta u = 0 \quad , \quad y > f(x) \quad . \tag{1.51}$$

Then, integration over the interior of the close path OSDBCO (Fig.1.2) and use of the second Green identity yields

$$\iint (u\Delta u' - u'\Delta u)dS = \int_{OSDBCO} \left(u \frac{du'}{dn} - u' \frac{du}{dn}\right)d\ell = 0 \quad ,$$

where du'/dn and du/dn are outward normal derivatives. When we calculate the contour integral, the contributions of segments OC and DB cancel because u and u' are pseudo-periodic functions; due to the boundary conditions, the contribution of the curve line is zero whatever the polarization. Finally we obtain

$$\int_0^d \left(u \frac{\partial u'}{\partial y} - u' \frac{\partial u}{\partial y}\right)dx = 0 \quad , \quad \text{for} \quad y > a \quad , \tag{1.52}$$

thus establishing the lemma.

1.2.5 The Reciprocity Theorem

This theorem comes out when we apply the lemma to a particular pair of functions. As usual, u will be the field when the grating G is illuminated under the incidence θ (see Fig.1.4a)

$$u = \exp(i\alpha x - i\beta y) + \sum_n B_n \exp(i\alpha_n x + i\beta_n y) \quad , \quad \text{for} \quad y > a \quad , \tag{1.53}$$

$$\frac{\partial u}{\partial y} = -i\beta \exp(i\alpha x - i\beta y) + \sum_n i\beta_n \exp(i\alpha_n x + i\beta_n y) \quad , \tag{1.54}$$

$$\alpha = k_1 \sin\theta \quad , \quad \beta = k_1 \cos\theta \quad , \quad \alpha_n = \alpha + nK \quad , \quad \beta_n^2 = k_1^2 - \alpha_n^2 \quad .$$

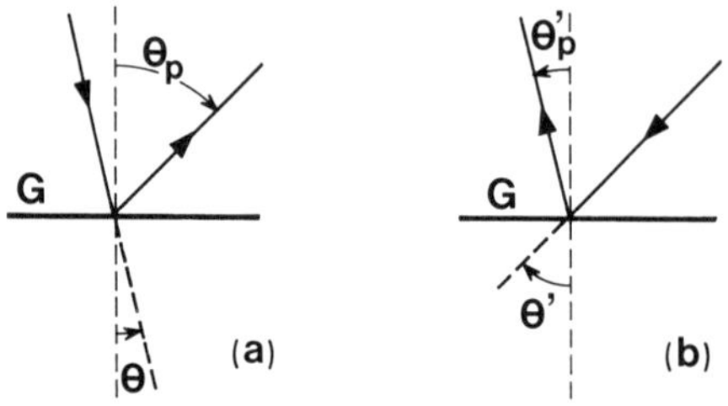

Fig. 1.4a,b. The reciprocity theorem. (a) θ is the angle of incidence and θ_p the angle of diffraction in the pth order; (b) when the angle of incidence is $\theta' = -\theta_p$, the pth diffracted wave propagates in the direction defined by $\theta_p' = -\theta$, as easily shown using the construction of Fig.1.3

As for u' (see Fig.1.4b) it will be the field when the grating is illuminated under the incidence $\theta' = -\theta_p$ $(p \in U)$

$$u' = \exp(-i\alpha_p x - i\beta_p y) + \sum_m B'_m \exp(i\alpha'_m x + i\beta'_m y) \quad , \quad \text{for} \quad y > a \quad , \tag{1.55}$$

$$\frac{\partial u'}{\partial y} = -i\beta_p \exp(-i\alpha_p x - i\beta_p y) + \sum_m i\beta'_m B'_m \exp(i\alpha'_m x + i\beta'_m y) \quad , \tag{1.56}$$

$$\alpha_p = \alpha + pK \quad , \quad \alpha'_m = -\alpha_p + mK \quad , \quad \beta'^2_m = k_1^2 - \alpha'^2_m \quad .$$

Inserting (1.53-56) in (1.52) one obtains without difficulty, but with a little patience, a very simple result, namely

$$\beta_p B_p = \beta B'_p \quad . \tag{1.57}$$

This tedious work is left as an exercise to the reader. [Note that $\alpha'_p = -\alpha$ and $\beta'_p = \beta$, do not forget that $\frac{1}{d}\int_0^d \exp(inKx) = \delta_{n,0}$.]
Equation (1.57) can be interpreted in terms of efficiencies. To do so, we first take the complex conjugate of its two members knowing that β and β_n are real numbers

$$\beta_p \overline{B_p} = \beta \overline{B'_p} \quad . \tag{1.58}$$

Then, from (1.57,58), we get after some evident manipulations

$$\frac{\beta_p}{\beta} B_p \overline{B_p} = \frac{\beta}{\beta_p} B'_p \overline{B'_p} \quad , \tag{1.59}$$

which means that (in Fig.1.4) *the efficiency* e_p *has the same value in situation (a) and in situation (b).*

Hereafter, we refer to this result as the reciprocity theorem for gratings. This theorem holds in particular for $p = 0$, i.e., $\theta_p = \theta$. In this case, a careful examination of Fig.1.4 leads to the equivalent and useful statement: *The efficiency in the zero order does not change when the grating G is rotated by 180° about an axis which is perpendicular to the plane on which it has been ruled* [1.2]. It must be emphasized that this theorem holds even if the grating profile *P* is not symmetrical with respect to the y axis.

1.2.6 The Conservation of Energy

Let us apply the lemma to another pair of functions, namely u (defined as in Sect. 1.2.5) and its complex conjugate $\overline{u}$ which, as does u, of course verifies the Helmholtz equation and the boundary condition. We get

$$\frac{1}{d}\int_0^d \left(u\frac{\partial \bar{u}}{\partial y} - \bar{u}\frac{\partial u}{\partial y}\right)dx = 0 \quad , \qquad \text{for} \quad y > a \quad ,$$

or

$$\mathrm{Im}\left\{\frac{1}{d}\int_0^d u\frac{\partial \bar{u}}{\partial u}dx\right\} = 0 \quad , \qquad \text{for} \quad y > a \quad . \tag{1.60}$$

We now have to perform this integral taking into account (1.53). It is again a question of patience. However we have to be careful when taking the complex conjugate of $\exp(i\alpha_n x + i\beta_n y)$; the right result is $\exp(-i\alpha_n x - i\beta_n y)$ if $n \in U$ and $\exp(-i\alpha_n x + i\beta_n y)$ if $n \notin U$ (because in the latter case $i\beta_n$ is real). Keeping this hint in mind, we write

$$u = \exp(i\alpha x - i\beta y) + \sum_{n \in U} B_n \exp(i\alpha_n x + i\beta_n y) + \sum_{n \notin U} B_n \exp(i\alpha_n x + i\beta_n y) \quad , \tag{1.61}$$

$$\bar{u} = \exp(-i\alpha x + i\beta y) + \sum_{n \in U} \overline{B_n} \exp(-i\alpha_n x - i\beta_n y) + \sum_{n \notin U} \overline{B_n} \exp(-i\alpha_n x + i\beta_n y) \quad . \tag{1.62}$$

Inserting (1.61,62) into (1.60), we find after cumbersome computations

$$\beta - \sum_{n \in U} \beta_n B_n \overline{B_n} = 0 \quad ,$$

$$1 = \sum_{n \in U} \frac{\beta_n}{\beta} B_n \overline{B_n} = \sum_{n \in U} e_n \quad . \tag{1.63}$$

This equality, which means that the *sum of the efficiencies is equal to unity*, is generally called the *energy balance criterion*. The physical interpretation is obvious: the incident energy is equal to the diffracted energy. It is worth noting that the B_n associated with evanescent waves are not involved in (1.63).

1.2.7 The Littrow Mounting

Let us recall that, in this usual mounting, there is a certain value of n for which the nth diffracted wave and the incident wave are propagating in opposite directions. As is easily shown by the geometrical construction (Fig.1.3) that can happen only with a negative value of n and if we have

$$2\sin\theta = -n\frac{\lambda}{d} \quad . \tag{1.64}$$

This relation explains why, in Littrow mounting, the efficiency can be expressed in terms of λ/d, instead of the incidence θ. As an exercise, let us consider the Littrow mounting in the -1 order ($2\sin\theta = \lambda/d$). Referring again to Fig.1.3, we can see that there are only two diffracted waves (the orders 0 and -1) provided that

$$2\frac{\lambda}{d} > 1 + \sin\theta \quad ,$$

that is to say, since $\sin\theta = \lambda/2d$

$$\frac{\lambda}{d} > \frac{2}{3} \quad . \tag{1.65}$$

As long as this inequality is verified, (1.63) implies

$$1 - e_0 = e_{-1} \quad ,$$

which proves that e_{-1} is not sensitive to a 180° rotation of the grating about Oy. Such a result has been observed experimentally for echelette gratings in the sixties and most opticists were very surprised because they were thinking that an echelette grating must be illuminated on its large facet to obtain high efficiencies.

1.2.8 The Determination of the Coefficients B_n by the Rayleigh Method

We know that outside the grooves, i.e., if $y > a$ [$a=\max f(x)$], the diffracted field can be written as a plane wave expansion

$$u^d(x,y) = \sum_n B_n\psi_n(x,y) = \sum_n B_n \exp(i\alpha_n x + i\beta_n y) \quad . \tag{1.44}$$

The validity of such an expansion has been established in Sect.1.2.3. When carefully rereading this section the reader will see that the proof fails in the groove region because, for a fixed y in the interval (0,a), the Helmholtz equation (1.32) is not verified everywhere but almost everywhere [not verified on P, i.e., for $y=f(x)$]. Indeed, on P, the existence of Δu^d itself can be questioned since we have no physical argument to justify the continuity of the first derivatives of u^d. It would be very dangerous to forget this restriction as shown by the following example: the derivative $H'(x)$ of the step function $H(x)$ [5] is zero almost everywhere (for $x\neq 0$); if we forget the word "almost" we find that $H(x)$ is a constant function, which of course is wrong.

Thus, at this stage of the game, we cannot say either that the Rayleigh expansion (1.44) is valid in the grooves region ($0<y<a$) or that it is not. It could be valid in spite of our inability to prove its validity! RAYLEIGH [1.1] made the assumption that, at least for shallow grooves, the plane wave expansion is valid everywhere in region 1 [i.e., for $y > f(x)$], and not only for $y > a$, as proved in Sect.1.2.3. He was followed by many others (FANO [1.3], MEECHAM [1.4], STROKE [1.5], BOUSQUET [1.6-7], PETIT [1.8], DELEUIL [1.9], JANOT and HADNI [1.10], YAKOVLEV [1.11]) who often began once again to take an interest in the theoretical study of gratings as soon as computer assistance became available for scientists. Under this assumption (usually

---

5 $H(x) = \begin{cases} 1 & \text{if } x > 0 \\ 0 & \text{if } x < 0 \end{cases}$.

called the Rayleigh assumption) the determination of the B_n coefficients is a simple matter. For example, in P polarization, they have to be chosen in order to satisfy the Dirichlet boundary condition

$$\sum_{n=-\infty}^{+\infty} B_n \exp[i\alpha_n x+i\beta_n f(x)] = -E^i[x,f(x)] \quad , \tag{1.66}$$

which can be written

$$\sum_{n=-\infty}^{+\infty} B_n \phi_n(x) = s(x) \quad ,$$

with

$$\phi_n(x) = \exp[i\alpha_n x+i\beta_n f(x)] \quad , \quad s(x) = -E^i[x,f(x)] \quad . \tag{1.67}$$

We recognize a very classical problem: the expansion of a function in a series of functions (the ϕ_n).

Of course, in numerical computations, we cannot deal with series but only with finite sums. N being a positive integer, we compute, for each B_n, an approximate value B_n^N which, we hope, tends to B_n when N tends to infinity. Three methods have been used to do this:

1) *Point Matching Method (PMM)*. One keeps only 2N+1 terms in the plane wave expansion and writes that the boundary condition is verified for 2N+1 values of x_p (p=-N,..., 0,...,N) chosen in the interval (0,d). We get

$$\sum_{n=-N}^{+N} B_n^N \phi_n(x_p) = s(x_p) \quad , \tag{1.68}$$

which is a linear system to be solved with the help of the computer.

2) *Fourier Series Method (FSM)*. Starting from (1.66) and multiplying its two members with $\exp(-i\alpha x)$ one obtains an equality between two periodic functions

$$\sum_{n=-\infty}^{+\infty} B_n \xi_n(x) = h(x) \quad , \tag{1.69}$$

with

$$\xi_n(x) = \exp[inKx+i\beta_n f(x)] \quad , \quad h(x) = -\exp[-i\beta f(x)] \quad . \tag{1.70}$$

This equality implies an equality between the Fourier coefficients. For any integer m, we have

$$\int_0^d \sum_{n=-\infty}^{+\infty} B_n \xi_n(x) \exp(-imKx)dx = \int_0^d h(x) \exp(-imKx)dx \quad ,$$

what we can write as

$$\sum_{n=-\infty}^{+\infty} a_{mn} B_n = h_m \quad , \tag{1.71}$$

with

$$a_{mn} = \int_0^d \xi_n(x) \exp(-imKx)dx \quad , \quad h_m = \int_0^d h(x) \exp(-imKx)dx \quad . \tag{1.72}$$

The B_n are therefore solutions of the infinite linear system (1.71) [an infinite number of equations ($m=-\infty,\ldots,0,\ldots,+\infty$) and an infinite number of unknowns ($n=-\infty,\ldots,0,\ldots,+\infty$)]. The B_n^N are solutions of the associated finite linear system

$$\sum_{n=-N}^{+N} a_{mn} B_n^N = h_m \quad , \quad (m \text{ or } n=-N,\ldots,0,\ldots,N) \quad . \tag{1.73}$$

Again, this system is treated by the computer.

3) *Variational Method.* Here, following MEECHAM [1.4], we introduce the integral

$$I = \int_0^d \left| s(x) - \sum_{n=-N}^{+N} B_n^N \phi_n(x) \right|^2 d\ell \quad ,$$

where $d\ell$ is the arc element on P. Then the B_n^N are chosen in such a way that this integral is minimized which, once more and as is well known, leads to the resolution of a linear system.

Of course we have to emphasize that a method is a good one only if, for small values of $|n|$ at least, B_n^N quickly reaches an asymptotic value when N increases (if not, the computation time would be prohibitive). Moreover, the calculated B_n (the asymptotic values of B_n^N) must be in agreement with the energy balance criterion (1.63) and the reciprocity theorem.

I personally worked up the PMM and the FSM around 1962 for triangular profile gratings. From a numerical point of view, the latter appeared to be the best one. The matrix coefficients a_{mn}, as well as the second members h_m, were expressed in closed form before being computed [which is possible whenever $f(x)$ is a piecewise linear function]. A detailed report on this work was published in 1963 [1.8] from which I extract the example illustrated in Fig.1.5. Very good results were obtained for $tg\alpha = 0.15$ [6]; the results are still acceptable for $tg\alpha = 0.30$ (the energy balance

6 Do not confuse the angle α with the first component of $\underline{k}^i$ which is designated by this greek letter.

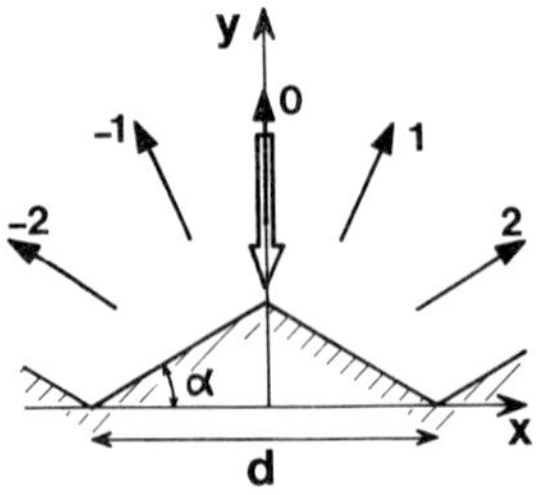

Fig. 1.5. The elementary profile is an isosceles triangle. The angle of incidence is zero; $d = 1.25$ μm; $\lambda = 0.546$ μm. There are five diffracted orders (-2, -1,0,+1,+2)

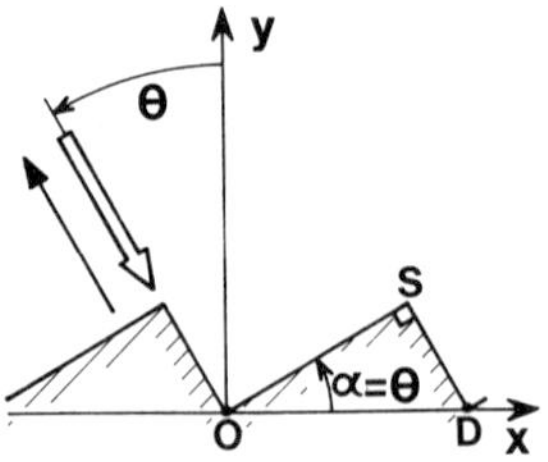

Fig. 1.6. The magnetic field is parallel to the z axis. The angle of incidence θ is equal to the blaze angle α. The Littrow condition ($\lambda=2d\ \sin\theta=2d\ \sin\alpha$) implies: $SD = \lambda/2$

is then satisfied within two percent) but the method fails completely for $\mathrm{tg}\alpha = 1$. Similar studies have been carried out for sinusoidal gratings [1.12,13]. Again the method fails drastically for sufficiently deep grooves. As a conclusion, we can retain that the Rayleigh assumption, as used in conjunction with the PMM or the FSM, does not allow us to predict the properties of all the gratings used in practice; the grooves of some of them are too deep.

To explain this failure one can, of course, question the validity of the Rayleigh assumption, which has been done as far back as the beginning of the fifties [1.14]; the topic soon provoked considerable controversy and heated discussions took place all through the sixties in most of the meetings devoted to the propagation of electromagnetic waves. What should we think today? The situation is clear for sinusoidal gratings [$f(x)=b\cos(Kx)$] in P polarization: for these, the Rayleigh assumption is untenable if $Kb > 0.448$ [1.12] and valid if $Kb < 0.448$ [1.15]; however, numerical experiments show that it is possible to obtain reliable results even if Kb is as much as twice the theoretical limit [1.12]. For an echelette grating, or more generally if $f(x)$ is not analytic, the Rayleigh assumption is, in general, not valid [1.16]. We say "in general" because there are in fact some exceptions: for instance, in the -1 order Littrow mounting and under incidence and polarization conditions described in Fig.1.6, the diffracted field is reduced to a single plane wave which goes back in the direction of propagation of the incident wave [1.17]. Lastly, for an arbitrary groove shape, we cannot say a priori whether the Rayleigh assumption is valid or not [1.18]. That is why the methods based upon the Rayleigh assumption have been progressively abandoned and replaced by more reliable methods which will be described later. Nevertheless the variational method has recently been brought back into fashion by Japanese workers at Kyushu University [1.19] and it seems that we have to agree with YASUURA when he claims that the method is exact whatever the groove shape. The reader will be perhaps surprised by this point of view which, at first sight, seems to be in contradiction with the fact that the Rayleigh hypothesis is not valid at least for certain profiles. This apparent paradox can be explained as follows:

Given a positive integer N, there exists a collection of complex numbers $Y_n(N)$ (let us call them Yasuura coefficients) which minimize the integral

$$\int_0^d \left| \sum_{n=-N}^{+N} Y_n(N)\phi_n(x)+E^i[x,f(x)] \right|^2 d\ell \quad .$$

When, for a fixed n, N tends to infinity, $Y_n(N)$ tends to a limit $Y_n(\infty)$. *This limit is nothing other than the coefficients* B_n *which appear in the Rayleigh expansion (1.44) valid for* $y > a$, but the series $\sum_n Y_n(\infty)\psi_n(x,y)$ is not necessarily convergent inside the grooves (for $0<y<a$). Nevertheless, when $N\to\infty$, the sequence $\sum_{n=-N}^{+N} Y_n(N)\cdot\psi_n(x,y)$ converges uniformly to $u^d(x,y)$ in any closed subset of the domain $y > f(x)$. I have to confess that all this is rather subtle. The interested reader should refer to a paper by MILLAR [1.20] in which he reviews and extends many important theoretical works. In any case, the reader will already find more information when reading Chap.2. For the present, let us return to numerical analysis.

As previously said, the variational method introduced by MEECHAM [7] around 1956 [1.4] was used again in 1972 by YASUURA and IKUNO [1.19] who gave it the name of improved point matching method (IPMM). This name comes from numerical considerations: the integral to be minimized is approximated with a Riemann sum of K terms ($K \geq 2N+1$) and it turns out that if $K = 2N+1$ the IPMM reduces to the PMM. Indeed the IPMM is simple and general but I agree with MILLAR [1.20] when he emphasized that it does not necessarily provide the most efficient means for solving a given problem. Up to now, most of the published results obtained with the IPMM concern shallow groove gratings and they probably could be obtained as quickly with the FSM. In my opinion, further computations using the IPMM would be valuable in order to compare their performances to those of other rigorous methods which have now proved reliable for deep groove gratings. A paper by KALHOR [1.21] is a first step in this direction; what he calls the Rayleigh method (in fact the IPMM) is compared with a modal expansion (Sect.1.2.12) for corrugated gratings in P polarization; he concludes that, when both methods give accurate and identical results, the IPMM is twice as fast.

1.2.9 An Integral Expression of u^d in P Polarization

Starting from (1.24) and keeping in mind that u(x,y) is zero in region 2, it is clear that u verifies a Helmholtz equation for any points except on the grating surface

$$\Delta u + k_1^2 u = 0 \quad , \quad \text{if} \quad y \neq f(x) \quad . \tag{1.74}$$

7 Apparently MEECHAM did not realize that the approximation he was using converges uniformly to the diffracted field even inside the grooves.

Because u and u^d are pseudo-periodic functions, they can be considered as distributions, i.e., as elements of the space R' defined in Appendix A. Doing so, we must not confuse Δu (the Laplacian in the sense of distributions) and $\{\Delta u\}$ (the Laplacian without precaution). Equation (1.74) must be written as

$$\{\Delta u\} + k_1^2 u = 0 \quad . \tag{1.75}$$

But, because u is continuous, its jump σ_u is zero and we have from (A.13)

$$\Delta u = \{\Delta u\} + \left(\frac{du}{dn} - 0\right)\delta_P \quad . \tag{1.76}$$

From (1.75,76), it follows that

$$\Delta u + k_1^2 u = \frac{du}{dn}\,\delta_P \quad .$$

It is easy to show that $du/dn = \hat{\underline{n}}_{21} \cdot \text{grad } u$ is closely related to the surface current $\underline{j}_S = j_S \hat{\underline{e}}_z$. One finds

$$\frac{du}{dn} = -i\omega\mu_0 j_S \quad ,$$

and finally, it turns out that u verifies the inhomogeneous Helmholtz equation

$$\Delta u + k_1^2 u = -i\omega\mu_0 j_S \delta_P \quad . \tag{1.77}$$

There is another way to obtain (1.77). It is founded on the fact that the three following equations are valid in the sense of distributions:

$$\text{div } \underline{E} = 0 \quad , \tag{1.78}$$

$$\text{curl } \underline{H} = \underline{j}_S \delta_P - i\omega\varepsilon_1 \underline{E} \quad , \tag{1.79}$$

$$\text{curl } \underline{E} = i\omega\mu_0 \underline{H} \quad . \tag{1.80}$$

Let us establish first (1.78). We know (A.11) that

$$\text{div } \underline{E} = \{\text{div } \underline{E}\} + \hat{\underline{n}}_{21} \cdot (\underline{E}_1 - 0)\delta_P \quad .$$

The two terms in the right-hand side are zero: the scalar product $\hat{\underline{n}}_{21} \cdot \underline{E}_1$ is zero because $\hat{\underline{n}}_{21}$ is perpendicular to $\underline{E}_1$; $\{\text{div } \underline{E}\}$ is zero because, in the sense of functions, $\text{div } \underline{E} = 0$ in region 1 (from Maxwell equations) and in region 2 (in which $\underline{E}=0$).

The proofs for (1.79,80) are obtained in a similar way. For instance, noting again $\underline{E}$ and $\underline{H}$ are null in region 2, it is clear, from Maxwell equations that

$$\{\text{curl } \underline{H}\} = -i\omega\varepsilon_1 \underline{E} \quad . \tag{1.81}$$

Moreover we know from (1.19) that

$$\hat{\underline{n}}_{21} \wedge (\underline{H}_1 - 0)\delta_P = \underline{j}_S \delta_P \quad . \tag{1.82}$$

Adding the corresponding members of (1.81,82) yields

$$\mathrm{rot}\,\underline{H} = \underline{j}_S\delta_P - i\omega\varepsilon_1\underline{E} \quad ,$$

which is (1.79). The proof for (1.80) is left as an exercise.
Let us now apply the curl operator to the two members of (1.80). Taking into account (1.78,79) we are led to the vector Helmholtz equation

$$\Delta\underline{E} + k_1^2\underline{E} = -i\omega\mu_0\underline{j}_S\delta_P \quad , \tag{1.83}$$

which proves $\underline{j}_S$ to be parallel to Oz ($\underline{j}_S = j_S\hat{\underline{z}}$).
Finally, in projection on the z axis, we again find (1.77)

$$\Delta u + k_1^2 u = -i\omega\mu_0 j_S\delta_P \quad .$$

The incident field u^i verifies the associate homogeneous equation

$$\Delta u^i + k_1^2 u^i = 0 \quad . \tag{1.84}$$

Consequently, u^d, as defined by (1.31), verifies also the inhomogeneous Helmholtz equation

$$\Delta u^d + k_1^2 u^d = -i\omega\mu_0 j_S\delta_P \quad . \tag{1.85}$$

It is worth noting that u^d satisfies an outgoing wave condition not only for $y\to\infty$ but also for $y\to-\infty$ [since for $y < f(x)$, $u^d = -u^i$]. This radiation condition ensures the uniqueness of the solution of (1.85) in which j_S is supposed to be known (see Chap.2). That is why the superficial current (or rather the associated distribution $j_S\delta_P$) is often looked upon as the source of the diffracted field u^d. To express $u^d(x,y)$ in terms of the surface current we first have to look for a function $G(x,y)$, belonging to R', verifying the outgoing wave condition for $y\to\pm\infty$, and so that

$$\Delta G + k_1^2 G = \delta_R \quad . \tag{1.86}$$

Then, as a consequence of (A.32,30,31) we shall write

$$u^d = -i\omega\mu_0 G * j_S\delta_P \quad ,$$

$$u^d(x,y) = -i\omega\mu_0\int G[x-x',y-f(x')]j_S(x')dl' \quad . \tag{1.87}$$

In this expression, the integral is to be taken over one period [8], (x,y) is an observation point, $[x',f(x')]$ is a point on the grating profile, and dl' is the arc element of P. Actually, $G[x-x',y-f(x')]$ plays the role of a Green function.

---

8 In this chapter, whenever the integration interval is not specified, the integration must be performed over one period.

It is possible to obtain G(x,y) in closed from by using its Fourier series expansion

$$G(x,y) = \sum_{n=-\infty}^{+\infty} G_n(y)\,\exp(i\alpha_n x) \quad . \tag{1.88}$$

Substitution in (1.86), taking into account the expansion (A.18) of δ_R, gives

$$\sum_{n=-\infty}^{+\infty} \left[G_n''+\left(k_1^2-\alpha_n^2\right)G_n\right] \exp(i\alpha_n x) = \sum_{n=-\infty}^{+\infty} \frac{1}{d}\,\delta(y)\,\exp(i\alpha_n x) \quad ,$$

which, in turn, implies, for any n

$$G_n'' + \left(k_1^2-\alpha_n^2\right)G_n = \frac{1}{d}\,\delta(y) \quad , \tag{1.89}$$

where $\delta(y)$ is the Dirac distribution (the familiar delta function). We can guess that $G_n(y)$ is a function which is twice differentiable in the complement of the origin. We rewrite therefore (1.89) in the form

$$\{G_n''\} + \left(k_1^2-\alpha_n^2\right)G_n + \sigma_{G_n'}\,\delta(y) + \sigma_{G_n}\,\delta'(y) = \frac{1}{d}\,\delta(y) \quad , \tag{1.90}$$

which proves that G_n is continuous at $y=0$ ($\sigma_{G_n}=0$) and that the jump of G_n' at $y=0$ is $1/d$ ($\sigma_{G_n'}=1/d$). Moreover, provided y is different from zero, we have, in the sense of functions

$$G_n'' + \left(k_1^2-\alpha_n^2\right)G_n = 0 \quad . \tag{1.91}$$

This differential equation can be integrated to give

$$G_n = \begin{cases} A_n \exp(-i\beta_n y) + B_n \exp(i\beta_n y) \quad , & \text{if } \; y > 0 \quad , \\ C_n \exp(-i\beta_n y) + D_n \exp(i\beta_n y) \quad , & \text{if } \; y < 0 \quad . \end{cases}$$

The outgoing wave condition implies $A_n = D_n = 0$ and, because of the continuity of G_n, necessarily $B_n = C_n$. The value of B_n is obtained by writing that the jump of G_n' at $y=0$ must be $1/d$. We are led to $B_n = 1/2id\beta_n$. Finally, we get

$$G_n(y) = \begin{cases} \dfrac{1}{2id\beta_n} \exp(i\beta_n y) \quad , & \text{if } \; y > 0 \\ \dfrac{1}{2id\beta_n} \exp(-i\beta_n y) \quad , & \text{if } \; y < 0 \end{cases}$$

which can be written more concisely

$$G_n(y) = \frac{1}{2id\beta_n} \exp(i\beta_n |y|) \quad . \tag{1.92}$$

Insertion of (1.92) in (1.88) yields the expression of G(x,y)

$$G(x,y) = \frac{1}{2id} \sum_{n=-\infty}^{+\infty} \frac{1}{\beta_n} \exp(i\alpha_n x + i\beta_n |y|) \quad , \tag{1.93}$$

which is a *fundamental result for the electromagnetic theory of gratings*. In anticipation of what follows, let us emphasize the presence of the *absolute value* of y in (1.93).

As a consequence of (1.87,93) we are now in a position to give an integral representation of $u^d(x,y)$

$$u^d(x,y) = -\int N(x,y,x') j_S(x') dl' \quad , \tag{1.94}$$

with

$$N(x,y,x') = \frac{\omega\mu_0}{2d} \sum_{n=-\infty}^{+\infty} \frac{1}{\beta_n} \exp\left[i\alpha_n (x-x') + i\beta_n |y-f(x')| \right] \quad . \tag{1.95}$$

Unlike the Rayleigh expansion, the integral representation (1.94) is valid at any points of region 1, in particular at the points for which $0 < y < a$. But let us suppose y to be greater than $a = \max f(x)$; then $y-f(x')$ is positive, we can replace $|y-f(x')|$ by $y-f(x')$, and (1.94) becomes

$$u^d(x,y) = \sum_{n=-\infty}^{+\infty} B_n \exp(i\alpha_n x + i\beta_n y) \quad ,$$

with

$$B_n = -\frac{\omega\mu_0}{2d} \int \frac{1}{\beta_n} j_S(x') \exp\left[-i\alpha_n x' - i\beta_n f(x') \right] dl' \quad . \tag{1.96}$$

Of course we recognize the Rayleigh expansion but, as a by-product, (1.96) gives the expression of B_n in terms of j_S. A similar simplification appears if y is supposed to be less than the minimum of f(x). Then $y-f(x')$ is negative and $|y-f(x')|$ can be replaced by $f(x')-y$. Again the field is given by a plane wave expansion

$$u^d(x,y) = \sum_{n=-\infty}^{+\infty} C_n \exp(i\alpha_n x - i\beta_n y) \quad , \tag{1.97}$$

with

$$C_n = -\frac{\omega\mu_0}{2d} \int \frac{1}{\beta_n} j_S(x') \exp\left[-i\alpha_n x' + i\beta_n f(x') \right] d\ell' \quad . \tag{1.98}$$

But, in region 2, we know u^d. We have, by definition

$$u^d(x,y) = -E^i(x,y) = -\exp(i\alpha x - i\beta y) \quad . \tag{1.99}$$

Equation (1.99), together with (1.98,39,42), leads to

$$-\frac{\omega\mu_0}{2d}\int\frac{1}{\beta_n}\,j_S(x')\,\exp\Big[-i\alpha_n x'+\beta_n f(x')\Big]d\ell' = -\delta_{n0} \quad . \tag{1.100}$$

This result will be very useful for checking the validity of numerical computations because it imposes strong relationships on the function $j_S(x)$.

1.2.10 The Integral Method in P Polarization

From (1.94,96) it turns out that the diffracted field is perfectly determined as soon as $j_S(x)$ is known. We can therefore choose this function as the unknown function in the grating problem.

Let $P(x,y)$ be a point in region 1 and $M[x,f(x)]$ a point located on the grating profile (Fig.1.7). From the boundary condition we know that

$$\lim_{P\to M} u^d(P) = -u^i(M) \quad , \tag{1.101}$$

so we write, using (1.94)

$$\int_0^d N[x,f(x),x']j_S(x')\,\frac{d\ell'}{dx'}\,dx' = \exp[i\alpha x-i\beta f(x)] \quad . \tag{1.102}$$

We have to be aware that some precautions are necessary when replacing (1.101) by (1.102). Actually we assume the existence of the integral which appears in (1.102) and the continuity of the function $u^d(x,y)$ as defined in (1.94). Both these statements are shown to be correct [1.22] but the existence of the integral is not a trivial matter: the kernel $K(x,x') = N[x,f(x),x']d\ell'/dx'$ is not bounded but integrable due to its weak singularity (a logarithmic singularity). It may be convenient to introduce a new unknown function, say $\tilde{j}_S$, defined as follows:

$$j_S(x) = \tilde{j}_S(x)\,\exp(i\alpha x) \quad . \tag{1.103}$$

Insertion of (1.103) in (1.102) gives

$$\int K_0(x,x')\tilde{j}_S(x')dx' = \exp[-i\beta f(x)] \quad , \tag{1.104}$$

with

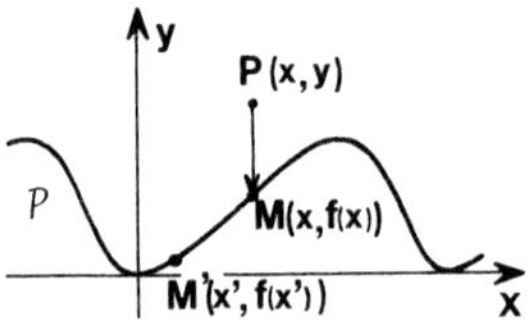

Fig. 1.7. The limiting process used in the integral method

$$K_0 = \frac{\omega\mu_0}{2d}[1+f'(x')]^{\frac{1}{2}} \sum_{n=-\infty}^{+\infty} \frac{1}{\beta_n} \exp[inK(x-x')+i\beta_n|f(x)-f(x')|] \quad . \tag{1.105}$$

Equation (1.104) is an integral equation of the first kind with a "weakly" singular kernel; the solution of this equation can be effected by the numerical methods described in Chap.3. Here, we will only say a few words about the Fourier series method we used in 1964 [1.23,24] so obtaining the first accurate and reliable results for deep grooves echelette gratings.

We assume that the periodic function $\tilde{j}_S$ is sufficiently well described by $M = 2N+1$ terms of its Fourier series

$$\tilde{j}_S = \sum_{n=-N}^{+N} x_n \exp(inKx) \quad . \tag{1.106}$$

We also replace the kernel K_0 by the truncated kernel $\hat{K}_0$ ($\hat{K}_0$ has $S = 2P+1$ terms; $n = -P,\dots,0,\dots,P$). Then we write that (1.104), which is an equality between two periodic functions, implies the equality between the corresponding Fourier coefficients. We thus obtain the linear system

$$\sum_{n=-N}^{+N} a_{mn} x_n(M) = b_m \quad , \tag{1.107}$$

with

$$b_m = \int_0^d \exp[-imKx - i\beta f(x)]dx \quad , \tag{1.108}$$

$$a_{mn} = \int_0^d \int_0^d \hat{K}_0(x,x') \exp(inKx' - imKx)dx\, dx' \quad . \tag{1.109}$$

Taking into account (1.105) we have to notice that a matrix coefficient a_{mn} is a sum of double integrals. Fortunately, for echelette gratings these integrals, as well as the ones that give the b_m, are obtained in closed form after cumbersome but easy algebraic computations. Of course the inversion of the linear system must be done with the help of the computer. Numerical experiments show that, in order to obtain reliable results, M must be about twice the number of the diffracted waves and that S must be twice as great again as M. Here is an example related to the symmetrical profile in Fig.1.5. For $\theta = 0$, $d = 1.25$ µm, $\lambda = 0.546$ µm, $\mathrm{tg}\alpha = 0.8$, $M = 11$ and $S = 23$, we obtained

$$B_0\bar{B}_0 = 0.2863 \quad , \qquad C_0\bar{C}_0 = 1{,}0075 \quad ,$$
$$B_1\bar{B}_1 = B_{-1}\bar{B}_{-1} = 0.0450 \quad , \quad C_1\bar{C}_1 = C_{-1}\bar{C}_{-1} = 1.3\ 10^{-4} \quad ,$$
$$B_2\bar{B}_2 = B_{-2}\bar{B}_{-2} = 0.6442 \quad , \quad C_2\bar{C}_2 = C_{-2}\bar{C}_{-2} = 9.9\ 10^{-4} \quad ,$$
$$\sum_{n\in U} B_n\bar{B}_n \cos\theta_n = 0.9942 \quad .$$

More information can be found in the original report [1.22] but today we have at our disposal more efficient computer programs (Chaps.3 and 4).

1.2.11 The Integral Method in S Polarization

We proceed in a way very similar to the one just used for P polarization but now u designates the z component of the magnetic field $\underline{H}$. Again we are able to express u^d in terms of the magnitude j_S of the surface current density. It turns out that we can exhibit a new kernel M(x,y,x') such as

$$u^d(x,y) = \int_0^d M(x,y,x')j_S(x')dx' \quad . \tag{1.110}$$

Then, using (1.110) we have to express a boundary condition. Starting from (1.19) and noting that $\underline{j}_S$ is tangent to P we write

$$\lim_{P\to M} \left[u^d(P)+u^i(P)\right] = j_S(M) \quad . \tag{1.111}$$

But here u^d, as given by (1.110), is discontinuous at the grating profile. As explained in Chap.3, the right integral equation is

$$\int_0^d M[x,f(x),x']j_S(x')dx' + u^i(x) = \frac{1}{2}\, j_S(x) \quad , \tag{1.112}$$

rather than

$$\int_0^d M[x,f(x),x']j_S(x')dx' + u^i(x) = j_S(x) \quad , \tag{1.113}$$

as expected at first glance. The use of (1.113) instead of (1.112) is a mistake which has been made by many people (including myself) and we had to wait until 1970 for the first rigorous solution of the grating problem in S polarization [1.25]. This proves that most opticists had probably never heard of what is called Maue's integral equation in microwave theory [Ref.1.26, p.354].

1.2.12 Modal Expansion Methods

The integral method which can be used for *any shape profile* involves rather complicated mathematics. On the other hand, for *certain profiles only*, the general diffraction problem can be treated in a particularly simple and clear fashion by matching a modal expansion in the grooves to the Rayleigh expansion. To illustrate such a possibility, let us first consider the lamellar grating [9] shown in Fig.1.8. Several recent papers have been devoted to this type of grating already extensively studied by WIRGIN in the early sixties [1.27-29].

9 Also called corrugated grating or rectangular-groove grating.

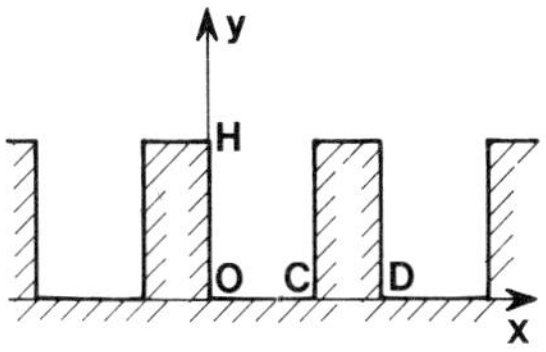

Fig. 1.8. The lamellar grating. $\overline{OH} = h$, $\overline{OC} = c$, $\overline{OD} = d$

Here, for brevity, we will only speak of normal incidence in P polarization, but the generality of the method is obvious.

We know that, for $y > h$, the total field $u(x,y)$ can be described by a Rayleigh expansion

$$u(x,y) = \exp(-ik_1y) + \sum_{n=-\infty}^{+\infty} B_n \exp(inKx+i\beta_n y) \quad , \tag{1.114}$$

with

$$\beta_n^2 = k_1^2 - n^2K^2 \quad ; \quad \beta_n \text{ or } \beta_n/i \text{ positive.}$$

Inside the first groove (i.e., for $0 < x < c$ and $0 < y < h$), $u(x,y)$, which is zero for $x = 0$ and $x = c$, can be represented by a Fourier sine series

$$u(x,y) = \sum_{n=1}^{\infty} u_n(y) \sin\left(\frac{n\pi x}{c}\right) \quad . \tag{1.115}$$

Insertion of (1.115) in the Helmholtz equation leads to

$$u_n(y) = a_n \cos(\mu_n y) + b_n \sin(\mu_n y) \quad ,$$

with

$$\mu_n^2 = k_1^2 - (n^2\pi^2/c^2) \quad , \quad \mu_n \text{ or } \mu_n/i \text{ positive.}$$

The boundary condition ($u=0$ if $y=0$) imposes $a_n = 0$ and finally, because u is zero inside the perfectly conducting metal, the periodic function $u(x,y)$ is given on the period interval by

$$u(x,y) = \begin{cases} \sum_{n=1}^{\infty} b_n \sin(\mu_n y) \sin\left(\frac{n\pi x}{c}\right) & \text{if } 0 \leq x \leq c \quad , \\ 0 & \text{if } c \leq x \leq d \quad . \end{cases} \tag{1.116}$$

In this way, the problem is reduced to the determination of two sequences of coefficients B_n and b_n, which can be made by matching the two expansions (1.114,116) at $y = h$.

We first write the continuity of u. In terms of Fourier coefficients, we get

$$B_n \exp(i\beta_n h) + \delta_{n0} \exp(-ik_1 h)$$
$$= \frac{1}{d} \int_0^c \left[\sum_{m=1}^{\infty} b_m \sin(\mu_m h) \sin\left(m\pi \frac{x}{c}\right) \right] \exp(-inKx) dx$$
$$= \sum_{m=1}^{\infty} P_{nm} b_m \quad , \tag{1.117}$$

i.e., in matrix notation

$$B + \mathcal{P} b = S \quad . \tag{1.118}$$

Obviously B and b are here unknown column matrices (whose elements are, respectively, B_n and b_n), S is a known column matrix [with elements $S_n = \delta_{n0} \exp(-2ik_1 h)$] and $\mathcal{P}$ is a square matrix [10] whose elements $\mathcal{P}_{nm}$ are obtained in closed form after some algebraic manipulations

$$\mathcal{P}_{nm} = \frac{1}{d} \exp(-i\beta_n h) \sin(\mu_m h) I_{nm} \quad , \tag{1.119}$$

with

$$I_{nm} = \begin{cases} \dfrac{m\pi c}{c^2 n^2 K^2 - m^2 \pi^2} [(-1)^m \exp(-inKc) - 1] & \text{if } \dfrac{4n^2}{d^2} \neq m^2 c^2 \\ \dfrac{c}{2} & \text{if } \dfrac{2n}{d} = mc \quad . \end{cases} \tag{1.120}$$

Let us call C(x) the periodic function defined on the period interval as follows:

$$C(x) = \begin{cases} 1 & \text{if } 0 < x < c \quad , \\ 0 & \text{if } c < x < d \quad . \end{cases} \tag{1.121}$$

Its Fourier coefficients C_n are easily obtained

$$C_n = \begin{cases} \dfrac{1}{2i\pi n} [1 - \exp(-inKc)] & \text{if } n \neq 0 \quad , \\ \dfrac{c}{d} & \text{if } n = 0 \quad . \end{cases} \tag{1.122}$$

In order to find another linear relation between B and b we consider the function $F(x,y) = u(x,y)C(x)$ which can be also described by a Fourier series

$$F(x,y) = \sum_{n=-\infty}^{+\infty} F_n(y) \exp(inKx) \quad . \tag{1.123}$$

---

10 There is no possible confusion between the matrix $\mathcal{P}$ used in this section and the grating profile.

It is clear that $\partial F/\partial y$ is continuous at $y = h$, which implies the continuity of $\partial F_n/\partial y$ With evident notations we write therefore

$$\left.\frac{\partial F_n}{\partial y}\right|_{y=h_+} = \left.\frac{\partial F_n}{\partial y}\right|_{y=h_-} . \tag{1.124}$$

For $y < h$, $F(x,y)$ is simply $u(x,y)$; F_n and $\partial F_n/\partial y$ are already known; we have

$$F_n(y) = \sum_{m=1}^{\partial} \frac{1}{d} \sin(\mu_n y) I_{nm} b_n \quad ; \quad \frac{\partial F_n}{\partial y} = \sum_{m=1}^{\infty} \frac{1}{d} \mu_n \cos(\mu_n y) I_{nm} b_m . \tag{1.125}$$

For the determination of $\partial F_n/\partial y$ for $y = h_+$, we make use of (1.114) and of a classical result related to the Fourier coefficients of a product:

$$\begin{aligned} F_n(y) &= \sum_{m=-\infty}^{+\infty} C_{n-m} E_m(y) \\ &= \sum_{m=-\infty}^{+\infty} C_{n-m} \left[B_m \exp(i\beta_m y) + \delta_{m0} \exp(-ik_1 y) \right] , \quad \text{for} \quad y > h , \end{aligned} \tag{1.126}$$

$$\frac{\partial F_n}{\partial y} = \sum_{m=-\infty}^{+\infty} C_{n-m} \left[i\beta_m B_m \exp(i\beta_m y) - ik_1 \delta_{m0} \exp(-ik_1 y) \right] . \tag{1.127}$$

Finally, if we rewrite (1.124) taking into account (1.125,126), we obtain the sought relation

$$\sum_m C_{n-m} \left[i\beta_m B_m \exp(i\beta_m h) - ik_1 \delta_{m0} \exp(-ik_1 h) \right] = \sum_m \frac{1}{d} \mu_n I_{nm} \cos(\mu_n h) b_m ,$$

or, in matrix language

$$\mathcal{Q}B + \mathcal{R}b = T , \tag{1.128}$$

with

$$\mathcal{Q}_{nm} = i\beta_m \exp(i\beta_m h) C_{n-m} ,$$

$$\mathcal{R}_{nm} = -\frac{1}{d} \mu_n \cos(\mu_n h) I_{nm} , \quad T_n = ik_1 C_n \exp(-ik_1 h) .$$

We are now in a position to get the column matrix B whose elements are the Rayleigh coefficients B_n. From (1.118,128) we get

$$b = (\mathcal{R} - \mathcal{Q}\mathcal{P})^{-1} (T - \mathcal{Q}S) , \tag{1.129}$$

then, from (1.118)

$$B = S - \mathcal{P}b . \tag{1.130}$$

In practice, for numerical computations, we have to truncate the expansions that appear in (1.114,115); in other words we have to assume that the field is sufficiently well described by N coefficients B_n and N coefficients b_n. Then, all the calculations are explicit except for the inversion of the square matrix $R - QP$ which is effected with a computer. The computation time τ is generally very low; typical values are $N = 15$ and $\tau = 0{,}3$ s on the UNIVAC 1108 computer [1.30]. It is worth noting that this method has been extended to the case of a grating composed of rectangular rods lying on a dielectric stack [1.31-33]. The study of this particular structure is of course closely linked with the fabrication of integrated circuits by masking [1.32].

The rectangular-groove grating is not the only one which can be treated by the modal expansion method. For instance, let us consider the groove profile of Fig.1.9, i.e., a triangular profile symmetrical with respect to the y axis. If $y > a$, the total field can be expressed in terms of a sum of propagating and evanescent plane waves (1.114) as we did for the lamellar grating. In the hatched region ($\alpha<\phi<\pi-\alpha$, $0<r<R$) the total field $u(r,\phi)$ which vanishes for $\phi = \alpha$ and $\phi = \pi - \alpha$ can be considered to be the restriction to the interval $(0,\pi-2\alpha)$ of an odd function of $\phi-\alpha$ defined on $(2\alpha-\pi,\pi-2\alpha)$. Therefore it can be expanded in a sine series (see footnote 6 on p.17)

$$u(r,\phi) = \sum_{n=1}^{\infty} u_n(r) \sin\left[\frac{n\pi}{\pi-2\alpha} (\phi-\alpha)\right] \quad . \tag{1.131}$$

If we write the Helmholtz equation using this series, we find that $u_n(r)$ is necessarily a Bessel function of the first kind (a Bessel function of the second kind [11] cannot occur because the hatched region encompasses point O for which $r = 0$)

$$u(r,\phi) = \sum_{n=1}^{\infty} A_n J_\nu(k_1 r) \sin\left[\tau_n(\phi-\alpha)\right] \quad , \tag{1.132}$$

where

$$\tau_n = \frac{n\pi}{\pi-2\alpha} \quad . \tag{1.133}$$

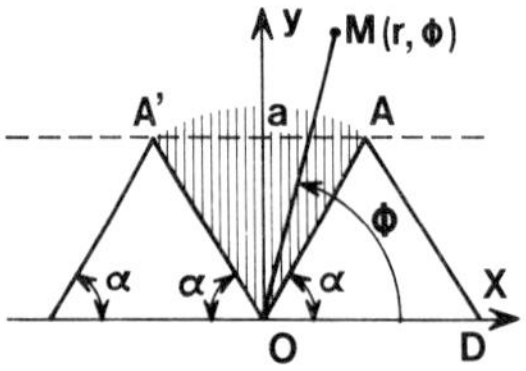

Fig. 1.9. A profile for which the method of JOVICEVIC and SESNIC [1.34] is rigorous; OA = OA' = R

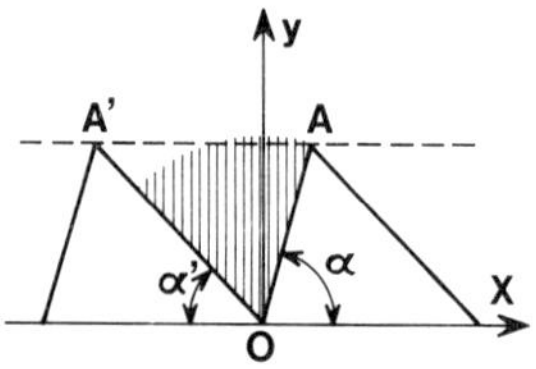

Fig. 1.10. A profile for which the method of JOVICEVIC and SESNIC seems questionable

11 Such a function is not bounded at the origin.

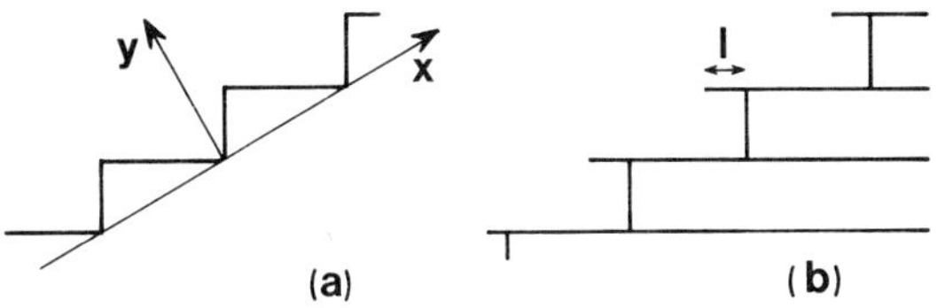

(a) (b)

Fig. 1.11. (a) The echelette grating; (b) the auxiliary diffracting structure: the plates and the shorting walls are infinitely thin and perfectly conducting

Thus the field is given by a plane wave expansion outside the grooves and by (1.132) in the groove region. We can determine the two sequences of unknown coefficients B_n and A_n by expressing the continuity of u and $\partial u/\partial y$ at $y = a$. This technique with which I agree for a symmetrical profile (Fig.1.9) has also been used [1.34] for any triangular profile (Fig.1.10). In this case, it seems that we cannot unreservedly put faith in the method because the hatched region, in which the validity of (1.132) [with now $\tau_n = n\pi/(\pi - \alpha - \alpha')$] is established by using the preceding arguments, does not include the whole groove region. In other terms, and to the author's knowledge, there is then no proof for the existence of the A_n coefficients. Nevertheless, and due to uniqueness, if we suppose that such coefficients have been found, the corresponding solution is, of course, the good one. This is not unreminiscent of the discussions about the validity of the Rayleigh assumption. No doubt, JOVICEVIC's method can often be very efficient but, because it implies an assumption, it seems dangerous to use it to write a computer program which purports to be general. Incidentally, if we look at [Ref.1.34, Fig.9] we note that neither the reciprocity theorem nor the energy balance criterion are very well satisfied. In our laboratory, we have drawn the same curve using a program based on an integral method; it turns out to be perfectly symmetrical and, for any incidence and polarization, the sum of the relative powers in all the orders is equal to unity within a precision always bettwe than 1%.

As a last example of an analytical method adapted to a particular profile, I would like to quote a very ingenious paper devoted to the echelette grating by ITOH and MITTRA [1.35]. This grating is viewed as a limiting case of an associate canonical problem for which an exact solution is known. The authors observed that the grating structure (Fig.1.11a) is derivable by the process of letting $\ell = 0$ in the geometry of Fig.1.11b. The latter, which of course is rather unusual in optics, is composed of an infinite set of staggered parallel plates with shorting walls which are recessed by the length ℓ into each opening. The way in which an exact solution is derived for this auxiliary structure is described in detail in [1.35]. It involves a Rayleigh expansion, a modal expansion between the plates, and a clever use of the two-dimensional version of the second Green identity.

1.2.13 Conical Diffraction

Throughout Sect.1.2, the incident wave vector has been supposed to be orthogonal to the grooves. What happens if this constraint is removed? A detailed answer will be given in Chap.3; here, without proof, are the results that everyone should know.

The incident plane wave can be described by the associated electric vector field that we call $\underline{E}^i$

$$\underline{E}^i = \underline{a}\exp[i(\alpha x-\beta y+\gamma z)] = \underline{a}\exp(i\underline{k}^i\cdot\underline{r}) \quad . \tag{1.134}$$

Without any loss of generality, the vector amplitude $\underline{a}$ can be assumed to be real and of unit modulus. Moreover, as is well known, $\underline{a}$ and $\underline{k}^i$ are perpendicular.

Outside the grooves [i.e., for $y > \max f(x)$], the diffracted field is given by a plane wave expansion [12]

$$\underline{E}^d = \sum_{n=-\infty}^{+\infty} \underline{B}_n \exp\left[i(\alpha_n x+\beta_n y+\gamma z)\right] = \sum_{n=-\infty}^{+\infty} \underline{B}_n \exp\left(i\underline{\chi}_n\cdot\underline{r}\right) \quad , \tag{1.135}$$

with

$$\alpha_n = \alpha + nK \quad \text{and} \quad \beta_n = \left(k_1^2-\alpha_n^2-\gamma^2\right)^{\frac{1}{2}} \quad . \tag{1.136}$$

Of course $\underline{B}_n$ is here a complex vector amplitude which generalizes the Rayleigh coefficient used until now in scalar problems. One verifies that for travelling diffracted waves

$$\underline{\chi}_n^2 = \alpha_n^2 + \beta_n^2 + \gamma^2 = k_1^2 = \underline{k}^{i2} \quad , \tag{1.137}$$

$$\underline{\chi}_n \cdot \hat{\underline{z}} = \underline{k}^i \cdot \hat{\underline{z}} = \gamma \quad . \tag{1.138}$$

From these two relations it turns out that the diffracted wave vectors lie on a cone whose axis is parallel to the grooves and whose half-angle is the angle between the incident wave vector and the ruling direction. That is why we speak of conical diffraction.

In order to determine the efficiencies, it is convenient to introduce an auxiliary problem in which the incident wave vector $\tilde{\underline{k}}^i$ is the projection of $\underline{k}^i$ on the xOy plane (the components of $\tilde{\underline{k}}^i$ are α, $-\beta$, 0). For this auxiliary problem let $e_n^{\shortparallel}$ and $e_n^{\perp}$ be the efficiencies, respectively, in P polarization ($\underline{E}$ parallel to the grooves) and in S polarization ($\underline{E}$ perpendicular to the grooves). Then it can be shown, that in conical diffraction (i.e., with the oblique wave vector $\underline{k}^i$), the nth order efficiency is given by

$$e_n = e_n^{\shortparallel}\sin^2\delta + e_n^{\perp}\cos^2\delta \quad , \tag{1.139}$$

---

12 In (1.135) $\underline{\chi}_n$ is the wave vector of the nth diffracted wave.

where δ is the angle between $\underline{a}$ and a certain unit vector $\underline{U}$. This vector $\underline{U}$ lies in the xOy plane and is perpendicular to $\tilde{\underline{k}}^i$. In other terms, and thanks to (1.139) *the use of the computer becomes unnecessary as soon as both the simple cases of polarization* (P and S) *have been solved*. That is why *they are said to be fundamental*.

1.3 The Dielectric or Metallic Grating

In this section we abandon the ideal situation of the perfectly conducting grating. Region 2 is now filled with a material of index ν_2. For the sake of simplicity, ν_2 is supposed to be real, the general case (ν_2 complex) being treated in Chap.4. Dealing with this dielectric grating, we illustrate the differential method, suggested in 1966 [1.36] and extensively developed in our laboratory since 1969 [1.37].

1.3.1 Generalities

The notations being still the same (see Sects.1.1.2 and 5), it is clear that function u verifies now a Helmholtz equation in each of the two regions 1 and 2

$$\Delta u + k_1^2 u = 0 \quad , \quad \text{if} \quad y > f(x) \quad , \tag{1.24}$$

$$\Delta u + k_2^2 u = 0 \quad , \quad \text{if} \quad y < f(x) \quad . \tag{1.140}$$

Outgoing wave conditions are verified by $u^d = u - u^i$ (for $y \to +\infty$) and by u (for $y \to -\infty$). As for the boundary conditions, we have to express the conservation of the tangential components of $\underline{E}$ and $\underline{H}$. After some elementary manipulations of Maxwell equations, this leads to the following conclusion: u is continuous whatever the polarization; du/dn (in P polarization) or ε^{-1}du/dn (in S polarization) have the same value whether calculated in region 1 or region 2.

1.3.2 The Diffracted Field Outside the Groove Region

If we assume the existence and uniqueness of the solution (see Chap.2) we can prove the pseudo-periodicity of the field exactly as we did in Sect.1.2.3

$$u(x,y) = \exp(i\alpha x) \sum_{n=-\infty}^{+\infty} v_n(y) \exp(inKx) = \sum_{n=-\infty}^{+\infty} v_n(y) \exp(i\alpha_n x) \quad .$$

Inserting this expansion in (1.24,140), we establish that, outside the groove region at least, the field can be described by Rayleigh expansions.

If $y > a$

$$u(x,y) = \exp(i\alpha x - i\beta y) + \sum_{n=-\infty}^{+\infty} R_n \exp(i\alpha_n x + i\beta_{n1} y) \quad , \tag{1.141}$$

with

$$\alpha_n = k_1 \sin\theta + nK \quad \text{and} \quad \beta_{n1}^2 = k_1^2 - \alpha_n^2 \quad . \tag{1.142}$$

If $y < 0$

$$u(x,y) = \sum_{n=-\infty}^{+\infty} T_n \exp(i\alpha_n x - i\beta_{n2} y) \quad , \tag{1.143}$$

with

$$\beta_n^2 = k_2^2 - \alpha_n^2 \quad . \tag{1.144}$$

Each of these two expansions contains propagating and evanescent waves depending on the value of n. Let us call U_1 (resp. U_2) the set of integers n for which β_{n1}^2 (resp. β_{n2}^2) is positive. The field associated with (1.141) has been described in Sect.1.2.3; we only have to replace U by U_1. If $n \in U_1$, that is to say if $\alpha_n^2 < k_1^2$, we can put

$$\alpha_n = k_1 \sin\theta + nK = k_1 \sin\theta_{n1} \quad , \quad -\frac{\pi}{2} < \theta_{n1} < \frac{\pi}{2} \quad , \tag{1.145}$$

$$\beta_{n1} = \left(k_1^2 - k_1^2 \sin^2\theta_{n1}\right)^{\frac{1}{2}} = k_1 \cos\theta_{n1} \quad , \tag{1.146}$$

and $R_n \exp(i\alpha_n x + i\beta_{n1} y)$ represents a plane wave propagating in the θ_{n1} direction (Fig.1.12). Similarly, if $n \in U_2$, we can put

$$\alpha_n = k_1 \sin\theta + nK = k_2 \sin\theta_{n2} \quad , \quad -\frac{\pi}{2} < \theta_{n2} < \frac{\pi}{2} \quad , \tag{1.147}$$

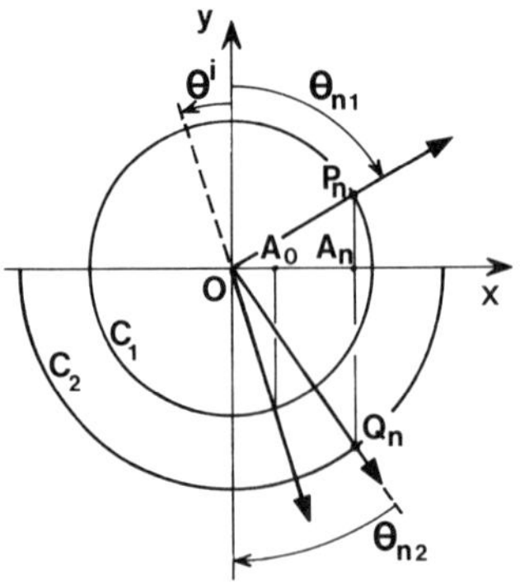

Fig. 1.12. The radius of circles C_1 and C_2 are, respectively, ν_1 and ν_2. If $\overline{A_0A_n} = n(\lambda_0/d)$, the nth reflected (resp. transmitted) diffracted wave is propagating along OP_n (resp. OQ_n)

$$\beta_{n2} = k_2^2 - k_2^2 \sin^2\theta_{n2} = k_2 \cos\theta_{n2} \quad , \tag{1.148}$$

and $T_n \exp(i\alpha_n x - i\beta_{n2} y)$ represents a transmitted plane wave propagating in the θ_{n2} direction.

Equations (1.145,147) are a generalization of the perfectly conducting grating formula (1.49). They immediately justify the geometrical construction in Fig. 1.12 insofar as they are rewritten under the equivalent form

$$\nu_1 \sin\theta_{n1} = \nu_1 \sin\theta + n \frac{\lambda_0}{d} \quad , \tag{1.149}$$

$$\nu_2 \sin\theta_{n2} = \nu_1 \sin\theta + n \frac{\lambda_0}{d} \quad , \tag{1.150}$$

where λ_0 is the wavelength in vacuum.

1.3.3 Maxwell Equations and Distributions

As a consequence of the field pseudo-periodicity we can consider $\underline{E}$, $\underline{D}$, $\underline{H}$, $\underline{B}$ as vector distributions whose components are elements of R' (see Appendix A.3) namely pseudo-periodic functions continuous and twice differentiable in the complement of the grating profile P. Exactly as we did in Sect.1.1.4 for an arbitrary surface S (and using then Schwartz distributions) we can say that Maxwell equations are valid in the sense of distributions, the differential operators curl and div being defined as in Appendix A.4. Because we have neither charge nor current, we have therefore in the sense of distributions

$$\text{curl } \underline{E} = i\omega \underline{B} \quad , \tag{1.151}$$

$$\text{curl } \underline{H} = -i\omega \underline{D} \quad , \tag{1.152}$$

$$\text{div } \underline{D} = 0 \quad , \tag{1.153}$$

$$\text{div } \underline{B} = 0 \quad . \tag{1.154}$$

Let us recall that *these four equations automatically take into account the boundary conditions on the grating surface*. The proof has already been given in Sect.1.1.4. For instance, (1.151) implies the conservation of the tangential component of $\underline{E}$.

1.3.4 The Principle of the Differential Method (in P Polarization)

Instead dealing with two Helmholtz equations (1.24,140), we can say that u (the z component of $\underline{E}$) verifies

$$\Delta u + k^2(x,y)u = 0 \quad , \tag{1.155}$$

with

$$k^2(x,y) = \begin{vmatrix} k_1^2 & \text{if} \quad y > f(x) \quad , \\ k_2^2 & \text{if} \quad y < f(x) \quad . \end{vmatrix} \tag{1.156}$$

It is worth noting that (1.155) holds in the sense of distributions, because both the jumps of u and du/dn being zero, the Laplacian in the sense of distribution is actually equal to the Laplacian without precaution ($\Delta u = \{\Delta u\}$). This result can alternatively be derived directly from (1.151,152) which imply

$$-\Delta \underline{E} + \text{grad div } \underline{E} = -i^2\omega^2\mu\varepsilon\underline{E} = k^2(x,y)\underline{E} \quad . \tag{1.157}$$

Moreover because $\underline{E}$ is supposed to be parallel to Oz, $\underline{n} \cdot \sigma_{\underline{E}}$ is necessarily zero. Consequently

$$\text{div } \underline{E} = \{\text{div } \underline{E}\} + \underline{n} \cdot \sigma_{\underline{E}}\delta_P = 0 + 0 \quad ,$$

and (1.157) becomes

$$\Delta \underline{E} + k^2(x,y)\underline{E} = 0 \quad ,$$

which implies (1.155) by projection on the z axis.

Let us now rewrite (1.155) using the series expansion of u and replacing the periodic function $k^2(x,y)$ by its Fourier series $\sum_{n=-\infty}^{+\infty} c_n(y) \exp(inKx)$. After some routine algebraic calculations we obtain

$$\sum_n \left[v_n'' - \alpha_n^2 v_n + \sum_m c_{n-m}(y) v_m \right] \exp(inKx) = 0 \quad .$$

The left-hand member is the expansion in Fourier series (with respect to x) of the null function. We have therefore, for any n

$$v_n'' - \alpha_n^2 v_n + \sum_{m=-\infty}^{+\infty} c_{n-m}(y) v_m = 0 \quad . \tag{1.158}$$

This means that the functions $v_n(y)$ are solutions of a differential system which can be written in matrix form

$$V'' = A(y)V \quad , \tag{1.159}$$

where V is a column matrix whose elements are the functions $v_n(y)$ and A is a known square matrix whose elements A_{nm} are defined by

$$A_{nm} = \alpha_n^2 \delta_{mn} - c_{n-m} \quad . \tag{1.160}$$

Furthermore, (1.158) implies the continuity of $v_n(y)$ and $v'_n(y)$ because, if these functions were discontinuous at some value of y (say y_0), the distribution $\delta(y-y_0)$ and its derivative would appear in the right-hand member.

Outside the groove region the Rayleigh expansions are valid and the $v_n(y)$ are exponential functions

$$v_n(y) = T_n \exp(-i\beta_{n2}y) \quad \text{if} \quad y < 0 \quad , \tag{1.161}$$

$$v_n(y) = \delta_{n0} \exp(-i\beta y) + R_n \exp(i\beta_{n1}y) \quad \text{if} \quad y > a \;[13] . \tag{1.162}$$

Differentiation of (1.161) leads to

$$v'_n(y) = -i\beta_{n2}T_n \exp(-i\beta_{n2}y) \quad ,$$

hence

$$\frac{v'_n(y)}{v_n(y)} = -i\beta_{n2} \quad , \quad \text{if} \quad y < 0 \quad .$$

Now, taking into account the continuity of v'_n/v_n we get

$$v'_n(0) + i\beta_{n2}v_n(0) = 0 \quad . \tag{1.163}$$

Similarly, starting from (1.162) instead of (1.161) we show that

$$v'_n(a) - i\beta_{n1}v_n(a) = -2i\beta\delta_{n0} \exp(-i\beta a) \quad . \tag{1.164}$$

Again, (1.163,164) can be written in matrix form as follows

$$V'(0) + L_0V(0) = 0 \quad , \tag{1.165}$$

$$V'(a) + L_aV(a) = S \quad . \tag{1.166}$$

L_0 and L_a are here diagonal known matrices whose elements are

$$[L_0]_{nm} = i\beta_{n2}\delta_{nm} \quad , \quad [L_a]_{nm} = -i\beta_{n1}\delta_{nm} \quad .$$

As for S, it is a column matrix which is in fact reduced to a number [$S_n = -2i\beta\delta_{n0}$ $\exp(-i\beta a)$].

Let us now notice that the field is perfectly determined when all the $v_n(y)$ are known on the interval (0,a). Actually (1.161,162) show that the knowledge of $v_n(0)$

---

13 β_{01} is simply $\beta = k_1 \cos\theta$.

and $v_n(a)$ allows us to get the coefficients R_n and T_n. *The grating problem is thus reduced to the resolution of the differential system (1.159) on the finite interval (0,a)* with the boundary conditions (1.165,166), which is a classical problem in applied mathematics. For numerical purposes, the series describing u must be replaced by a finite sum with $P = 2N+1$ terms and consequently all the matrices are finite matrices with P rows.

Numerical considerations will be extensively developed in Chap.4. Here we will only describe briefly the so-called "shooting method". This method is applicable provided a computer program is available that gives V(a) and V'(a) from V(0) and V'(0) taking into account (1.159). Starting with an arbitrary value of V(0) [say V(0) = C] we get successively V'(0) [from (1.165)], then V(a) and V'(a) (by numerical integration) and finally $U = V'(a) + L_a V(a)$ by simple operations of matrix algebra. The key step is to notice that, due to the linearity of (1.165,159), there exists a matrix M such that

$$MC = U \quad . \tag{1.167}$$

This matrix M can be determined numerically: its pth column is nothing else than the result obtained for U when the column matrix C has elements $C_n = \delta_{np}$, as is immediately verifiable. When M is known [which needs P numerical integrations of (1.159)] V(0) is given by the matrix equation

$$MV(0) = S \quad , \tag{1.168}$$

and, once more, the problem is reduced to a matrix inversion. The knowledge of V(0) allows us to compute the T_n coefficients and a further numerical integration gives V(a) and consequently the R_n coefficients. Of course this method works provided that the field is sufficiently well described by a few terms of its series expansion in order that the infinite differential system can be replaced by not too large a system of equations. Numerical experiments show that this is always the case when λ/d is greater than 0.2. As a check of accuracy we must observe the convergence of successive approximations when P increases; we can also use reciprocity considerations [1.38] and verify the energy balance criterion which here takes the intuitive form

$$\cos\theta = \sum_{n \in U_1} R_n \bar{R}_n \cos\theta_{n1} + \sum_{n \in U_2} T_n \bar{T}_n \cos\theta_{n2} \quad . \tag{1.169}$$

We have to emphasize that, as explained in Chap.4, the differential method is applicable to structures more general than the ordinary grating; we can for instance suppose that an inhomogeneous domain D, in which the optical index $\nu(x,y)$ is a periodic function with respect to x, lies on several homogeneous layers which themselves lie on a homogeneous substrate. The other case of polarization (S polarization) can be treated although some numerical difficulties have been encountered when dealing with aluminum gratings (see Chap.4). A computer program is also available to solve the general case of conical diffraction which cannot be reduced to the two simple cases of polarization (S and P) as was possible for perfectly conducting grating. *It must therefore be noted that, for dielectric or metallic gratings, we had better not speak of fundamental cases but rather of usual cases of polarization.*

1.4 Miscellaneous

In the previous sections, we have successively used two types of gratings (the perfectly conducting grating and the dielectric grating) to give the fundamental ideas on which the methods are based which have up to now been the most employed (Rayleigh's method, integral and differential methods, modal methods). It is possible that this way of doing things might not give an overall picture of all the work which has been done on the subject. That is why I would like, in this last section, to describe briefly some methods of study of which we have not yet had the opportunity to speak.

In Sect.1.3 we have used the dielectric grating to illustrate the powerful differential method, which does not mean that an integral method cannot be utilized for this type of grating. Actually, in our laboratory, several computer programs based upon integral methods are available for dealing with gratings of finite conductivity even if they are coated with several dielectric layers (see Chap.3). By the way, it is conversely possible to employ differential methods when dealing with perfectly conducting gratings. The key step is to perform a change in coordinates that maps the grating profile P onto the x axis, which complicates the partial differential equations but facilitates the expression of the boundary condition. Then the initial problem turns into another one which can be reduced to the resolution of a system of ordinary differential equations. The use of a conformal mapping was first successfully investigated around 1971 by NEVIERE et al. [1.39] as is explained in Chap.4. More recently good results have been obtained by CHANDEZON [1.40] using the simple coordinate transformation

$$x_1 = x \quad , \quad x_2 = y - f(x) \quad , \quad x_3 = z \quad . \tag{1.170}$$

The x_i being nonorthogonal curvilinear coordinates, the basic equations are a little more complicated but, unexpectedly, the problem finally reduces to a differential system with constant coefficients. This allows interesting analytical considerations and the author, using perturbation techniques and Padé approximants, proposes several quite simple formulae to represent the variations of the efficiencies [1.40].

Coming back to the dielectric grating, another method is worth quoting which needs only the resolution of an algebraic linear system [1.41]. The functions $v_n(y)$ which appears in Sect.1.3.4 are represented on the interval (0,a) by a Fourier series corresponding to a period b greater than $a = \max f(x)$. In other words, the field is expressed in the groove region (i.e., for $0 < y < a$) as a double Fourier series

$$u(x,y) = \exp(i\alpha x) \sum_n \sum_m B_{nm} \exp(inKx) \exp\left(im \frac{2\pi}{b} y\right) \quad . \tag{1.171}$$

Rayleigh expansions are still used for $y < 0$ or $y > a$. The unknow coefficients R_n, T_n, B_{nm} entering the various expansions are determined by the requirement that the field must be a solution of (1.155) and by enforcing the proper boundary conditions along the straight lines $y = 0$ and $y = a$. The method has been exploited numerically for echelette gratings in P polarization and we have verified that all the efficiency curves reproduced in the quoted paper are in agreement with the ones obtained in our laboratory using integral or differential methods. This systematic confrontation has been undertaken because we questioned a priori the rapidity of convergence of the double series in (1.171). MARCUSE [1.41] did in fact relate some difficulties observed when increasing the groove depth; nevertheless all the usual profiles seem to be accessible.

Lastly, to end this rapid survey, we should perhaps draw attention to some numerical studies of finite gratings with only a very small number of grooves (about ten at the most). Some data have already been obtained both by a differential method [1.42] and by an integral method [1.43,44]. In the latter case, despite the fact that one is led to solve large systems of linear equations, computations are made in relatively short computations times, with reasonable storage and good precision, by using the AKAIKE-ROBIN inversion algorithm [1.45] which is a generalization of earlier work by TRENCH [1.46].

References

1.1 J.W.S. Rayleigh: Proc. Roy. Soc. A *79*, 399 (1907)
1.2 R. Petit: Opt. Acta *14*, 301-310 (1967)
1.3 U. Fanò: J. Opt. Soc. Am. *31*, 213 (1941)
1.4 W.C. Meecham: J. Appl. Phys. *27*, 361 (1956)
1.5 G.W. Stroke: Rev. Opt. *39*, 350 (1960)
1.6 P. Bousquet: C. R. Acad. Sci. *256*, 3422 (1963)
1.7 P. Bousquet: C. R. Acad. Sci. *257*, 80 (1963)
1.8 R. Petit: Rev. Opt. *42*, 263 (1963)
1.9 R. Deleuil: C. R. Acad. Sci. *258*, 506 (1963)
1.10 C. Janot, A. Hadni: J. Phys. *24*, 1073 (1963)
1.11 E.A. Yakovlev, Opt. Spektr. *19*, 417 (1965)
1.12 R. Petit, M. Cadilhac: C. R. Acad. Sci., Ser. A-B, *262*, 468 (1966)
1.13 G.R. Jiracek: IEEE Trans. AP-*21*, 393 (1973)
1.14 B.A. Lippmann: J. Opt. Soc. Am. *43*, 408 (1953)
1.15 R.F. Millar: Proc. Cambridge Phil. Soc. *69*, 217 (1971)
1.16 M. Nevière, M. Cadilhac: Opt. Commun. *2*, 235 (1970)

1.17 A. Marechal, G.W. Stroke: C. R. Acad. Sci. *249*, 2042 (1959)
1.18 R.F. Millar: Proc. Cambridge Phil. Soc. *69*, 175 (1971)
1.19 H. Ikuno, K. Yasuura: IEEE Trans. AP-*2*, 657 (1973)
1.20 R.F. Millar: Radio Science *8*, 785 (1973)
1.21 H.A. Kalhor: IEEE Trans. AP-*24*, 884 (1976)
1.22 R. Petit: Rev. Opt. *45*, 249 (1966)
1.23 R. Petit, M. Cadilhac: C. R. Acad. Sci. *259*, 2077 (1964)
1.24 R. Petit: C. R. Acad. Sci. *260*, 4454 (1965)
1.25 J. Pavageau, J. Bousquet: Opt. Acta *17*, 469 (1970)
1.26 J. Van Bladel: *Electromagnetic Fields* (McGraw-Hill, New York 1964)
1.27 A. Wirgin: Alta frequenza, Selected papers from the URSI Symp. *38*, 327 (1969)
1.28 A. Wirgin: Thèse A.O. 1429, Faculté des Sciences d'Orsay (1967)
1.29 A. Wirgin, R. Deleuil: J. Opt. Soc. Am. *59*, 1348 (1969)
1.30 D. Maystre, R. Petit: Opt. Commun. *5*, 90 (1972)
1.31 J.L. Roumiguières, D. Maystre, R. Petit: Opt. Commun. *7*, 402 (1973)
1.32 J.L. Roumiguières: Thèse de 3ème cycle, Université d'Aix-Marseille III, Centre de St-Jérôme (1976)
1.33 J.L. Roumiguières, D. Maystre, R. Petit: *Proc. fifth Colloquium on Microwave Communication*, Budapest (1974)
1.34 S. Jovicevic, S. Sesnic: J. Opt. Soc. Am. *62*, 865 (1972)
1.35 T. Itoh, R. Mittra: IEEE Trans. MTT-*17*, 319 (1969)
1.36 R. Petit: Rev. Opt. *45*, 353 (1966)
1.37 G. Cerutti-Maori, R. Petit, M. Cadilhac: C. R. Acad. Sci., Ser. B *268*, 1060 (1969)
1.38 M. Nevière, P. Vincent: Opt. Acta *23*, 557 (1976)
1.39 M. Nevière, M. Cadilhac, R. Petit: IEEE Trans. AP-*21*, 37 (1973)
1.40 J. Chandezon: Thèse d'Etat, Université de Clermont-Ferrand II (1979)
1.41 D. Marcuse: Bell Syst. Tech. J. *55*, 1295 (1976)
1.42 J.P. Hugonin, R. Petit: Opt. Commun. *20*, 360 (1977)
1.43 P. Facq: Ann. Telecommunic. *31*, 99 (1976)
1.44 P. Facq: Thèse A.O. 12470, Université de Limoges (1977)
1.45 H. Akaïké: SIAM J. Appl. Math. *24*, 234 (1973)
1.46 W.F. Trench: SIAM J. Appl. Math. *12*, 515 (1964)
1.47 B.W. Ross: *Analytic Functions and Distributions in Physics and Engineering* (Wiley, New York 1969) pp.289-360
1.48 E.M. de Jager: In *Mathematics Applied to Physics*, ed. by E. Roubine (Springer, Berlin, Heidelberg, New York 1970) pp.52-109
1.49 L. Schwartz: *Théorie des Distrbutions* (Hermann, Paris 1966)
1.50 L. Schwartz: *Mathematics for Physical Sciences* (Addison-Wesley, London 1967)

Appendix A: The Distributions or Generalized Functions

There is probably no miraculous way to teach in depth the theory of distributions in a few pages... The purpose of this appendix is only to make it unnecessary for physicists to have to consult specialized mathematics books to understand certain considerations related to grating theories. We seek here therefore utility and brevity rather than the rigor and the elegance which would characterize a treatment for mathematicians. We suppose, of course, that the reader has some ideas on Fourier series, convolution and also on generalized functions (another name for distributions) at least in the case of one variable [1.47,48].

A.1 Preliminaries

It is known that a functional is an application of a vector space of functions F on the field C of complex numbers. In other terms, we say that we have defined a functional T on F if we are able to associate a complex number c with any function ϕ of F; we write

$$\forall\phi \in F \quad , \quad <T,\phi> = c \quad .$$

A functional T is said to be linear if, for any pair of complex numbers λ_1 and λ_2, and for any pair of functions ϕ_1 and ϕ_2 belonging to F, we have

$$<T,\lambda_1\phi_1+\lambda_2\phi_2> = \lambda_1<T,\phi_1> + \lambda_2<T,\phi_2> \quad .$$

A functional is said to be continuous if

$$\phi_j \to \phi \quad \text{implies:} \quad <T,\phi_j> \to <T,\phi> \quad .$$

We must note that this definition supposes the understanding of the expression "ϕ_j tends to ϕ", that is to say the choice of a notion of convergence in F. (We have to make precise what we mean when we say that a sequence of functions tends to a given function.) Conversely, because both $<T,\phi_j>$ and $<T,\phi>$ are numbers, subtle comments are not necessary to understand the meaning of "$<T,\phi_j>$ tends to $<T,\phi>$": the reader is supposed to know the theory of numerical sequences. The set of all the linear and continuous functionals defined on the space F is called F'.

Further if λT and the sum of two functionals T_1 and T_2 are, respectively, defined by

$$\forall\lambda \in C \quad , \quad \forall\phi \in F \quad , \quad <\lambda T,\phi> \overset{\text{def}}{=} \lambda<T,\phi> \quad ,$$

$$\forall\phi \in F \quad , \quad <T_1+T_2,\phi> \overset{\text{def}}{=} <T_1,\phi> + <T_2,\phi> \quad ,$$

it turns out that F' is also a vector space of functions. This space F' is called the dual topological space of F and when we say that a sequence T_j of functionals (belonging to F') tends to a functional T, we mean that

$$\forall\phi \in F \quad , \quad <T_j,\phi> \to <T,\phi> \quad .$$

Following SCHWARTZ [1.49,50] people generally speak of distributions as linear and continuous functionals on the space $\mathcal{D}$ of functions differentiable any number of times and whose support is bounded [14]. The distributions are therefore the elements of $\mathcal{D}'$. On the other hand, very often in this book, we call distributions the linear and continuous functionals on a certain space $\mathcal{R}$ well adapted to grating problems ($\mathcal{R}$ recalls "rêseau", the french word for "grating") and, when they appear, the elements of $\mathcal{D}'$ are called Schwartz distributions to avoid confusion.

A.2 The Function Space $\mathcal{R}$

We can say that this space is associated with a periodic structure (period d with respect to x), an angle θ (which represents generally the angle of incidence of a plane wave) and a wavelength $\lambda = 2\pi/k$. It would be better to speak of $\mathcal{R}_{d,\theta,\lambda}$ but we will use the notation $\mathcal{R}$ for simplicity.

$\mathcal{R}$ is the set of all the complex valued and infinitely differentiable functions $\psi(x,y)$, with bounded support in y, and such that

$$\psi(x+d,y) = \psi(x,y) \exp(-ikd \sin\theta) \quad . \tag{A.1}$$

One says that a sequence of $\psi_j(x,y)$ belonging to $\mathcal{R}$ (j=1,2,...) converges to zero in the sense of $\mathcal{R}$ if ψ_j and all its derivatives converge to zero (for $j\to\infty$) uniformly. Of course, ψ_j tending to ψ means that $\psi - \psi_j$ tends to zero.

A.3 The Space $\mathcal{R}'$

A.3.1 Definitions

$\mathcal{R}'$ is the dual topological space of $\mathcal{R}$ and its elements are the distributions. A distribution is therefore a linear and continuous functional on $\mathcal{R}$. It is generally easy to decide if a given functional on $\mathcal{R}$ is linear but more difficult to decide if it is continuous. Fortunately (and mathematicians have good reasons for that) we can assume that all the linear functionals on $\mathcal{R}$ that we encounter in practice are also continuous. Consequently the reader (who is supposed to have only a pragmatic point of view) is allowed to forget the rather subtle notion of convergence defined at the end of the previous paragraph. He can, for $\mathcal{R}$, confuse, without danger, the dual topological space with the dual space (the set of all the linear functionals) and retain only that a distribution is a linear functional on $\mathcal{R}$.

A.3.2 Examples of Distributions

Very often, in grating theory, we deal with functions u(x,y) such that

$$\forall y \quad , \quad u(x+d,y) = u(x,y) \exp(ikd \sin\theta) \quad . \tag{A.2}$$

---

14 Such a function vanishes outside some bounded interval.

With any function of this type (called a pseudo-periodic function), which is moreover supposed to be a locally Lebesgue integrable function, we associate the functional T_u defined by

$$\forall\psi \in R \quad , \quad \langle T_u,\phi \rangle = \iint_{-\infty}^{+\infty} u(x,y)\psi(x,y)dx\, dy \quad , \tag{A.3}$$

where the integration in y is, in fact, performed for a fixed x, on the finite support of $\psi(x,y)$ and the integration in x is performed over one period [15]. Because we use Lebesgue integrals, T_u is actually linked with a class of functions (the class of all functions equal to ψ almost everywhere). Very often in physics we write u instead of T_u, confusing thus the function itself and the associated functional. This is an incorrect but usual notation which leads us to write

$$\forall\psi \in R \quad , \quad \langle u,\phi \rangle = \iint_{-\infty}^{+\infty} u(x,y)\psi(x,y)dx\, dy \quad . \tag{A.4}$$

Sometimes we deal with physical magnitudes, such as a surface current, which are only defined on the line P which represents the profile of the grating [i.e., the graph of y=f(x)]. Let ρ be such a magnitude and let $d\ell$ be the arc element of P; the distribution $\rho\delta_P$ is then defined by

$$\forall\psi(x,y) \in R \quad , \quad \langle \rho\delta_P,\psi \rangle = \int \rho\psi d\ell \quad ,$$

$$= \int \rho(x)\psi[x,f(x)]\, \frac{d\ell}{dx}\, dx \quad . \tag{A.5}$$

The distribution of Dirac on R is the functional denoted δ_R and defined by

$$\forall\psi(x,y) \in R \quad , \quad \langle \delta_R,\psi(x,y) \rangle = \psi(0,0) \quad . \tag{A.6}$$

We have to admit that our notations are not perfectly clear: here the subscript R refers to the space on which δ_R is defined; conversely in $\rho\delta_P$ the subscript P refers to the line on which ρ is defined. Of course $\rho\delta_P$ is also a functional on R and by the way should not be regarded as a product of ρ by δ_P. In our opinion, it would be better to write $T_{\rho,P}$ instead of $\rho\delta_P$, but experience shows this latter notation to be more flexible and we retain it... .

A.4 Derivative of a Distribution

If $T \in R'$, we call the partial derivative of T with respect to x, the distribution denoted $\partial T/\partial x$ and defined by

15 Recall that, whenever the integration interval is not specified, the integration must be performed over one period.

$$\forall\psi \in R \quad , \quad \left\langle \frac{\partial T}{\partial x}, \psi \right\rangle = -\left\langle T, \frac{\partial \psi}{\partial x} \right\rangle \quad . \tag{A.7}$$

The other partial derivatives are defined in the same way

$$\left\langle \frac{\partial T}{\partial y}, \psi \right\rangle = -\left\langle T, \frac{\partial \psi}{\partial y} \right\rangle \quad ,$$

$$\left\langle \frac{\partial^2 T}{\partial x^2}, \psi \right\rangle = -\left\langle \frac{\partial T}{\partial x}, \frac{\partial \psi}{\partial x} \right\rangle = \left\langle T, \frac{\partial^2 \psi}{\partial x^2} \right\rangle \quad ,$$

$$\left\langle \frac{\partial^2 T}{\partial y^2}, \psi \right\rangle = \left\langle T, \frac{\partial^2 \psi}{\partial y^2} \right\rangle \quad ,$$

$$< \Delta T, \psi > = \left\langle \frac{\partial^2 T}{\partial x^2} + \frac{\partial^2 T}{\partial y^2}, \psi \right\rangle = \left\langle \frac{\partial^2 T}{\partial x^2}, \psi \right\rangle + \left\langle \frac{\partial^2 T}{\partial y^2}, \psi \right\rangle \quad ,$$

$$= \left\langle T, \frac{\partial^2 \psi}{\partial x^2} \right\rangle + \left\langle T, \frac{\partial^2 \psi}{\partial y^2} \right\rangle = < T, \Delta\psi > \quad . \tag{A.8}$$

Due to these definitions, any distribution is infinitely differentiable. In particular, any locally integrable and pseudo-periodic function has first and second derivatives; but these derivatives, which are elements of R', are not necessarily functions. For example, let us consider a pseudo-periodic function u(x,y) which is regular in the complement of P. This function defines a distribution T_u (Sect.A.3.2) whose derivative $\partial(T_u)/\partial x$ is the distribution which maps any function ψ of R, onto the number $-< T_u, \partial\psi/\partial x >$. It is interesting to compare $\partial(T_u)/\partial x$ with the distribution $T_{\partial u/\partial x}$ associated with the pseudo-periodic function $\partial u/\partial x$ (the derivative of u at the usual sense of the theory of functions) defined in the complement of P and supposed to be locally integrable.

Let us refer to Fig.1.13 in which I_1 and I_2 are, respectively, the interiors of the contours OSDBAO and OSDB'A'O.

$$\forall\psi \in R \quad , \quad -\left\langle \frac{\partial}{\partial x} (T_u), \psi \right\rangle = \iint_{-\infty}^{+\infty} u \frac{\partial \psi}{\partial x} \, dx \, dy$$

$$= \iint_{I_1} u \frac{\partial \psi}{\partial x} \, dx \, dy + \iint_{I_2} u \frac{\partial \psi}{\partial x} \, dx \, dy \quad .$$

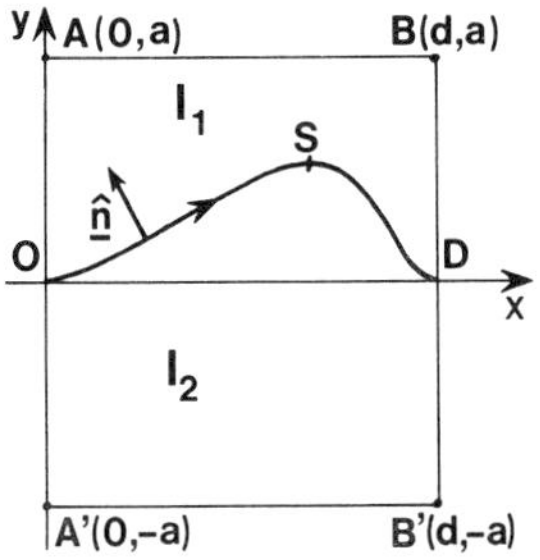

Fig. 1.13. For a given function $\psi(x,y)$, it is possible to choose the ordinate a of A in order to have $\psi(x,a) = 0$. $\hat{n}$ is the normal unit vector; its components are denoted in the text by n_x and n_y

$$\iint_{I_1} u\,\frac{\partial\psi}{\partial x}\,dx\,dy = \iint_{I_1} \frac{\partial}{\partial x}(u\psi)dx\,dy - \iint_{I_1} \frac{\partial u}{\partial x}\,\psi\,dx\,dy \quad .$$

If we introduce the vector function $\underline{V}$ with components $(u\psi,0,0)$ we have, using the Green theorem

$$\iint_{I_1} \frac{\partial}{\partial x}(u\psi)dx\,dy = \iint_{I_1} \mathrm{div}\,\underline{V}\,dx\,dy = -\int_{C_1} \underline{V}\cdot\hat{\underline{n}}\,d\ell \quad ,$$

where C_1 is the closed path OSDBAO. But because ψ vanishes on AB and because u and ψ are pseudo-periodic, the integral on C_1 is reduced to the integral on OSD (the contributions of DB and AO cancel)

$$\iint_{I_1} u\,\frac{\partial\psi}{\partial x}\,dx\,dy = -\int_{OSD} u\,\psi\,n_x\,d\ell - \iint_{I_1} \frac{\partial u}{\partial x}\,\psi\,dx\,dy \quad .$$

Likewise

$$\iint_{I_2} u\,\frac{\partial\psi}{\partial x}\,dx\,dy = \int_{OSD} u\,\psi\,n_x\,d\ell - \iint_{I_2} \frac{\partial u}{\partial x}\,\psi\,dx\,dy \quad .$$

Finally, if we denote by σ_u the jump of the function when crossing the line P in the sense indicated by $\hat{\underline{n}}$, we get

$$\forall\psi\in R \quad , \quad \left\langle \frac{\partial}{\partial x}(T_u),\psi\right\rangle = \int_{OSD} \sigma_u\,\psi\,n_x\,d\ell + \iint_{-\infty}^{+\infty} \frac{\partial u}{\partial x}\,\psi\,dx\,dy \quad ,$$

$$= \left\langle \sigma_u\,n_x\,\delta_P,\psi\right\rangle + \left\langle T_{\partial u/\partial x},\psi\right\rangle \quad .$$

This means that

$$\frac{\partial}{\partial x}(T_u) = n_x\,\sigma_u\,\delta_P + T_{\partial u/\partial x} \quad .$$

If, as is generally the case in physics, we write u instead of T_u, the first term becomes $\partial u/\partial x$. However we cannot use $\partial u/\partial x$ instead of $T_{\partial u/\partial x}$ in the second member, because this would be completely confusing. Consequently, and following L. Schwartz, we decide that hereafter $T_{\partial u/\partial x}$ will be denoted by $\{\partial u/\partial x\}$ and we write our fundamental result as

$$\frac{\partial u}{\partial x} = \left\{\frac{\partial u}{\partial x}\right\} + n_x\,\sigma_u\,\delta_P \quad .$$

In words, the derivative in the sense of distributions $(\partial u/\partial x)$ is the derivative without precaution $(\{\partial u/\partial x\})$ plus a singular distribution whose support is P.
If now, we call a vector distribution a set of three distributions (just as a vector function is a set of three functions), we can speak of gradient, divergence,

curl, Laplacian in distribution theory. In cartesian rectangular coordinates, the definitions of these operators are the same as in function theory but all the derivatives must be understood as being derivatives in the sense of distributions. If σ_u, $\sigma_{du/dn}$, $\sigma_{\underline{u}}$ denote, respectively, the jumps of u, du/dn and $\underline{u}$ when crossing P in the sense indicated by $\hat{\underline{n}}$, tedious but elementary calculations show that

$$\text{grad } u = \{\text{grad } u\} + \hat{\underline{n}}\, \sigma_u\, \delta_P \quad , \tag{A.10}$$

$$\text{div } \underline{u} = \{\text{div } \underline{u}\} + \hat{\underline{n}} \cdot \sigma_{\underline{u}}\, \delta_P \quad , \tag{A.11}$$

$$\text{curl } \underline{u} = \{\text{curl } \underline{u}\} + \hat{\underline{n}} \wedge \sigma_{\underline{u}}\, \delta_P \quad , \tag{A.12}$$

$$\Delta u = \{\Delta u\} + \sigma_{du/dn}\, \delta_P + \text{div}(\hat{\underline{n}}\cdot\sigma_u \delta_P) \quad . \tag{A.13}$$

These formulae are fundamental [16] for grating theory. It is recommended to learn the first one by heart; it is easy to deduce from it the second and the third rules (we have only to change a product into a scalar product, then into a vector product); deducing the last one is a good exercise starting with $\Delta u = \text{div}(\text{grad } u)$. Of course all the notations must be perfectly understood; to be as clear as possible let us emphasize for instance that

{div u} is the distribution $\{\partial u_x/\partial x\} + \{\partial u_y/\partial y\}$,

{grad u} is the vector distribution with components $\{\partial u/\partial x\}$ and $\{\partial u/\partial y\}$, etc...

A.5 Expansion with Respect to the Basis $e_j(x) = \exp[i(nK+k \sin\theta)x] = \exp(i\alpha_n x)$

A.5.1 Theorem

Any distribution T (element of R') can be expanded in terms of the $e_j(x)$; the coefficients $t_n(y)$ of this expansion are Schwartz distributions (i.e., linear functionals acting on the functions $\phi(y)$ which are infinitely differentiable and vanish outside a bounded interval). More precisely

$$\text{if} \quad T \in R' \quad , \quad T = \sum_{n=-\infty}^{+\infty} t_n(y) e_n(x) \quad , \tag{A.14}$$

with

$$\forall\phi \in \mathcal{D} \quad , \quad < t_n(y), \phi(y) > \overset{\text{def}}{=} \frac{1}{d} < T, \phi(y)\overline{e_n}(x) > \quad , \tag{A.15}$$

$$\forall\psi \in R \quad , \quad < t_n(y) e_n(x), \psi(x,y) > \overset{\text{def}}{=} < t_n(y), \int e_n(x)\psi(x,y)dx > \quad . \tag{A.16}$$

---

16 They are also true in $\mathcal{D}'$ (i.e., for Schwartz distributions) but P is then any surface in the three-dimensional space.

Of course, it can be verified that if T is the distribution associated with a pseudo-periodic function u(x,y), the coefficients $t_n(y)$ are given by the formula

$$t_n(y) = u_n(y) = \frac{1}{d}\int u(x,y)\overline{e_n}(x)dx \quad , \qquad \text{(A.17)}$$

which is obtained easily by expanding the periodic function u(x,y) exp(-ikx sinθ) in a Fourier series.

A.5.2 Proof

We have to establish that, provided ψ belong to R

$$<T,\psi> = \left\langle \sum_{n=-\infty}^{+\infty} t_n(y)e_n(x),\psi(x,y)\right\rangle \quad ,$$

$$= \sum_{n=-\infty}^{+\infty} \left\langle t_n(y),\int e_n(x')\psi(x',y)dx'\right\rangle \quad ,$$

$$= \sum_{n=-\infty}^{+\infty} \left\langle T,\frac{1}{d}\overline{e_n}(x)\int e_n(x')\psi(x',y)dx'\right\rangle \quad .$$

We may proceed as follows

a) Expanding in a Fourier series, with respect to x, the function ψ(x,y) exp(ikx sinθ) it turns out that

$$\text{if} \quad \psi \in R \quad , \quad \psi(x,y) = \sum_{n=-\infty}^{+\infty} \psi_n(y)\overline{e_n}(x) \quad ,$$

with

$$\psi_n(y) = \frac{1}{d}\int \psi(x',y)e_n(x')dx' \quad .$$

b) Noting that the nth Fourier coefficient of an infinitely differentiable function tends to zero more rapidly than any power of 1/n, it is possible to show that the equality

$$\psi(x,y) = \lim_{N\to\infty} \sum_{n=-N}^{+N} \psi_n(y)\overline{e_n}(x)$$

is still true in R (i.e., using the notion of convergence associated with R).

c) We only have to apply T to both members of this equality to obtain the expected result.

A.5.3 Application to $\delta_{\mathcal{R}}$ [also denoted $\delta_{\mathcal{R}}(x,y)$]

We have to retain that

$$\delta_{\mathcal{R}} = \sum_{n=-\infty}^{+\infty} \frac{1}{d}\,\delta(y)e_n(x) \quad . \tag{A.18}$$

The demonstration is easy: we know that $\delta_{\mathcal{R}} = \sum_n t_n(y)e_n(x)$ with

$$\forall\phi(y) \in \mathcal{D} \quad , \quad <t_n(y),\phi(y)> = \frac{1}{d} < \delta_{\mathcal{R}},\phi(y)\overline{e_n}(x) > \quad ,$$
$$= \frac{1}{d}\,\phi(0) = \left\langle \frac{1}{d}\,\delta(y),\phi \right\rangle \quad .$$

This implies $t_n(y) = \frac{1}{d}\,\delta(y)$, where $\delta(y)$ is the Dirac distribution. Likewise, if we define $\delta_{\mathcal{R}}(x-a,y-b)$ by

$$\forall\psi(x,y) \in \mathcal{R} \quad , \quad <\delta_{\mathcal{R}}(x-a,y-b),\psi(x,y)> = \psi(a,b) \quad , \tag{A.19}$$

it can be shown that

$$\delta_{\mathcal{R}}(x-a,y-b) = \sum_{n=-\infty}^{+\infty} \frac{1}{d}\,\delta(y-b)\overline{e_n}(a)e_n(x) \quad . \tag{A.20}$$

A.6 Convolution

A.6.1 Memoranda on the Product of Convolution in $\mathcal{D}_1'$

By $\mathcal{D}_1'$ we mean the subspace of $\mathcal{D}'$ whose elements are functions of *one* variable. The reader is supposed to have heard of the convolution in this space and we will recall here some important results without proof.

If S and T are distributions of $\mathcal{D}_1'$ their convolution product $S * T$ is defined, subject to its existence, by

$$\forall\phi(y) \in \mathcal{D} \quad , \quad <S*T,\phi> = <S(x)T(y),\phi(x+y)> \quad , \tag{A.21}$$

where $S(x)T(y)$ is a tensorial product

$$<S(x)T(y),\phi(x+y)> = \left\langle S(x),<T(y),\phi(x+y)> \right\rangle = \left\langle T(y),<S(x),\phi(x+y)> \right\rangle \quad . \tag{A.22}$$

Likewise

$$\forall\phi(y) \in \mathcal{D} \quad , \quad <S*T*Y,\phi> = <S(x)T(y)U(z),\phi(x,y,z)> \quad . \tag{A.23}$$

Even if these definitions are forgotten the following properties must be known:

a) $S*T$ exists if at least one of the two distributions has a bounded support; then the product is commutative.

b) $S*T*U$ exists if at least two of the three distributions have a bounded support; then the product is associative.

c) The Dirac distribution is the unit of convolution

$$\forall T \in \mathcal{D}'_1 \quad , \quad \delta * T = T \quad . \tag{A.24}$$

d) The convolution by a derivative of δ is a differentiation

$$\delta' * T = T' \quad , \quad \delta'' * T = T'' \quad . \tag{A.25}$$

e) The convolution by $\delta(x-a)$ is a translation

$$\delta(y-a) * T(y) = T(y-a) \tag{A.26}$$

with

$$\forall \phi(y) \in \mathcal{D} \quad , \quad \langle T(y-a), \phi(y) \rangle \overset{\text{def}}{=} \langle T(y), \phi(y+a) \rangle \quad . \tag{A.27}$$

f) If $f(y)$ and $g(y)$ are locally integrable functions, $f*g$ is the function $h(y)$ defined by

$$h(y) = \int_{-\infty}^{+\infty} f(y')g(y-y')dy' = \int_{-\infty}^{+\infty} g(y')f(y-y')dy' \quad . \tag{A.28}$$

It must be noted as a consequence that to differentiate (or translate) a product of convolution we have only to differentiate (or translate) one or the other of the two factors.

A.6.2 Convolution in $\mathcal{R}'$

Definition.

Let $S = \sum_{n=-\infty}^{+\infty} s_n(y)e_n(x)$ and $T = \sum_{n=-\infty}^{+\infty} t_n(y)e_n(x)$ be two distributions of $\mathcal{R}'$, we define, subject to its existence, their convolution product by

$$S * T = d \sum_{n=-\infty}^{+\infty} s_n(y) * t_n(y)e_n(x) \quad . \tag{A.29}$$

$S*T$ is therefore also an element of $\mathcal{R}'$. Using this definition, which probably would appear as a property in a mathematics book [17], we are able to establish all the results we need for grating theory.

---

17 In mathematics, the product of convolution is generally defined using the tensorial product as recalled for $\mathcal{D}'_1$ in Sect.A.6.1.

Properties.

a) $\delta_{\mathcal{R}}$ is the unit of convolution in $\mathcal{R}'$

$$\forall T \in \mathcal{R}' \quad , \quad \delta_{\mathcal{R}} * T = d \sum_{n=-\infty}^{+\infty} \frac{1}{d}\,\delta(y) * t_n(y)e_n(x) = \sum_{n=-\infty}^{+\infty} t_n(y)e_n(x) = T \quad . \qquad (A.30)$$

b) If $G(x,y)$ is a function of $\mathcal{R}'$ and $\rho\delta_P$ the distribution defined previously, the product $\rho\delta_P * G$ is a function of x and y defined by

$$\rho\delta_P * G = \int \rho(x')G[x-x',y-f(x')]d\ell' \quad , \qquad (A.31)$$

where $M'[x',f(x')]$ is a point on P and $d\ell'$ the associated length element.

Here is a brief and formal demonstration of this important result. We first expand $\rho\delta_P$ and G as explained in Sect.A.5.1. From (A.14,15,17) we get

$$\rho\delta_P = \sum_n t_n(y)e_n(x) \quad , \quad \text{with} \quad < t_n(y),\phi(y) > = \frac{1}{d}\int \rho(x')\phi[f(x')]\overline{e_n}(x')d\ell' \quad ,$$

$$G(x,y) = \sum_n g_n(y)e_n(x) \quad , \quad \text{with} \quad g_n(y) = \frac{1}{d}\int G(x,y)\overline{e_n}(x)dx \quad .$$

The sought result is a consequence of these two expansions and of definitions previously given:

$$\forall\psi \in \mathcal{R}\,, < G * \rho\delta_P,\psi > = \left\langle d\sum_n t_n(y) * g_n(y)e_n(x),\psi\right\rangle \quad , \qquad \text{from (A.29)},$$

$$= d\sum_n \left\langle t_n(y) * g_n(y), \int e_n(x)\psi(x,y)dx\right\rangle \quad , \qquad \text{from (A.16)},$$

$$= d\sum_n \left\langle t_n(z)g_n(y),\int e_n(x)\psi(x,y+z)dx\right\rangle \quad , \qquad \text{from (A.21)},$$

$$= d\sum_n \left\langle t_n(z),\int_{-\infty}^{+\infty}\int g_n(y)e_n(x)\psi(x,y+z)dy\,dx\right\rangle \quad , \qquad \text{from (A.22)},$$

$$= d\sum_n \left\langle t_n(z),\int_{-\infty}^{+\infty}\int g_n(Y-z)e_n(x)\psi(x,Y)dY\,dx\right\rangle \quad ,$$

$$= d\sum_n \frac{1}{d}\int \rho(x')\left[e_n(x')\int_{-\infty}^{+\infty}\int g_n[Y-f(x')]e_n(x)\psi(x,Y)dY\,dx\right]d\ell' \quad , \qquad \text{from (A.15)},$$

$$= \left\langle \int \rho(x')\sum_n e_n(x-x')g_n[Y-f(x')]d\ell',\psi(x,Y)\right\rangle \quad ,$$

$$= \left\langle \int \rho(x')G[x-x',y-f(x')]d\ell',\psi(x,y)\right\rangle \quad .$$

c) In order to differentiate a convolution product in $\mathcal{R}'$ we have only to differentiate one or the other of the two factors. This is obvious for the differentiation with respect to x

$$\frac{\partial S}{\partial x} = \sum_n \frac{\partial}{\partial x}\left[s_n(y)e_n(x)\right] = \sum_n ik\alpha_n s_n(y)e_n(x) \quad ,$$

$$\frac{\partial}{\partial x}(S*T) = d\sum_n \frac{\partial}{\partial x}\left[s_n(y)*t_n(y)e_n(x)\right] \quad ,$$

$$= d\sum_n ik\alpha_n s_n(y) * t_n(y)e_n(x) = \frac{\partial S}{\partial x} * T \quad .$$

For the case of the differentiation with respect to y, we have only to note that the result holds in $\mathcal{D}_1'$ as recalled in Sect.A.6.1

$$\frac{\partial S}{\partial y} = \sum_n \frac{ds_n}{dy} e_n(x) \quad ,$$

$$\frac{\partial}{\partial y}(S*T) = d\sum_n \frac{\partial}{\partial y}\left[s_n(y)*t_n(y)e_n(x)\right] \quad ,$$

$$= d\sum_n \frac{\partial}{\partial y}\left[s_n(y)*t_n(y)\right]e_n(x) \quad ,$$

$$= \sum_n \frac{ds_n}{dy} * t_n e_n(x) = \frac{\partial S}{\partial y} * T \quad .$$

d) The previous result can be extended easily to the more general case when a differential linear operator L with constant coefficients is applied to a convolution product

$$L(S*T) = LS * T = S * LT \quad . \tag{A.32}$$

e) As a consequence of c) it turns out that if G is a function, the convolution product $G*\mathrm{div}(\rho\hat{\underline{n}}\delta_P)$ is also given by a curvilinear integral on P

$$G * \mathrm{div}(\rho\,\hat{\underline{n}}\,\delta_P) = G * \frac{\partial}{\partial x}(\rho\, n_x\,\delta_P) + G * \frac{\partial}{\partial y}(\rho\, n_y\,\delta_P) \quad ,$$

$$= \frac{\partial G}{\partial x} * \rho\, n_x\,\delta_P + \frac{\partial G}{\partial y} * \rho\, n_y\,\delta_P \quad ,$$

$$= \int \rho(x')\hat{\underline{n}}(x') \cdot \mathrm{grad}\, G[x-x', y-f(x')]\,\frac{d\ell'}{dx'}\,dx' \quad ,$$

$$= \int \rho(M')\,\frac{dG}{dn}(M-M')\,d\ell' \quad , \tag{A.33}$$

where M(x,y) is the field point and M'[x',f(x')] is a point on P. We have to emphasize that the notation grad G[x-x',y-f(x')] is ambiguous; in fact we have first to compute grad G (this is clear!) and then to replace x by x-x' and y by y-f(x'). The same remark can be made about the normal derivative dG(M-M')/dn.

2. Some Mathematical Aspects of the Grating Theory

M. Cadilhac

This chapter deals with mathematical considerations needed for the understanding of grating problems: a survey of some properties of the solutions of the Helmholtz equation, of uniqueness and reciprocity theorems, and of the foundations of Yasuura's improved point-matching technique.

Many aspects of these questions rest upon hard functional analysis, and for this reason they are beyond our scope. Our only purpose is to say what a physicist has to know about these topics and give some simple proofs for a selected number of situations. Moreover, we shall only consider Dirichlet boundary conditions (corresponding to the $E_{\parallel}$ polarization).

2.1 Some Classical Properties of the Helmholtz Equation

The Helmholtz equation (steady-state wave equation)

$$\Delta u + k^2 u = 0 \tag{2.1}$$

is a linear elliptic partial differential equation. The main consequence is that if k^2 is constant (or analytic) in a domain, every continuous solution is *analytic* in this domain (an analytic function of two variables x and y is not to be confused with an analytic function of $z = x + iy$), so that if $u = 0$ in a subdomain, it can be inferred that $u = 0$ in the whole domain [2.1-4].

The most classical problem associated with the Helmholtz equation is the following boundary value problem: To find a function u which satisfies the Helmholtz equation in a domain Ω, and takes preassigned values on the boundary $\partial\Omega$ (Dirichlet problem). k^2 is supposed constant in Ω, the boundary values are supposed to be sufficiently regular. The solution is sought in a so-called Sobolev space: the space of square integrable functions, with generalized derivatives of order 1 and 2, which are also square integrable (in Ω).

If Ω is bounded, the result is simple: there is one (existence) and only one (uniqueness) solution, if k^2 does not belong to a discrete infinite set of values, i.e., the elements of the spectrum of $-\Delta$ (which is an unbounded operator with dense domain).

But the scattering problems correspond to physical contexts where Ω is not bounded. The situation is well understood if the boundary $\partial\Omega$ is bounded, [2.5]. In this case, the spectrum of $-\Delta$ is the set of all positive numbers, so that if k^2 is negative or strictly complex, there is one and only one solution (in the Sobolev space). In actual problems, k^2 is generally positive and the situation is more complicated. Yet, one can arrive at an existence and uniqueness theorem for the solution, on two conditions:

a) Replace the square integrability in Ω by the local square integrability in $\bar{\Omega}$ for the function and its derivatives ($\bar{\Omega} = \Omega + \partial\Omega$).

b) Impose a restriction on the behavior of the solution at infinity, namely the famous Sommerfeld radiation condition [2.5].

2.2 The Radiation Condition for the Grating Problem

In the grating problem, the domain Ω where the Helmholtz equation has to be solved is defined by $y > f(x)$, if $y = f(x)$ as usual defines the grating profile $\mathscr{P}$. We consider the case of infinite conductivity and $E_{\parallel}$. Ω and $\partial\Omega$ (that is $\mathscr{P}$) are now both unbounded, and this accounts for serious complications. For a general analysis of this type of problems, see [2.7,8].

The first difficulty encountered is to express the condition at infinity properly: since we can now reach infinity without leaving the vicinity of the boundary, it is no longer possible to keep the conventional Sommerfeld condition.

Let us consider, in the half plane $y > \max f(x)$, a function u with the three assumptions:

a) u is bounded for large y.

b) u is a solution of Helmholtz equation with real k^2: $\Delta u + k^2 u = 0$.

c) For fixed y, u can be Fourier-analyzed as a function of x, in some sense (u is for example square integrable in x; more generally, it is a tempered distribution).

Let us write, at least formally

$$y > \max(f) \qquad u(x,y) = (2\pi)^{-1/2} \int_{-\infty}^{+\infty} U(\alpha,y) \exp(i\alpha x)dx \quad . \tag{2.2}$$

From the Helmholtz equation we have

$$\frac{\partial^2 U(\alpha,y)}{\partial y^2} + (k^2 - \alpha^2)\, U(\alpha,y) = 0 \quad ,$$

so that $U(\alpha,y)$ has the form

$$U(\alpha,y) = U_+(\alpha) \exp[i\beta(\alpha)y] + U_-(\alpha) \exp[-i\beta(\alpha)y] \quad , \tag{2.3}$$

where $\beta^2(\alpha) = k^2 - \alpha^2$, with $\beta(\alpha)$ or $\beta(\alpha)/i$ positive. Since u is bounded, we must have $U_-(\alpha) = 0$ for $|\alpha| > k$ and u(x,y) is a sum of three terms which can be interpreted as an incoming wave, an outgoing wave and an evanescent wave

$$u = u_{in} + u_{out} + u_{ev} \quad , \tag{2.4}$$

where

$$u_{in}(x,y) = (2\pi)^{-1/2} \int_{-k}^{+k} U_-(\alpha) \exp[i\alpha x - i\beta(\alpha)y]d\alpha \quad , \tag{2.5}$$

$$u_{out}(x,y) = (2\pi)^{-1/2} \int_{-k}^{+k} U_+(\alpha) \exp[i\alpha x + i\beta(\alpha)y]d\alpha \quad , \tag{2.6}$$

$$u_{ev}(x,y) = (2\pi)^{-1/2} \int_{|\alpha| > k} U_+(\alpha) \exp[i\alpha x + i\beta(\alpha)y]d\alpha \quad . \tag{2.7}$$

We shall say that u satisfies the radiation condition (or the outgoing wave condition) if $u_{in} = 0$ or, equivalently, $U_-(\alpha) = 0$.

Of course, the indices + and - have to be exchanged in the preceding definitions if the domain considered is below the profile.

2.3 A Lemma

Let us consider two functions u and v defined in the domain $y > f(x)$, with the following properties:

a) They are bounded for large y.
b) They are solutions of a Helmholtz equation, with real k^2.
c) They are square integrable in x and locally square integrable in y.
d) They assume, almost evereywhere on the profile $\mathscr{P}$, square integrable boundary values and normal derivatives.

Let us define the following sesquilinear functional:

$$W_{\mathscr{P}}(u,v) = \frac{i}{2} \int_{\mathscr{P}} \left(u \frac{d\bar{v}}{dn} - \bar{v} \frac{du}{dn}\right)ds \quad , \tag{2.8}$$

where the bar indicates the complex conjugate.

Then we also have (this is our fundamental lemma)

$$W_{\mathscr{P}}(u,v) = \int_{-k}^{+k} \beta(\alpha) \, [U_+(\alpha)\overline{V_+(\alpha)} - U_-(\alpha)\overline{V_-(\alpha)}]d\alpha \quad , \tag{2.9}$$

where U_+; V_+, U_-, V_- are the outgoing and incoming amplitudes of U and V, such as they are defined in the preceeding section.

Proof: take $y_1 > \max f(x)$ and transform

$$\iint\limits_{f(x) < y < y_1} (u \, \Delta\bar{v} - \bar{v} \, \Delta u) \, dx \, dy$$

(which is zero according to the Helmholtz equation), using the Green formula[1]. We get

$$W_{\mathscr{P}}(u,v) = \frac{i}{2} \int_{-\infty}^{+\infty} \left(u \frac{\partial \bar{v}}{\partial y} - \bar{v} \frac{\partial u}{\partial y}\right) dx \qquad \text{for} \quad y = y_1 \quad .$$

Then the Plancherel-Parseval theorem gives the wanted result [if we take the conditions $U_-(\alpha) = V_-(\alpha) = 0$ for $|\alpha| > k$] into account.

2.4 Uniqueness Theorems

We shall not discuss the existence of the solution of the grating problem, although it must not be considered as a futile matter: mathematical existence is not a consequence of physical evidence. But it is not possible to discuss the existence problems by using only simple arguments. See for example [2.5,6] for the "limiting absorption method".

We simply intend to give an elementary proof of uniqueness in two particular cases, both with $E_{\parallel}$. The other polarization case is not necessarily a straightforward extension.

2.4.1 Metallic Grating, with Infinite Conductivity

If two solutions of the Helmholtz equation for $y > f(x)$ assume the same boundary value for $y = f(x)$ and both satisfy the radiation condition, they are identical.

It is sufficient to prove that a solution u of the Helmholtz equation, a) with null incoming amplitude, b) vanishing on the boundary, vanishes evereywhere. Take first $v = u$ and apply the fundamental lemma, Sect.2.3. This gives

$$\int_{-k}^{+k} \beta(\alpha) \, |U_+(\alpha)|^2 \, d\alpha = 0 \quad ,$$

so that $U_+(\alpha) = 0$ for $|\alpha| < k$, at least almost everywhere. Take now $v = \partial u/\partial y$ and apply the lemma again. The result is

$$W_{\mathscr{P}}\left(u, \frac{\partial u}{\partial y}\right) = -\frac{1}{2} \int_{\mathscr{P}} n_y \left|\frac{du}{dn}\right|^2 ds = 0 \quad ,$$

where n_y is the y component of the normal unit vector on the profile, oriented upward. With $n_y > 0$, $du/dn = 0$. Having $u = 0$ and $du/dn = 0$ on the profile, we can

1 From c) and d), the Green formula is valid, although the domain is not bounded.

conclude that $u = 0$ in some neighborhood of $\mathscr{P}$ (theorem of Holmgren [2.1]). Since u is analytic, $u = 0$ everywhere in the domain $y > f(x)$. Q.E.D.

It is important to note that the proof is valid if the lemma holds; in particular, the local square integrability is a necessary assumption. This is related to the edge conditions in the case of profiles with angular points [2.5,9], i.e., the famous Meixner conditions.

2.4.2 Dielectric Grating

The uniqueness problem can be given the following form:

u satisfies the equations

a) $\Delta u + k_1^2 u = 0$ above the profile $y > f(x)$,

b) $\Delta u + k_2^2 u = 0$ below the profile $y < f(x)$, (k_1, k_2 real different).

c) u and du/dn are continuous across the profile.

d) u has only outgoing and evanescent components with amplitudes $U_+^{(1)}(\alpha)$ and $U_-^{(2)}(\alpha)$, respectively, above and below $\mathscr{P}$.

To show that $u = 0$.

Using the lemma 2.3 with $v = u$, we get

$$W_{\mathscr{P}}(u,u) = \int_{-k_1}^{+k_1} \beta_1(\alpha)\, |U_+^{(1)}(\alpha)|^2\, d\alpha \quad \text{with} \quad \beta_1(\alpha) = \sqrt{k_1^2 - \alpha^2} \quad ,$$

and also

$$W_{\mathscr{P}}(u,u) = -\int_{-k_2}^{+k_2} \beta_2(\alpha)\, |U_-^{(2)}(\alpha)|^2\, d\alpha \quad \text{with} \quad \beta_2(\alpha) = \sqrt{k_2^2 - \alpha^2} \quad ;$$

so that $U_+^{(1)}(\alpha) = 0 \quad \text{for} \quad |\alpha| < k_1$,

$$U_-^{(2)}(\alpha) = 0 \quad \text{for} \quad |\alpha| < k_2 \quad .$$

Let us now take $v = \partial u/\partial y$. Applying the lemma again, we obtain $W_{\mathscr{P}}(u,v) = 0$. We note that the value of $W_{\mathscr{P}}$ does not depend on the side of $\mathscr{P}$ considered, although dv/dn has a jump

$$\left(\frac{dv}{dn}\right)_+ - \left(\frac{dv}{dn}\right)_- = \left(k_2^2 - k_1^2\right) u n_y$$

as can be seen for example by tyking the y derivative of the Helmholtz equation in the sense of distributions. Considering the corresponding jump of $W_{\mathscr{P}}$, we obtain

$$\int_{\mathscr{P}} |u[x,f(x)]|^2 dx = 0 \quad .$$

Hence $u = 0$ on $\mathscr{P}$ and we are left with the conditions of Sect.2.4.1. Therefore $u = 0$ everywhere. Q.E.D.

2.5 Reciprocity Relations

The lemma 2.3 can also be used to establish the very important reciprocity relations. To illustrate this point, let us consider the case of a dielectric grating, and two functions u and v satisfying

a) the Helmholtz equation

$$\Delta u + k_1^2 u = 0 \qquad \Delta v + k_1^2 v = 0 \qquad \text{for} \quad y > f(x) \quad ,$$

$$\Delta u + k_2^2 u = 0 \qquad \Delta v + k_2^2 v = 0 \qquad \text{for} \quad y < f(x) \quad ,$$

b) u, v, du/dn, dv/dn are continuous across the profile y = f(x).

Let, for $|\alpha| < k$, $U_-^{(1)}(\alpha)$, $U_+^{(2)}(\alpha)$ be the incoming amplitudes corresponding to the function u in each medium, $U_+^{(1)}(\alpha)$, $U_-^{(2)}(\alpha)$ the outgoing amplitudes of u, and $V_-^{(1)}(\alpha)$, $V_+^{(2)}(\alpha)$, $V_+^{(1)}(\alpha)$, $V_-^{(2)}(\alpha)$ the analogous amplitudes of v. Then the lemma gives two expressions of $W_{\mathscr{P}}(u,\bar{v})$

$$W_{\mathscr{P}}(u,\bar{v}) = \int_{-k_1}^{+k_1} \beta_1(\alpha)\, [U_+^{(1)}(\alpha)\, V_-^{(1)}(-\alpha) - U_-^{(1)}(\alpha)\, V_+^{(1)}(-\alpha)]d\alpha \tag{2.10}$$

and the same expression with the index 2 instead of 1.

To obtain these expressions, we have to point out that the amplitudes of $\bar{v}$ are deduced from those of v by successively

a) taking the complex conjugate,

b) changing α into $-\alpha$,

c) changing the index + into - and vice versa.

Therefore, the general reciprocity relation can be written

$$\begin{aligned} &\int_{-k_1}^{+k_1} \beta_1(\alpha)\, [U_+^{(1)}(\alpha)\, V_-^{(1)}(-\alpha) - U_-^{(1)}(\alpha)\, V_+^{(1)}(-\alpha)]d\alpha \\ &= \int_{-k_2}^{+k_2} \beta_2(\alpha)\, [U_+^{(2)}(\alpha)\, V_-^{(2)}(-\alpha) - U_-^{(2)}(\alpha)\, V_+^{(2)}(-\alpha)]d\alpha \quad . \end{aligned} \tag{2.11}$$

Let us give a particular application, namely the reciprocity by transmission. Let us suppose that u has only one incoming amplitude $U_{in}(\alpha) = U_-^{(1)}(\alpha)$ so that $U_+^{(2)}(\alpha) = 0$. Similarly, let us put $V_{in}(\alpha) = V_+^{(2)}(\alpha)$ and $V_-^{(1)}(\alpha) = 0$, and also $U_{out}(\alpha) = U_-^{(2)}(\alpha)$ and $V_{out}(\alpha) = V_+^{(1)}(\alpha)$. Then, the reciprocity relation is

$$\int_{-k_2}^{+k_1} \beta_1(\alpha)\, U_{in}(\alpha)\, V_{out}(-\alpha)\, d\alpha = \int_{-k_2}^{+k_2} \beta_2(\alpha)\, U_{out}(\alpha)\, V_{in}(-\alpha)\, d\alpha \quad . \tag{2.12}$$

In the case of unit plane incident waves

$$U_{in}(\alpha) = \delta(\alpha - \alpha_1)$$

$$V_{in}(\alpha) = \delta(\alpha - \alpha_2)$$

the relation becomes

$$\beta_1(\alpha_1)\, V_{out}(-\alpha_1) = \beta_2(\alpha_2)\, U_{out}(-\alpha_2) \quad . \tag{2.13}$$

Under this simple form, the physical interpretation is clear. Other particular applications of the general reciprocity relation could be easily derived. See also [2.13,14].

It is interesting to emphasize that for the couple $(u,\bar{v})$ the lemma is valid even if k_2^2 is complex, so that the *reciprocity theorem* can be applied to the case of *absorbing media* as well.

2.6 Foundation of the Yasuura Improved Point-Matching Method [2.15]

It has been shown in Sect.1.2.8 that the Yasuura method is based on the convergence of least square approximants of $s(x) = -E^i[x,f(x)]$ by linear combinations of $\phi_n(x) = \exp[i\alpha_n x + i\beta_n f(x)]$ (the Rayleigh functions). More precisely, if

$$s_N(x) = \sum_{-N}^{+N} Y_n(N)\, \phi_n(x)$$

is the best least square approximant of order N of s(x), the principles of Yasuura's method are

$$\text{a)} \lim_{N\to\infty} \int_0^d |s(x) - s_N(x)|^2\, d\ell = 0 \quad , \tag{2.14}$$

and

$$\text{b)} \lim_{N\to\infty} Y_n(N) = B_n \text{ (the wanted quantities)} \quad . \tag{2.15}$$

This happens because the ϕ_n have a very particular property, namely they are members of a topological basis. This notion is wellknown for orthogonal systems. But the situation is much more complicated in our case, where the ϕ_n are not orthogonal in general.

2.6.1 Definition of a Topological Basis

Let us consider in general a topological vector space (that is an abstract set which is linear and in which some convenient convergence notion has been defined). A family of vectors $\underline{e}_n$ (elements of the set) is said to be *total* (or *complete*) if every vector $\underline{x}$ can be represented as a limit of linear combinations of the $\underline{e}_n$, i.e., if there exist sets of coefficients $C_n(N)$ such that

$$\underline{x} = \lim_{N\to\infty} \sum_{n=0}^{N} C_n(N)\, \underline{e}_n \quad . \tag{2.16}$$

As an example, in the set of all complex continuous functions on the interval [0,1], with the topology of the uniform convergence, the family of the $\{x^n\}$ (n = 0, 1, 2, etc...) is total (this is a famous theorem of Weierstrass').

If one also has the property that the $C_n(N)$ have a limit C_n when $N\to\infty$, the family is said to be a topological basis. It can be shown that, in the preceding example, the $\{x^n\}$ are *not* such a topological basis.

It is also important to know that, in general, for a topological basis, one has *not*

$$\underline{x} = \sum C_n \underline{e}_n \quad ,$$

the $\sum$ indicating a series. The special case of orthogonal bases in Hilbert space is an exception.

2.6.2 The System of Rayleigh Functions is a Topological Basis

We first intend to establish that the family of Rayleigh functions $\phi_n(x) = \exp[i\alpha_n x + i\beta_n f(x)]$ is total in the space of square integrable pseudo-periodic functions for the L_2 topology. It is easily shown, from the theory of Hilbert spaces, that this is equivalent to the following property: no function, except the null function, is orthogonal to all ϕ_n. We first assert that

if, for some function ψ, we put

$$J(\alpha) = \int_{-\infty}^{+\infty} \overline{\psi(x)} \exp[i\alpha x + i\beta(\alpha)f(x)]dx \tag{2.17}$$

and if $J(\alpha) = 0$ for any real α, then $\psi(x) = 0$, at least almost everywhere.

Formal proof: The function

$$g(x,y) = \int_{-\infty}^{+\infty} [4i\pi\beta(\alpha)]^{-1} \exp[i\alpha x + i\beta(\alpha)|y|]d\alpha \tag{2.18}$$

is the solution of $(\Delta + k^2)g(x,y) = \delta(x)\delta(y)$ and satisfies the outgoing wave condition [in fact, $g(x,y) = (4i)^{-1} H_0^+(kr)$].
If we now consider the function

$$\Phi(x,y) = \int_{-\infty}^{+\infty} \overline{\psi(x')}\, g[x - x', y - f(x')]dx' \tag{2.19}$$

we have

$$(\Delta + k^2)\Phi(x,y) = \overline{\psi(x)}\, \delta[y - f(x)] \tag{2.20}$$

so that ϕ is continuous everywhere.

On the other hand, we can easily see that, for $y < \min f(x)$,

$$\phi(x,y) = \int_{-\infty}^{+\infty} [4i\pi\beta(\alpha)]^{-1} \exp[-i\alpha x - i\beta(\alpha)y]\, J(\alpha)\, d\alpha \tag{2.21}$$

and, from the hypothesis $J(\alpha) = 0$, we get $\phi(x,y) = 0$ there. Since $\phi(x,y)$ is analytic in the domain $y < f(x)$ (see Sect.2.1), $\phi(x,y) = 0$ in this domain, and, from the continuity $\phi[x,f(x)] = 0$. Since ϕ satisfies the radiation condition, we can use the uniqueness theorem of Sect.2.4.1 to deduce that $\phi(x,y) = 0$ for $y > f(x)$. Finally, $\phi = 0$ everywhere. By applying the operator $\Delta + k^2$, we get $\psi = 0$ almost everywhere.

If ψ is pseudo-periodic, $J(\alpha)$ can be written

$$J(\alpha) = K \sum_n \delta(\alpha - \alpha_n) \int_0^d \overline{\psi(x)} \exp[i\alpha_n x + i\beta_n f(x)]dx \quad . \tag{2.22}$$

If we assume that, for each n,

$$\int_0^d \overline{\psi(x)} \, \phi_n(x) \, dx = 0 \quad ,$$

we have $J(\alpha) = 0$ and therefore $\psi = 0$ Q.E.D.

That the $Y_n(N)$ tend to B_n when s_N tends to s is a direct consequence of the fact that the B_n depend continuously on s (for the L_2) topology). Although this property seems very natural from a physical point of view, it is in close relation with the existence theorems and will not be further discussed. See also [2.11].

2.6.3 The Convergence of the Rayleigh Series; A Counterexample

The Rayleigh series $\sum B_n \phi_n(x)$ cannot be expected in general to be convergent for x fixed, even for regular profiles. PETIT and CADILHAC [2.10] have shown that if $f(x) = h \cos(Kx)$ and $Kh > 0.448$ the Rayleigh series as a matter of fact is not convergent. The proof is carried out by assuming the contrary, showing that the sum could be extended to complex values of x in a domain containing points with an arbitrarily large imaginary part, and by exhibiting a contradiction at infinity. The details are the following: Let us assume that

$$\sum_{-\infty}^{+\infty} B_n \exp[i\alpha_n x + i\beta_n h \cos(Kx)] = -\exp[-i\beta_0 h \cos(Kx)] \tag{2.23}$$

for any x. Then, considering the case $Kx = \pi$, we get

$$\lim_{|n|\to\infty} |B_n| \exp(nu) = 0 \quad ,$$

with $u = hK$. Hence the Rayleigh series converges for complex $Kx = \xi + i\eta$ to analytic function in the open set defined by

$$u - |\eta| + u \cos(\xi) \cosh(\eta) > 0 \quad . \tag{2.24}$$

This set is connected if $u > 0.448$, and in this case the limit must be equal to $-\exp[-ikh \cos(Kx)]$ in the whole set. But this appears to be in contradiction with the asymptotic behavior for $\eta \to \infty$.

It has been shown by MILLAR [2.11] that the Rayleigh series converges when this contradiction does not hold. VAN DEN BERG and FOKKEMA have analytically investigated the validity of the Rayleigh hypothesis for general analytical profiles [2.12].

References

2.1 R. Courant, D. Hilbert: *Methods of Mathematical Physics*, Vol. 2 (Interscience, New York 1962)
2.2 C. Miranda: *Partial Differential Equations of Elliptic Type* (Springer, Berlin, Heidelberg, New York 1970)
2.3 L. Hörmander: *Linear Partial Differential Equations* (Springer, Berlin, Heidelberg, New York 1964)
2.4 S. Agmon: *Lectures on Elliptic Boundary Value Problems* (Van Nostrand Reinhold, New York 1965)
2.5 C.H. Wilcox: *Scattering Theory for the d'Alembert Equation in Exterior Domains*, Lecture Notes in Methematics, Vol. 442 (Springer, Berlin, Heidelberg, New York 1975)
2.6 B.R. Vainberg: Russ. Math. Surveys 21, No. 3 (May/June 1966)
2.7 J.C. Guillot, C.H. Wilcox: C.R. Acad. Sci. Ser. A 282, 1171 (17 May 1876)
2.8 J.C. Guillot, C.H. Wilcox: Math. Z. 160, 89 (1978)
2.9 J. Meixner: IEEE Trans. AP-20 (4), 442 (1972)
2.10 R. Petit, M. Cadilhac: C.R. Acad. Sci. 262, 468 (14 Feb. 1966)
2.11 R.F. Millar: Radio Sci. 8, (8,9) 785 (Aug./Sept. 1973)
2.12 P.M. Van den Berg, J.T. Fokkema: J. Opt. Soc. Am. 69, 27 (1979)
2.13 D. Maystre, R.C. MacPhedran: Opt. Commun. 12, 164 (1974)
2.14 M. Nevière, P. Vincent: Opt. Acta 23, 557 (1976)
2.15 K. Yassura, T. Itakura: Kyushu Univ. Tech. Rpt. 38, (1) 72 (1965); 38, (4) 378 (1965); 39, (1) 51 (1966)

3. Integral Methods

D. Maystre

With 8 Figures

In this chapter, we class as an integral method any theory able to reduce in a rigorous manner a problem of diffraction by a grating to the resolution of a linear integral equation or a system of coupled linear integral equations. According to the mathematicians [3.1], an integral equation of the second kind may be written in the form

$$u(s) = v(s) + \lambda \int_b^a W(s,t)\, u(t)\, dt \quad , \tag{3.1}$$

where the kernel W(s,t) is a function of the two variables s and t, defined and continuous in the interval (a,b), λ is a parameter, u(s) and v(s) are two functions continuous in the same interval. The problem is to determine u(s) when W(s,t) and v(s) are known. Note that, under some conditions, this definition may be extended to the case where W(s,t), u(s) and v(s) are piecewise continuous or even singular. An integral equation of the first kind may be derived from (3.1) by replacing the left-hand member by zero. The reader interested in the study of this type of equation may consult [3.1] for the theory or [3.2] for the applications in physics. Our purpose is essentially to show that the use of applied mathematics makes it possible to concentrate in an integral equation a problem of diffraction by a grating. We shall indicate also how such an equation may be numerically solved by computer. The notations are those of Chap. 1 and we frequently employ the elements of the theory of distributions of SCHWARTZ [3.3,4] adapted to the grating problem as is explained in Appendix A.

3.1 Development of the Integral Method

The integral methods were the first rigorous methods used for solving the problem of diffraction by a perfectly conducting grating. They were proposed practically at the same time by PETIT and CADILHAC [3.5], WIRGIN [3.6], and URETSKI [3.7]. To our knowledge, the first numerical implementation of the theory was achieved by PETIT [3.8,9] in TE polarization, i.e., with the incident electric field parallel to the grooves[1]. However, in the other fundamental case of polarization, the non-integrability of the kernel of the integral equation necessitated a "renormalization" unjustified from a mathematical point of view. In fact, it was not until 1967 that the first rigorous integral method for the TM case of polarization was proposed by PAVAGEAU et al. [3.10].

After these early studies, one might have thought that all the questions of practical interest were solved. Indeed, since the gratings were essentially used in the visible region, research people generally thought that the aluminum coating on the grating surface might be accurately considered as perfectly conducting. This is certainly one of the reasons why the later studies devoted to the more complicated problem of the dielectric or metallic grating were few in number. The first relevant works, proposed by WIRGIN [3.11], NEUREUTHER and ZAKI [3.12] and VAN DEN BERG [3.13], led these authors to the resolution of two coupled integral equations, containing two unknown functions. Bearing in mind limited memory storage of computers, the solving of such a system involves substantial new difficulties. As far as we know, the relevant published results were practically limited to the case of dielectric gratings. At the same time, the development of spatial optics and the rapid progress of the grating technology involved the use of gratings in the far and extreme ultraviolet. In these regions where the metals are poorly conducting, the model of perfectly conducting grating must obviously be abandoned. It is the reason why a new approach of the problem of dielectric or metallic grating was proposed in 1972 by MAYSTRE [3.14,15]. The feature of the new method was to lead to only one integral equation, containing one unknown function. After adoption of some new numerical methods for summation and integration of the kernel, the resulting computer program proved able to cover the whole spectrum, for any polarization and for any shape of the grating surface [3.16]. The first numerical results obtained in the visible region with this program clearly showed the inadequacy of the model of the perfectly conducting grating [3.17].

This surprising fact is certainly at the origin of the further development of the integral method. Indeed, the simultaneous development of the differential method by NEVIERE et al. [3.18] led many research people to the opinion that this last approach, considerably simpler from a theoretical point of view and more easily implemented numerically, would be better adapted and more powerful than the integral method. In fact, the differential approach which also proved to be of immense value is plagued by considerable numerical difficulties in the visible and infrared regions, because of the great conductivity of metals such aluminum in these regions. This explains the recent efforts to study conical diffraction mountings [3.16] and especially to deal with dielectric coated gratings [3.19-21], buried

---

1 Also called S or $\underline{E}_{\parallel}$ polarization in Chap. 1.

gratings and bimetallic gratings [3.22], using integral theories. Nowadays, the computer programs issued from the integral methods are able to deal with virtually any problem of practical interest in the field of diffraction gratings. They are therefore widely utilized and moreover, they are the only ones which work in some cases of great practical importance.

3.2 Presentation of the Problem and Intuitive Description of an Integral Approach

3.2.1 Presentation of the Problem

Figure 3.1 represents the most general periodic structure studied in this chapter, which is called multiprofile grating. We consider a rectangular coordinate system Oxyz and Q cylindrical surfaces P_q of arbitrary shape having the same period d. For simplicity's sake, we suppose that P_q is given by a function $y = f_q(x)$. In the simplest case where Q = 1, we denote the only profile by P, whose equation is $y = f(x)$. Assuming a time dependence in $\exp(-i\omega t)$, we call ν_q the complex index of the material filling the region M_q which lies between P_{q-1} and P_q. The index ν_1 is generally equal to 1 and the lowest material filling M_{Q+1} which is in practice a metal or a dielectric, may be perfectly conducting.

In M_1, an electromagnetic monochromatic plane wave with unit amplitude strikes the grating with a wave vector $\underline{k}_1$ ($|\underline{k}_1| = k_1 = \nu_1 k_0 = 2\pi\nu_1/\lambda_0$, where λ_0 is the wavelength in the vacuum). Throughout this chapter, except in Sect.3.7, $\underline{k}_1$ is assumed to lie in the Oxy plane.

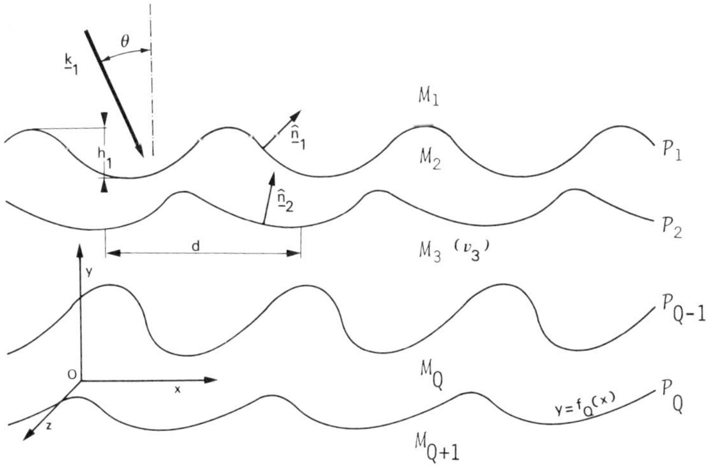

Fig. 3.1. Presentation of the general problem. The interface P_q, whose equation is $y = f_q(x)$, has a period d and a modulation depth h_q, ν_q denotes the complex index of the material filling M_q

The problem is to find the electromagnetic field (depending on the polarization of the incident wave) at any point of the space.

3.2.2 Intuitive Description of an Integral Approach

The first step of an integral approach consists of expressing the fields at any point P of the space in terms of integrals on P_q. These integrals contain a finite set of L well-chosen unknown functions $\phi_\ell (\ell = 1,\ldots,L)$ defined on P_q. Thus, the problem of the determination of the fields at any point of the space is reduced to the determination of L functions of x. It is worth noting that the fields so obtained must satisfy both the Maxwell equations in any region M_q and the outgoing wave conditions (OWC). Throughout this chapter, we assume also that these fields satisfy the integrability condition of MEIXNER [3.23] when P_q has edges[2]. The OWC, defined in Chaps. 1 and 2, states first that in the region of M_1 located above the grooves, the diffracted field, i.e., the difference between total and incident field, must go away from P_1 and remain finite when $y \to \infty$. Secondly, a similar condition is needed in M_{Q+1} for the total field under the grooves of P_Q. It must be noted that, depending on the different authors, the mathematical tools used to establish such integral expressions are of great variety: formalism of Green's functions [3.6,12], formalism of potentials [3.24], and so one. We consider that the theory of distributions, which combines rigor and versatility, permits us to describe in a synthetic manner all the integral theories [3.3,25].

After expressing the field in terms of the L unknown functions ϕ_ℓ, the last step is to determine all these functions. To this aim, one must first use the integral expressions of the field in order to express in terms of the functions ϕ_ℓ the limit value of the field and its normal derivative on both sides of any surface P_q. Then, writing the boundary conditions on these interfaces provides a set of coupled integral equations whose unknown are the functions ϕ_ℓ. Naturally, the determination of the L unknown functions requires L integral equations.

Thus, this intuitive presentation leads us to a basic remark: the most fundamental originality of an integral theory lies in the astute choice of the functions ϕ_ℓ. Roughly, the less the number L, the more the integral apporach is well adapted. In the simplest case of the uncoated perfectly conducting grating, the physicist's intuition leads one to consider the surface current density on the profile as a good unknown, since it clearly permits one to represent the diffracted field at any point of the space. Unfortunately, this indication no longer holds in the more complicated cases.

2 Local square integrability as explained in Sect.2.4.1.

3.3 Notations, Mathematical Problem and Fundamental Formulae

We will now state the mathematical formulation of the problem schematized in Fig. 3.1. Moreover, in order to facilitate the understanding of what follows, we concentrate in some basic formulae the results derived from the theory of distributions which will be used throughout the chapter. The reader not interested in the mathematical demonstration of these last formulae, contained in Sect.3.3.2, may be content to refer to the conclusion of this section.

3.3.1 Notations and Mathematical Formulation

We deal successively with the two fundamental cases of polarization and, assuming the uniqueness of the solution (see Chap.2), it can be shown that the total electric (or magnetic) field $\underline{E}(x,y)$ [or $\underline{H}(x,y)$] denoted by $\underline{F}(x,y) = F(x,y) \cdot \hat{\underline{z}}$ remains parallel to Oz in the TE (or TM) case of polarization. Furthermore, $F(x,y)$ is "pseudo-periodic" (Sect.1.2.2)

$$\forall (x,y), \quad F(x+d,y) = F(x,y)\exp(ik_1 x \sin\theta) \quad . \tag{3.2}$$

Then, by using the Helmholtz equation in the region M_q

$$\forall q, \quad \forall P \in M_q \, , \quad F(P) + k_q^2 \, F(P) = 0 \quad , \tag{3.3}$$

developing $F\exp(-ik_1 x \sin\theta)$ and taking into account the OWC, we deduce simply that *outside the grooves*, the field may be expressed as a plane wave expansion (see Sect. 1.2.3), in such a way that

$\forall j, \quad \forall P \in M_j$ and outside the grooves

$$F(x,y) = \sum_{n=-\infty}^{+\infty} A_n^{(j)} \exp\left[i(\alpha_n x - \beta_n^{(j)} y)\right] + \sum_{n=-\infty}^{+\infty} B_n^{(j)} \exp\left[i(\alpha_n x + \beta_n^{(j)} y)\right] \quad , \tag{3.4}$$

with

$$A_n^{(1)} = \delta_{n0} \quad \forall n \quad , \tag{3.5}$$

$$B_n^{(Q+1)} = 0 \quad \forall n \quad , \tag{3.6}$$

$$\alpha_n = k_1 \sin\theta + nK \quad , \tag{3.7}$$

$$\beta_n^{(j)} = \sqrt{k_j^2 - \alpha_n^2} \quad \text{or} \quad i\sqrt{\alpha_n^2 - k_j^2} \quad , \tag{3.8}$$

$$K = 2\pi/d \quad .$$

When k_j is real[3], if we denote by $\mathcal{U}^{(j)}$ the finite set of integers n such that β_n^j is

---

3 Let us note that when k_j is complex, the definition of $\beta_n^{(j)}$ by (3.8) is not sufficient since it requires us to take the square root of a complex number. In order to avoid difficulties, we must specify the determination of $\beta_n^{(j)}$. In the following, we assume that $\mathrm{Re}\{\beta_n^{(j)}\} + \mathrm{Im}\{\beta_n^{(j)}\}$ is positive (see Chap. 5).

real, it must be observed that $A_n^{(j)}$ and $B_n^{(j)}$ are the complex coefficients of plane wave propagating, respectively, downwards and upwards. In order to distinguish the two fundamental cases of polarization, we will denote by $A_n'^{(j)}$ and $B_n'^{(j)}$ the coefficients of the plane wave expansions for TM polarization.

Thus, to state the mathematical problem, we must add to the Helmholtz equation (3.3) and to the OWC (3.4-6) the boundary conditions on the interfaces $\mathcal{P}_i$. From the continuity conditions for the fields, it can easily be seen that, when M_{Q+1} is not perfectly conducting

$$\forall i \in (1,Q), \quad F_+[x,F_i(x)] = F_-[x,f_i(x)] \quad , \tag{3.9a}$$

$$C_i \frac{dF}{dn_+}[x,f_i(x)] = C_{i+1} \frac{dF}{dn_-}[x,f_i(x)] \quad , \tag{3.10a}$$

with

$$C_i = \begin{cases} \mu_0/\mu_i & \text{in TE case} \quad , \\ \varepsilon_0/\varepsilon_i & \text{in TM case} \quad . \end{cases}$$

The function $F_\pm(x,f_i(x))$ denotes the limit of $F(x,y)$, when P, lying on $\Gamma_\pm$ (Fig.3.2) tends towards $M \in \mathcal{P}_i$. Similarly

$$\frac{dF}{dn_\pm}[x,f_i(x)] = \underline{\hat{n}}_i(x) \cdot \lim_{P\to M} [\mathit{grad}\ F] \quad .$$

The constants μ_0, μ_i, ε_0 and ε_i are the magnetic permeability and the electric permittivity of the vacuum and of the material filling M_i.

When M_{Q+1} is perfectly conducting, (3.9a) and (3.10a) must be replaced for $i = Q$ by only one equation

$$F_+[x,f_Q(x)] = 0 \quad \text{in TE case} \quad , \tag{3.9b}$$

$$\frac{dF}{dn_+}[x,f_Q(x)] = 0 \quad \text{in TM case} \quad . \tag{3.10b}$$

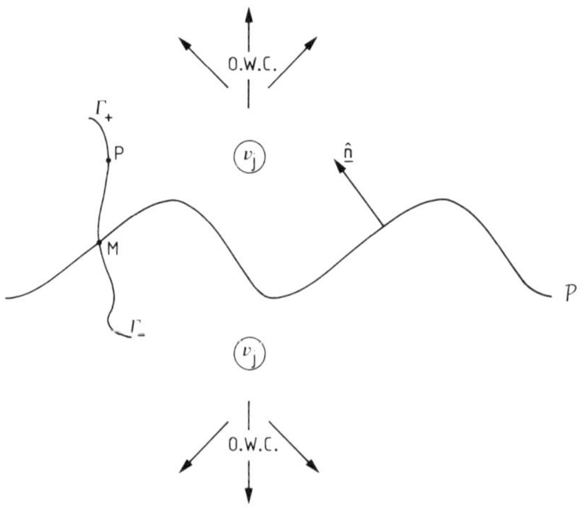

Fig. 3.2. The lines Γ_+ and Γ_- which have arbitrary shapes contain the point M of $\mathcal{P}$ and are, respectively, located above and below $\mathcal{P}$; the normal unit $\underline{\hat{n}}$ is oriented upwards

Since the magnetic permeability $\mu_i = \mu_0$, the boundary conditions (3.9a) and (3.10a) for TE polarization state finally that the field and its normal derivative are continuous. It is not the case for TM polarization where the normal derivative is not continuous since $\varepsilon_i = \nu_i^2$.

If the uniqueness of the solution of the problem defined by (3.3-6), (3.9,10) is ensured when Q = 1 (Chap.2), it is not the case otherwise. Indeed, we know that in particular conditions, modes may propagate in the structure of Fig.3.1. So, hereafter, we assume that this is not the case.

3.3.2 Basic Formulae of the Integral Approach

We will deal now with a nonphysical problem whose resolution with the help of the theory of distributions will allow us to point out some basic formulae whose utilization considerably simplifies the resolution of the various problems.

Let us suppose that a profile P separates two semi-infinite media, *both having the same index* ν_j (Fig.3.2). We consider a function U(x,y) having the pseudo-periodicity of the field [see (3.2)] and satisfying the two following conditions:

$$\Delta U + k_j^2 U = 0 \quad \text{everywhere, except on } P \quad , \tag{3.11}$$

$$\text{OWC for } y \to \pm\infty \quad . \tag{3.12}$$

We denote, respectively, by $\tau(x)$ and $\eta(x)$ the jumps of U and its normal derivative when P is crossed in the direction of the normal

$$\tau(x) = U_+[x,f(x)] - U_-[x,f(x)] \quad , \tag{3.13}$$

$$\eta(x) = \frac{dU}{dn_+}[x,f(x)] - \frac{dU}{dn_-}[x,f(x)] \quad . \tag{3.14}$$

Then, referring to the Appendix A, we can express the equation satisfied by U in the sense of distributions. For this purpose, we define the elementary solution of the Helmholtz equation [see (1.93)]

$$G^{(j)}(x,y) = \frac{1}{2id} \sum_{n=-\infty}^{+\infty} \frac{1}{\beta_n^{(j)}} \exp\left[i\,\beta_n^{(j)}|y| + i\alpha_n x\right] \quad . \tag{3.15}$$

This enables us to find the expression of U everywhere (see Appendix A)

$$U(x,y) = \int_0^d G^{(j)}[x - x', y - f(x')]\,\eta(x')\,d\ell' + \int_0^d \frac{dG^{(j)}}{dn}[x - x', y - f(x')]\,\tau(x')\,d\ell' \quad . \tag{3.16}$$

Inversely, it is worth noting that this expression satisfies (3.11-14). The above equation makes it possible to calculate the limits of U and its normal derivative on both sides of P. Indeed, it must be noted that when y is fixed, U(x,y) given by (3.16) is the product of a continuous function by a Fourier series in x. Hence, choosing Γ_+ and Γ_- parallel to Ox in Fig.3.2 and applying the classical theorem about the value of a discontinuous Fourier series on its discontinuity, yields

$$\frac{1}{2}\{U_+[x,f(x)] + U_-[x,f(x)]\} = U[x,f(x)] \quad , \tag{3.17}$$

where U[x,f(x)] is obtained by replacing y by f(x) in the right-hand member of (3.16). Eliminating $U_+(x)$ or $U_-(x)$ between (3.13) and (3.17), we can derive

$$\tilde{U}_\pm[x,f(x)] = \pm\frac{\tilde{\tau}}{2} + \int_0^d G^{(j)}(x,x')\,\tilde{\eta}(x')dx' + \int_0^d N'^{(j)}(x,x')\,\tilde{\tau}(x')dx' \quad , \tag{3.18}$$

with

$$\tilde{U}_\pm[x,f(x)] = U_\pm[x,f(x)]\exp(-ik_1\, x\sin\theta) \quad ,$$

$$\tilde{\tau}(x) = \tau(x)\exp(-ik_1 x\sin\theta) \quad ,$$

$$\tilde{\eta}(x) = \eta(x)\sqrt{1+f'^2(x)}\exp(-ik_1 x\sin\theta) \quad ,$$

$$G^{(j)}(x,x') = \frac{1}{2id}\sum_{n=-\infty}^{+\infty}\frac{1}{\beta_n^{(j)}}\exp\left[i\beta_n^{(j)}|f(x)-f(x')|+inK(x-x')\right] \quad , \tag{3.19}$$

$$N'^{(j)}(x,x') = \frac{1}{2d}\sum_{n=-\infty}^{+\infty}\left\{\mathrm{sgn}[f(x)-f(x')] - \frac{\alpha_n}{\beta_n^{(j)}}f'(x')\right\} \times \exp\left[i\beta_n^{(j)}|f(x) - f(x')| + inK(x-x')\right] \quad , \tag{3.20}$$

f'(x) denoting the derivative of f(x) and sgn(a) being, respectively, equal to +1 or -1 according to whether a is positive or negative. It is worth noting that the functions $\tilde{U}$, $\tilde{\tau}$, $\tilde{\eta}$, G^j and $N'^{(j)}$ are periodic. For simplicity, it is interesting to represent (3.18) in operator form[4]

$$\tilde{U}_\pm = \pm\tilde{\tau}/2 + G^{(j)}\{\tilde{\eta}\} + N'^{(j)}\{\tilde{\tau}\} \quad . \tag{3.21}$$

In the same way, the above considerations on Fourier series applied to *grad* U show that, provided that $\tau(x) = 0$

$$\frac{d\tilde{U}}{dn_\pm} = \pm\tilde{\eta}/2 + N^{(j)}\{\tilde{\eta}\} \quad , \tag{3.22}$$

with

$$N^{(j)}(x,x') = \frac{1}{2d}\sum_n\left\{\mathrm{sgn}|f(x)-f(x')| - \frac{\alpha_n}{\beta_n^{(j)}}f'(x)\right\} \times \exp\left[i\beta_n^{(j)}|f(x)-f(x')| + inK(x-x')\right] \quad , \tag{3.23}$$

$$\frac{d\tilde{U}}{dn_\pm}[x,f(x)] = \frac{dU}{dn_\pm}[x,f(x)]\sqrt{1+f'^2(x)}\exp(-ik_1 x\sin\theta) \quad .$$

---

4 For simplicity, the operator $G^{(j)}$ is denoted by the same symbol as the function of two variables, but $G^{(j)}\{\tilde{\eta}\}$, which is the function of one variable obtained when the operator $G^{(j)}$ is applied to the function $\tilde{\eta}$, must not be confused with the function $G^{(j)}(x,x')$ of the two variables x and x'.

Now, noting in (3.15,16) that the term $|y-f(x')|$ is, respectively, equal to $y - f(x')$ or $f(x') - y$ according to whether $y > \sup[f(x')]$ or $y < \inf[f(x')]$, it turns out that outside the grooves $U(x,y)$ is described by plane wave expansions similar to that of the right-hand member of (3.4)

$$\forall y > \sup[f(x)] : U(x,y) = \sum_{n=-\infty}^{+\infty} D_n^+ \exp(i\alpha_n x + i\beta_n^{(j)} y) \quad ,$$

$$\forall y < \inf[f(x)] : U(x,y) = \sum_{n=-\infty}^{+\infty} D_n^- \exp(i\alpha_n x - i\beta^{(j)} y) \quad ,$$

with

$$D_n^{\pm} = [s_n^{(j)\pm}, \tilde{\eta}] + [t_n^{(j)\pm}, \tilde{\tau}] \quad , \tag{3.24}$$

$$s_n^{(j)\pm}(x) = \frac{1}{2id\beta_n^{(j)}} \exp[\pm\, i\beta_n^{(j)} f(x) - inKx] \quad , \tag{3.25}$$

$$t_n^{(j)\pm}(x) = \frac{1}{2d}\left[\pm 1 - \frac{\alpha_n}{\beta_n^{(j)}} f'(x)\right] \exp\left[\pm\, i\beta_n^{(j)} f(x) - inKx\right] \quad , \tag{3.26}$$

the symbol $[u,v]$ being defined by

$$[u,v] = \int_0^d u(x)\, v(x)\, dx \quad .$$

Finally, in order to bring together all the notations in this section, let us *return to the physical problem* of Fig.3.1 to give the expression of the incident field F^i and its normal derivative on P_1, the region M_1 now having the index ν_1

$$\tilde{F}^i(x,f_1(x)) = F^i[x,f_1(x)] \exp(-ik_1 x \sin\theta) = \exp\left[-i\beta_0^{(1)} f_1(x)\right] \quad , \tag{3.27}$$

$$\frac{d\tilde{F}^i}{dn}[x,f_1(x)] = \frac{dF^i}{dn}[x,f_1(x)] \sqrt{1 + f_1'^2(x)} \, \exp(-ik_1 x \sin\theta)$$

$$= -i \left[\beta_0^{(1)} + \alpha_0 f_1'(x)\right] \exp\left[-i\beta_0^{(1)} f_1(x)\right] \quad . \tag{3.28}$$

To conclude this section, *when a pseudo-periodic function* $U(x,y)$ *everywhere continuous, except on* P, *satisfies (3.11,12), it can be described by (3.16) in terms of the jumps* τ *and* η *of* U *and its normal derivative across* P. *Moreover, (3.21) provides the limits* $U_{\pm}[x,f(x)]$ *of* U *on both sides of* P *and (3.22) the limits* $dU/dn_{\pm}[x,f(x)]$ *of its normal derivative in the particular case where* $\tau(x) \equiv 0$. *In the general case,* U *may be expressed outside the grooves of* P *by plane wave expansions whose coefficients are given by (3.24).*

3.4 The Uncoated Perfectly Conducting Grating

In this problem, the profile P separates the region M_1 filled by a dielectric and a region M_2 which is perfectly conducting. After the pioneering works described in [3.5-10], many papers have been published by the same authors and many others.

Today, many integral treatments are available in published form [3.13,24,26-32]. These differ either in the exact form of the integral equation or in its numerical treatment. A review of this matter may be found in [3.33]. McCLELLAN and STROKE have also given a method for the TM case of polarization but no numerical results supported the theory given in this paper [3.34]; in our opinion their fundamental equation appears to be false since it implies the continuity of the tangential component of the magnetic field at the grating surface.

3.4.1 The TE Case of Polarization

It has been established in Chap. 1 that this problem can be reduced to the resolution of a singular integral equation of the first kind. Our purpose is now to describe another mathematical treatment, which leads to an integral equation of the second kind, with a non singular kernel. This formulation, described by PAVAGEAU et al. [3.10] is inspired by the ideas of MAUE [3.35] for a two-dimensional problem of diffraction.

Let us remember that the mathematical problem, stated by (3.3,5,9b) is to find the electric field $F(x,y)$ which vanishes in M_2, and is defined in M_1 by the three following conditions:

$$\forall P \in M_1, \ \Delta F(P) + k_1^2 F(P) = 0 \quad , \tag{3.29}$$

$$\text{OWC for } F^d(P) = F(P) - F^i(P) \text{ when } y \to +\infty \quad , \tag{3.30}$$

$$\forall x, \ F_+[x,f(x)] = F^i[x,f(x)] + F_+^d[x,d(x)] = 0 \quad . \tag{3.31}$$

It must be observed that since $F(x,y) = 0$ in M_2, the boundary condition (3.31) involves the continuity of $F(x,y)$ on P.

The start of the Pavageau formalism is identical to that proposed by Petit. Indeed, in order to express the field $F(x,y)$ everywhere, we must define at any point of the space an intermediate function $U(x,y)$, identical to the diffracted field in M_1

$$\forall P, \ U(x,y) = F(x,y) - F^i(x,y) \quad . \tag{3.32}$$

The fundamental interest of this continuous "generalized diffracted field" is that it satisfies both the Helmholtz equation (3.29) in M_1 and M_2, and the OWC when $y \to \pm\infty$. $U(x,y)$ obviously thus satisfies (3.11,12) with $j = 1$. Defining the jump $\phi(x)$ of the normal derivative of U across P allows us to obtain an integral representation $U_s(x,y)$ of $U(x,y)$ everywhere by changing $\eta(x)$ in $\phi(x)$ and $\tau(x)$ in 0 in (3.16)

$$U_s(x,y) = \int_0^d G^{(1)}[x-x',y-f(x')] \ \phi(x') \ d\ell' \quad .$$

We have to point out that if ϕ is now arbitrarily given, $U_s(x,y)$ still satisfies (3.29,31). In other words, $U_s(x,y)$ defines a set of functions depending on ϕ. By adding F^i to U_s, we obtain a new set F_s of functions which satisfy (3.29,30), but generally do not satisfy the boundary condition (3.31)[5]. The distinction between the actual field F and its integral representation F_s, which perhaps look superfluous here, will be very useful (and even necessary) for the understanding of the more complicated theories.

The integral equation of Petit is obtained by stating directly that F_{s+} must satisfy (3.31), the expression of $U_{s+}[x,f(x)]$ being provided by (3.21)

$$G^{(1)}\{\tilde{\phi}\} = -\tilde{F}^i \quad , \text{ with } \tilde{\phi}(x) = \phi(x)\sqrt{1+f'^2(x)}\exp(-ik_1 x \sin\theta) \tag{3.33}$$

where the expressions of $G^{(1)}$ and $\tilde{F}^i$ are given by (3.19,27). From (3.24), we derive the coefficients B_n of the plane wave expansion in M_1[6]

$$\forall n,\ B_n = [s_n^{(1)+},\tilde{\phi}] \quad , \tag{3.34}$$

and a relation which will provide a verification of the correct resolution of (3.33)

$$\forall n,\ \delta_{n0} = [s_n^{(1)-},\tilde{\phi}] \quad . \tag{3.35}$$

The alternative boundary condition proposed by Pavageau can be expressed in the following form:

$$\frac{dF_s}{dn_-} = \frac{dU_s}{dn_-}[x,f(x)] + \frac{dF^i}{dn}[x,f(x)] = 0 \quad . \tag{3.36}$$

First, let us show that this equation is equivalent to (3.31). In other words, we have to prove that the limit F_{s+} of the field above P vanishes as soon as (3.36) is satisfied. Let us first note that F_s already satisfies (3.29,30) and, in addition, is continuous on P and satisfies also the Helmholtz equation in M_2 and an OWC for $y \to -\infty$[7]. According to the considerations of uniqueness developed in Chap. 2, F_s which satisfies the Helmholtz equation in M_2, an OWC when $y \to -\infty$, and whose normal derivative below P vanishes [according to (3.36)] vanishes everywhere in M_2. Now, it is a continuous function, therefore its limit above P is zero, that is to say the boundary condition (3.31).

The integral equation of Pavageau is obtained by introducing in (3.36) the expression of $dU_s/dn_-\ [x,f(x)]$ derived from (3.22)

---

5 It is worth noting that, for the moment, the boundary condition has been used to establish that $U_s(x,y)$ must be continuous and therefore that $\tau = 0$, but inversely the continuity of U_s does not imply the boundary condition.

6 Since the field vanishes outside M_1, we eliminate now the superscript of $B_n^{(1)}$.

7 Since U_s and F^i both satisfy the Helmholtz equation in M_2 and the OWC when $y \to -\infty$, the same conditions are verified by $F_s = U_s + F^i$.

$$\frac{\tilde{\phi}}{2} = \frac{d\tilde{F}^i}{dn} + N^{(1)}\{\tilde{\phi}\} \quad , \tag{3.37}$$

where $d\tilde{F}^i/dn[x,f(x)]$ and $N^{(1)}(x,x')$ are given by (3.28,23). It will be seen in Sect.3.8 that the kernel $N^{(1)}$ is not singular. Thus, the integral equations of Petit or of Pavageau may be used to determine the same unknown function $\tilde{\phi}$.

It must be noted that the theory used by Pavageau to obtain (3.37) was quite different from the above formalism. Indeed, Pavageau invoked the simple relation between the unknown function $\phi(x)$ and the surface current density $j(x)$, the above equation being derived from the boundary condition for the magnetic field $\underline{H}$

$$\hat{\underline{n}} \wedge \underline{H}[x,f(x)] = j(x)\ \hat{\underline{z}}$$

Without doubt, the presentation made by Pavageau has a greater significance from a physical point of view. Nevertheless, in our opinion, the above presentation allows the clear proof of the equivalence between the two formalisms proposed by Petit and Pavageau.

3.4.2 The TM Case of Polarization

The mathematical problem is now to determine the magnetic field $F(P)$ vanishing in M_2 and defined in M_1 by (3.29,30) and a new boundary condition stated by (3.10b)

$$\frac{dF}{dn_+}[x,f(x)] = \frac{dF^i}{dn}[x,f(x)] + \frac{dF^d}{dn_+}[x,f(x)] = 0 \quad . \tag{3.38}$$

As with TE polarization, it is interesting to define by (3.32) a generalized diffracted field satisfying both (3.29) in M_1 and M_2 and the OWC when $y \to -\infty$. Now, this function which satisfies (3.11,12) is not continuous but, according to (3.38), its normal derivative is continuous. Hence, denoting the jump of the function by $\psi(x)$ enables us to describe $U(x,y)$ in an integral form derived from (3.16) by replacing $\eta(x)$ with 0 and $\tau(x)$ with $\psi(x)$. The integral expression F_s obtained by adding F^i satisfies (3.29,30). Moreover its normal derivative is the same above and below P, and F_s satisfies in M_2 the Helmholtz equation and the OWC when $y \to -\infty$. Taking into account all these above properties, the boundary condition (3.38) may be written in the following equivalent form:

$$U_{s-}[x,f(x)] + F^i[x,f(x)] = 0 \quad . \tag{3.39}$$

Indeed, according to the theorems of uniqueness of Chap. 2, $U_s + F^i$ which satisfies an Helmholtz equation in M_2, an OWC when $y \to -\infty$ and which vanishes below P according to (3.39), vanishes everywhere in M_2. Hence, its normal derivative below P is zero and since $\eta = 0$, it is also the case for the normal derivative above P; (3.38) is then satisfied.

Making explicit in (3.39) the value of $U_{s-}[x,f(x)]$ from (3.21), we finally derive the integral equation of the second kind

$$\frac{\tilde{\psi}}{2} = \tilde{F}^i + N'^{(1)}\{\tilde{\psi}\} \quad , \tag{3.40}$$

with $\tilde{\psi}(x) = \psi(x)\ \exp(-ik_i x\ \sin\theta)$, $\tilde{F}^i$ and $N'^{(1)}$ being given by (3.27,20).

After resolution of (3.40), the amplitudes B'_n of the plane waves diffracted in M_1 can be obtained using (3.24) by

$$B'_n = [t_n^{(1)+}, \tilde{\psi}] \quad , \tag{3.41}$$

and furthermore

$$\delta_{n0} = [t_n^{(1)-}, \tilde{\psi}] \quad . \tag{3.42}$$

It is interesting to note that the kernels $N'^{(1)}$ and $N^{(1)}$ of the Pavageau integral equations for TM and TE cases may be deduced from each other only by replacing $f'(x')$ by $f'(x)$.

Other integral equations may be found in TE and TM cases. We will briefly describe one of these equations, relative to TM, because it will be encountered below. This equation, which to our knowledge cannot be found in the literature of gratings, is obtained by defining a new and peculiar function $U(x,y)$ by

$$U(x,y) = F^d \text{ in } M_1 \quad , \tag{3.43}$$

$$\Delta U(P) + k_1^2 U(P) = 0 \text{ in } M_2 \quad , \tag{3.44}$$

$$U_-[x,f(x)] = U_+[x,f(x)] = F_+^d[x,f(x)] \quad , \tag{3.45}$$

$$U \text{ satisfies an OWC when } y \to -\infty \quad . \tag{3.46}$$

According to Chap. 2, $U(x,y)$ is unique, even in M_2, because it satisfies (3.44, 46) and its value on P is fixed by (3.45). It is worth noting that this new function U for TM polarization, which has not physical significance in M_2, is continuous everywhere. Obviously, U satisfies (3.11,12) with $j = 1$. Thus, since the jump of U across P is zero, $dU_s/dn_+[x,f(x)]$ can be deduced from (3.22) by replacing $\tilde{\eta}$ by $\tilde{\phi}$, the jump of the normal derivative of U on P. According to the boundary condition (3.38) and keeping in mind (3.43), it turns out

$$\frac{\tilde{\phi}}{2} = -\frac{d\tilde{F}^i}{dn} - N^{(1)}\{\tilde{\phi}\} \quad , \tag{3.47}$$

which is an integral equation of the second kind whose kernel is that of the integral equation of Pavageau in TE polarization.

From (3.24), the complex amplitudes B'_n are now given by

$$B'_n = \left[s_n^{(1)+}, \tilde{\phi}\right] \quad .$$

It is interesting to note that the function $\tilde{\phi}$ is not equal to the surface current density and has no physical significance.

3.5 The Uncoated Dielectric or Metallic Grating

Without doubt, it is nowadays fully recognized that an accurate theoretical knowledge of the performances of gratings used in the ultraviolet, visible (and sometimes infrared) regions requires taking into account the finite conductivity of the metal. The numerical programs obtained from the theories of finite conductivity gratings are therefore the most commonly used today. The first rigorous theoretical works in this domain proposed the solving of two coupled integral equations [3.11-13]. Because of its fundamental numerical advantage, and since it has been thoroughly tested by comparison with experiment [3.36-39] and with the numerical results of NEVIERE's differential method [3.18], we present here the approach where only one unknown function is needed and so only one integral equation has to be numerically solved [3.14-16]. We will show why this method, as opposed to the differential one, is able to deal with highly conducting metals.

Another method needing only one unknown function has been proposed by WIRGIN [3.40] but it does not seem to have been numerically implemented.

3.5.1 The Mathematical Boundary Problem

It is stated by (3.2-8,9a,10a) with Q = 1, hence we must determine a pseudo-periodic total field F such that

$$\forall P \in M_1, \ \Delta F(P) + k_1^2 F(P) = 0 \quad , \tag{3.48}$$

$$\forall P \in M_2, \ \Delta F(P) + k_2^2 F(P) = 0 \quad , \tag{3.49}$$

$$\text{in } M_1, \ F^d(P) = F(P) - F^i(P) \text{ satisfies an OWC when } y \to +\infty \quad , \tag{3.50}$$

$$\text{in } M_2, \ F(P) \text{ satisfies an OWC when } y \to -\infty \quad , \tag{3.51}$$

$$F(P) \text{ is continuous across } P \quad , \tag{3.52}$$

$$\frac{dF}{dn_+}[x,f(x)] = \frac{C_2}{C_1}\frac{dF}{dn_-}[x,f(x)] \quad , \tag{3.53}$$

$$\text{with } \frac{C_2}{C_1} = \begin{cases} \dfrac{\mu_1}{\mu_2} = 1 \text{ in TE case} \quad , & (3.54a) \\[2ex] \dfrac{\varepsilon_1}{\varepsilon_2} = \dfrac{\nu_1^2}{\nu_2^2} \text{ in TM case} \quad . & (3.54b) \end{cases}$$

It turns out that the equations are the same in TE and TM polarizations: only the value of C_1/C_2 is modified. So these two cases will be treated together.

3.5.2 Vital Importance of the Choice of a Well-Adapted Unknown Function

Previously, in the case of an infinitely conducting metal, physical intuition and mathematical convenience suggest generally that the field must be described by a function linked to the surface current density on P. It is not the case any longer and we must therefore make a fundamental choice. The first idea that comes in mind is to adopt as the unknown function the value of the field F(M) on P, because of its continuity. Unfortunately, it turns out that this given function is not enough to describe the field everywhere in an integral form. In fact, it is also necessary to know the value of the normal derivative dF/dn_+ on P. Thus, with two unknown functions, it becomes indispensable to obtain two integral equations [3.11-13].

We will show now that a well-chosen function can describe the field everywhere, but before definit it mathematically, it seems useful to give its physical meaning. Let us imagine that the material of M_1, having electrical constants ε_1 and μ_1, fills all the space, including M_2. We also suppose that one can impose on P any surface current density $j(x)\hat{\underline{z}}$. One can imagine for example that a large number of infinite (with respect to z), very fine, conducting wires are located on P, the current in each of these wires being chosen arbitrarily (Fig.3.3). Let us assume that there exists a particular surface current density $j_d(M)$ which gives rise *at any point of* M_1 to an electric field $E'^{(d)}(P)$ equal to the actual diffracted field $F^d(P)$ of the physical problem. Naturally, following these considerations, it turns out that, in the TM case, the fictitious electric field E'^d must be equal to the actual magnetic field H^d in M_1. Of course, the fictitious field E'^d created by $j_d(M)$ in M_2 satisfies the Helmholtz equation (3.48) and is not related in any way to the actual field in M_2, one of the reasons being that the latter satisfies (3.49). Now, if $j_d(M)$ exists, it is obvious that is knowledge will make it possible to describe the diffracted field $F^d = E'^d$ (and hence the total field F) at any point of M_1. We thus know the field $F_+[x,f(x)]$ and its normal derivative $dF/dn_+[x,f(x)]$ on P. As pointed out at the beginning of this section, these two functions are sufficient to describe the field everywhere (including M_2) using an integral representation. In conclusion of this section, it must be emphasized that the originality of this approach consists in showing that the two unknown functions of the previous integral methods, i.e., the field and its normal derivative above P, can both be derived from the knowledge of only one well-chosen function $j_d(x)$ defined on P.

3.5.3 Mathematical Definition of the Unknown Function and Determination of the Field and Its Normal Derivative Above P

Figure 3.4 shows in outline the different steps of the theoretical method. Let us consider the function U(P) defined by the following conditions

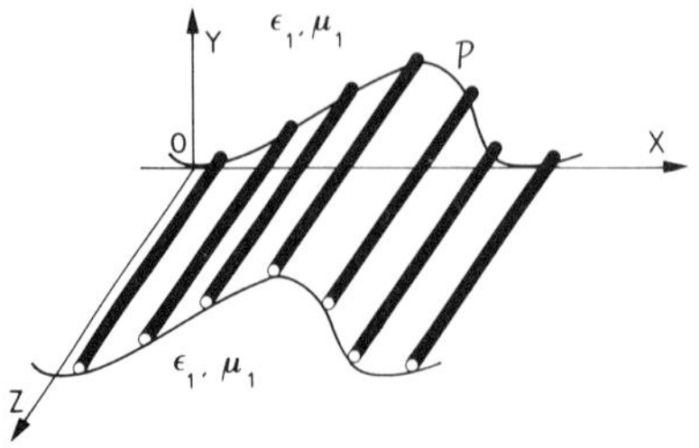

Fig. 3.3. Intuitive definition of a fictitious current density $j(x)\hat{z}$: the material filling M_1 in the physical problem fills now all the space, including M_2, and the current in each of the wires is chosen in order to generate in M_1 the diffracted field of the physical problem

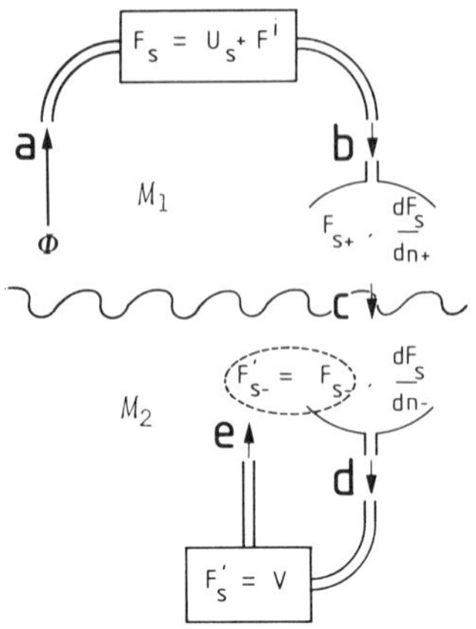

Fig. 3.4. Successive steps of an integral method used for metallic gratings

$$U(P) = F^d(P) \text{ in } M_1 \quad , \tag{3.55}$$

$$\Delta U(P) + k_1^2 U(P) = 0 \text{ in } M_2 \quad , \tag{3.56}$$

$$U_-[x,f(x)] = U_+[x,f(x)] = F_+^d[x,f(x)] \quad , \tag{3.57}$$

$$U \text{ satisfies an OWC for } y \to -\infty \quad . \tag{3.58}$$

It is interesting to notice that the problem defined by (3.55-58) is exactly the same as the problem previously defined by (3.43-46) to treat the TM case of polarization. Therefore, according to Sect.3.4.2, we deduce that U(P) so defined is unique and that its integral expression $U_s(P)$ may be deduced from (3.16) by replacing $\tau(x)$ by 0 and $\eta(x)$ by $\phi(x)$, the jump of the normal derivative of U across *P* (arrow a). According to Sect.3.2.2, this function ϕ is our unknown function and we know (arrow b) that the values of the limits of U_s and its normal derivative above *P* are given by (3.21,22)

$$\tilde{U}_{s+} = G^{(1)}\{\tilde{\phi}\} \quad , \tag{3.59}$$

$$\frac{d\tilde{U}_s}{dn_+} = \tilde{\phi}/2 + N^{(1)}\{\tilde{\phi}\} \quad . \tag{3.60}$$

Furthermore, from (3.24), the amplitudes $B_n^{(1)}$ of the plane wave expansion of U (i.e., of the diffracted field F^d) in M_1 will be given from ϕ by

$$B_n^{(1)} = [s_n^{(1)+}, \tilde{\phi}] \quad . \tag{3.61}$$

3.5.4 Expression of the Field in M_2 as a Function of ϕ

Keeping in mind that U(P) is equal to the diffracted field in M_1, and taking into account the boundary conditions (3.52,53), we can express the limits of the field and its normal derivative below P (arrow c)

$$\tilde{F}_{s-} = \tilde{F}_{s+} = \tilde{F}^i + G^{(1)}\{\tilde{\phi}\} \quad , \tag{3.62}$$

$$\frac{d\tilde{F}_s}{dn_-} = \frac{C_1}{C_2}\left[\frac{d\tilde{F}^i}{dn} + \tilde{\phi}/2 + N^{(1)}\{\tilde{\phi}\}\right] \quad . \tag{3.63}$$

In order to describe the field in M_2 in terms of ϕ, we define a function V(P) by the two following relations:

$$V(P) = \begin{cases} 0 \text{ in } M_1 \quad , & (3.64) \\ F(x,y) \text{ in } M_2 \quad . & (3.65) \end{cases}$$

Since it vanishes in M_1 and it is equal to the actual field in M_2, V(P), according to (3.49,51), satisfies (3.11,12) with j = 2. Now, the jump of V and its normal derivative across P are the opposite of F_- and dF/dn_-, therefore we derive the integral expression $V_s(P)$ of V(P) by changing η into $-(dF_s/dn_-)$ and τ into $-F_{s-}$ in (3.16) (arrow d). In the same way, the limit $F'_{s-} = V_{s-}$ of this expression below P (arrow e) can now be deduced from (3.21)

$$\begin{aligned} \tilde{F}'_{s-} &= -\left(-\tilde{F}_{s-}\right)/2 + G^{(2)}\left\{-\frac{d\tilde{F}_s}{dn_-}\right\} + N'^{(2)}\left\{-\tilde{F}_{s-}\right\} \\ &= \left(\tilde{F}_i + G^{(1)}\{\tilde{\phi}\}\right)/2 + G^{(2)}\left\{-\frac{C_1}{C_2}\left(\frac{d\tilde{F}^i}{dn} + \tilde{\phi}/2 + N^{(1)}\{\tilde{\phi}\}\right)\right\} \\ &\quad + N'^{(2)}\left\{-\tilde{F}_i - G^{(1)}(\tilde{\phi})\right\} \quad . \end{aligned} \tag{3.66}$$

3.5.5 Integral Equation

It may appear very curious to obtain with (3.62,66) two different integral expressions for the field below P. To explain this apparent contradiction, it must be remembered that $V_s(P)$ satisfies the Helmholtz equation (3.49) in M_1 and M_2, an OWC when $y \to -\infty$, and that *the jumps of V and its normal derivative* are, respectively, equal to $-F_{s-}$ and $-(dF_s/dn_-)$. Obviously this is not sufficient to prove that the expression of V_s so obtained satisfies (3.64,65). In other words, introducing an arbitrary function $\tilde{\phi}$ into (3.62,66) will lead us to find two different functions $\tilde{F}_{s-}$ and $\tilde{F}'_{s-}$. The integral equation is obtained simply by writing the equality of these two expressions, and finally we get

$$\left[G^{(1)}/2 + N'^{(2)}G^{(1)} + \frac{C_1}{C_2}\left(G^{(2)}/2 + G^{(2)}N^{(1)}\right)\right]\{\tilde{\phi}\}$$

$$+ \tilde{F}^i/2 + N'^{(2)}\{\tilde{F}^i\} + \frac{C_1}{C_2}\,G^{(2)}\left\{\frac{d\tilde{F}^i}{dn}\right\} = 0 \quad , \qquad (3.67)$$

which is a singular integral equation of the first kind.

Let us now briefly establish that the resolution of this equation will provide the actual field F(P). To this end, let us consider the integral expression $F_S''(P)$ of the field everywhere, defined by

$$F_S''(P) = \begin{cases} U_S(P) + F^i(P) \text{ in } M_1 & , \\ V_S(P) \text{ in } M_2 & . \end{cases}$$

Obviously, the function so defined satisfies (3.48-51), whatever the function ϕ. This conclusion does not hold any longer for the boundary conditions (3.52,53). However, taking into account (3.59,62), it can easily be seen that the integral equation states the continuity of F_S'' on P and implies (3.52), but apparently not (3.53). Let us remember that the jump of V_S is equal to $-F_{S-}$. Thus, since the integral equation (3.67) states that the limit of F_S'' below P is also equal to $-F_{S-}$, we conclude that its limit above P is zero. The function $V_S(P)$ satisfies in M_1 an Helmholtz equation, an OWC when $y \to -\infty$, therefore, since its limit above P is zero, it vanishes everywhere in M_1, as in Chap. 2. We can thus deduce that its normal derivative above P vanishes and since the jump of this normal derivative is equal to $-(dF_S/dn_-)$, the normal derivative below P is $+(dF_S/dn_-)$. Taking into account (3.60,63), this shows clearly that F_S'' satisfies (3.53). We must conclude that F_S'' identifies to the actual field F, provided that the integral equation (3.67) is satisfied.

The amplitudes $B_n^{(1)}$ of the diffracted field in M_1 are given by (3.61). In the same manner, according to (3.24), (3.62-65), the amplitudes $A_n^{(2)}$ of the plane waves in M_2 may be obtained from

$$A_n^{(2)} = -\left[t_n^{(2)-}, \tilde{F}^i + G^{(1)}\{\tilde{\phi}\}\right] - \left[s_n^{(2)-}, \frac{C_1}{C_2}\left(\frac{d\tilde{F}^i}{dn}\right) + \frac{\tilde{\phi}}{2} + N^{(1)}\{\tilde{\phi}\}\right] \quad . \quad (3.68)$$

3.5.6 Limit of the Equation when the Metal Becomes Perfectly Conducting

One of the most important features of the integral method described above is that it permits us to establish a continuity between the metallic grating and the perfectly conducting one. We show here that, thanks to a simple property of $G^{(2)}$ and $N'^{(2)}$, the integral equation (3.67) tends, when the grating index becomes infinite (i.e., when the metal becomes perfectly conducting) towards the integral equations (3.33,47), encountered for the perfectly conducting gratings. Indeed, it can be proved that, when $|\nu_2| \to \infty$, the functions $G^{(2)}(x,x')$ and $N'^{(2)}(x,x')$ tend towards "delta distributions" in such a way that, for any continuous periodic function p(x)

$$\int_0^d G^{(2)}(x,x')\,p(x')\,dx' \to \frac{A_1(x)}{\nu_2}\,p(x) \quad ,$$

$$\int_0^d N'^{(2)}(x,x')\,p(x')\,dx' \to \frac{A_2(x)}{\nu_2}\,p(x) \quad ,$$

$A_1(x)$ and $A_2(x)$ being bounded functions related to the function $f(x)$ and its derivatives. A physical meaning of this phenomenon may be found by noting that $G^{(2)}(x,y)$ represents the value at any point P of the field diffracted in an infinite medium of index ν_2 by an elementary linear current located on the Oz axis. It is obvious that, when $|\nu_2|\to\infty$ (i.e., when the conductivity of the metal increases), the field tends towards zero, except in the points very close to the Oz axis: $G^{(2)}$ tends towards a delta distribution.

Furthermore, if we remember that C_1/C_2 is equal to 1 in the TE case, and to $(\nu_2/\nu_1)^2$ in the TM case, it follows elementarily that (3.67) tends towards (3.33) in the TE case and towards (3.47) in the TM case. From these theoretical considerations, it appears as one of the fundamental advantages of the integral method. Indeed, the integral equation contains in a closed form all the conditions of the mathematical problem (Helmholtz equation, OWC, boundary conditions). Using relevant mathematical analysis, it is possible to predict the behavior of the kernel. This study makes it possible to prepare efficiently the numerical application and even sometimes to investigate fundamental properties of the solution.

3.6 The Multiprofile Grating

We will deal now with the general type of grating described in Fig.3.1. In practice, the lower material filling M_{Q+1} is a metal and the above regions are filled with dielectrics. The use of dielectric coatings can have various justifications. For example, a thin dielectric film is sometimes used to prevent the metal from oxidizing. On the other hand, in recent years, convenient dielectric coatings have been more and more used to reduce the energy absorbed by the gratings. For instance, gratings with more than twenty dielectric coatings are used as beam sampling mirrors for high power lasers.

To our knowledge, the first numerical results in the field were published by VAN DEN BERG [3.13] for an infinitely conducting grating overcoated with only one dielectric layer filling a region between the metal surface and a plane parallel to xOz. In this particular case, where only one interface is modulated, the theory proposed by VAN DEN BERG resulted in a single integral equation of the second kind. Since the work of VAN DEN BERG, numerous calculations have been performed by NEVIERE et al., using a differential formalism [3.18,41]. Because of the difficulties encountered by NEVIERE in the visible and infrared regions, two integral methods were recently proposed. The first [3.19] is theoretically able to deal with the arbitrary grating in Fig.3.1., without limitations concerning the shape of the profiles P_i or the conductivity of M_i. The case where M_{Q+1} is perfectly conducting is included in the possibilities of the numerical program which, because of time computation and memory storage, is limited to gratings covered with a number of dielectric layers smaller than 3. The second, more recently elaborated, is devoted to the study of a particular type of multiprofile gratings: those for which two consecutive profiles have no interpenetration [3.21]. With this limitation, this method requires lower computation times and is particularly well adapted to the case where many interfaces are plane.

At the same time, BOTTEN has solved the problem of a multilayer transmission grating [3.20,22], when its structure consists of a single grating surface and a series of plane inteference films placed above and below it. The originality of this method, which leads to the computation of only one unknown function, is to define a new "Green function", well adapted to the problem. The same author has devoted numerous theoretical and numerical studies to the bimetallic grating[8], involving the resolution of a single equation.

It is not possible to review here these theories which are generally very complicated. Thus, because of its generality, we have chosen to outline briefly the approach described in [3.19]. The mathematical problem is described by (3.3-10).

The successive steps of the method are shown in Fig.3.5. First, let us consider, in Fig.3.6a, a function $U_{Q+1}(x,y)$ satisfying the following conditions:

$U_{Q+1}(x,y) = F(x,y)$ in M_{Q+1} (below P_Q) ,

$\Delta U_{Q+1} + k^2_{Q+1}\, U_{Q+1} = 0$, above P_Q ,

U_{Q+1} is continuous on P_Q ,

U_{Q+1} satisfies an OWC when $y \to +\infty$.

Such a function satisfies (3.11,12). Thus $U_{Q+1}(x,y)$ may be expressed as a function of the unknown ϕ_Q, the jump of the normal derivative of U_{Q+1} on P_Q(arrow a of Fig. 3.5). Since U_{Q+1} is identical to the field F in P_{Q+1}, the arrow b is, in this first step, an identity and we can derive the limits of the field and its normal derivative below P_{Q+1} (arrow c). By using the boundary conditions (3.9a,10a) on P_Q, we deduce the limits of the field and its normal derivative above P_Q in terms of ϕ_Q (arrow d).

---

8 For example, such a structure may be composed of an array of cylinders, alternating in conductivity, deposited on a perfectly conducting substrate.

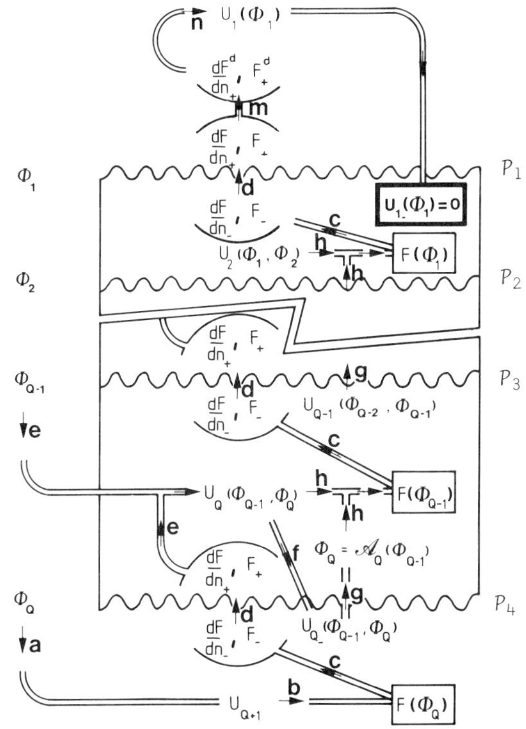

Fig. 3.5. Successive steps of an integral method used for a multi-profile grating

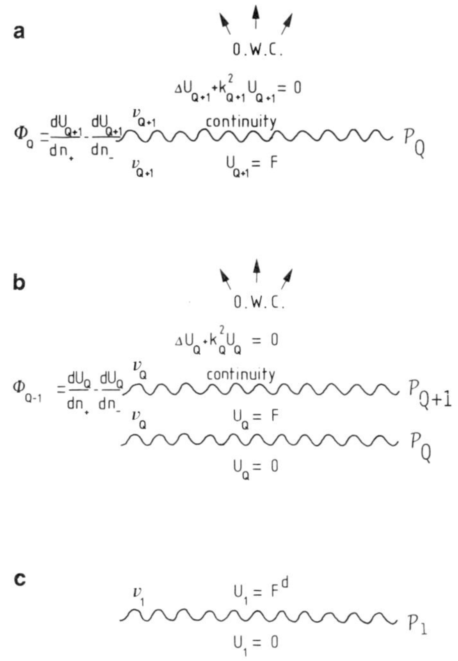

Fig. 3.6a-c. Definition of the unknown functions and calculation of the field. (a) Definition of ϕ_Q and calculation of the field below P_Q; (b) Definition of ϕ_{Q-1} and calculation of the field between P_{Q-1} and P_Q in terms of ϕ_{Q-1} and ϕ_Q (the limits of the field and its normal derivative above P_Q are known in terms of ϕ_Q); (c) Calculation of the diffracted field above P_1 in terms of ϕ_1 (the limits of the field and its normal derivative above P_1 are known in terms of ϕ_1)

Let us now consider in Fig.3.6b a new function $U_Q(x,y)$ satisfying the following conditions:

$U_Q = 0$ in M_{Q+1} (below P_Q) ,

$U_Q = F(x,y)$ in M_Q (between P_{Q-1} and P_Q) ,

$\Delta U_Q + k_Q^2 U_Q = 0$ above P_{Q-1} ,

U_Q is continuous on P_{Q-1} ,

U_Q satisfies an OWC when $y \to +\infty$.

Two fundamental properties of U_Q must be noted. First, since U_Q vanishes in M_{Q+1} and is equal to the actual field in M_Q, it satisfies everywhere (except on P_Q and P_{Q+1}) the Helmholtz equation with the coefficient k_Q^2. Thus, it is possible to write for U_Q the Helmholtz equation in the sense of the distributions with a right-hand member, similar to (1.85) but this time with two lines of discontinuity. In this equation will appear the jump of U_Q and its normal derivative when P_Q is crossed: these terms are already known as functions of ϕ_Q (arrow d). In the right-hand member, also appears the jump of the normal derivative of U_Q on P_{Q-1}, that we call ϕ_{Q-1}. Thus, by using the elementary solution $G^{(Q)}(x,y)$ of the Helmholtz equation, U_Q may be expressed in an integral form as a function of ϕ_Q and ϕ_{Q-1} (two arrows e). Now, the jump of $U_Q(\phi_{Q-1},\phi_Q)$ on P_Q is equal to $F_+(\phi_Q)$, however, we have to ensure that the limit of U_Q above P_Q is equal to $F_+(\phi_Q)$. This can be obtained by writing that the limit of U_Q below P_Q vanishes (arrow f). This condition gives rise to a linear relation between ϕ_Q and ϕ_{Q-1}. The inversion of a linear operator then leads to an evaluation of ϕ_Q in terms of ϕ_{Q-1} (arrow g). The use of this relation now permits us to know U_Q in terms of the single function ϕ_{Q-1}, then to express the field in M_Q in terms of ϕ_{Q-1} (arrows h). By taking the limit below P_{Q-1}, we deduce the values of the field and its normal derivative below P_{Q-1} (new arrow c); thus the boundary conditions on P_{Q-1} allows us to calculate the limit of the field and its normal derivative above P_{Q-1}. This makes it possible to define a new function $U_{Q-1}(x,y)$. In the same way, we can successively express the field in the region M_i as a function of a single unknown ϕ_{i-1}. Finally, the limit of the field and its normal derivative above M_1 may be written in terms of the single unknown ϕ_1.

Now, we have to find an integral equation in order to compute ϕ_1 and to deduce elementarily all the functions ϕ_i (i = 1, Q). To this aim, we now easily derive the limits of the diffracted field $F_+^d(\phi_1)$ and its normal derivative $dF^d(\phi_1)/dn_+$ above P_1 from the limits of the total field and its normal derivative (arrow m). So, let us now consider the function $U_1(x,y)$ defined in the following manner (Fig. 3.6c):

$U_1(x,y) = F^d$ in M_1 ,

$U_1(x,y) = 0$ below P_1 .

This function satisfies (3.11,12), thus U_1 can be expressed as a function of its jump $F^d_+(\phi_1)$ and of the jump $dF^d(\phi_1)/dn_+$ of its normal derivative (arrow n). We now have to ensure that the limit of U_1 above P_1 is equal to $F^d_+(\phi_1)$. This can be obtained by writing that the limit $U_{1-}(\phi_1)$ below P_1 vanishes. This condition gives rise to a single integral equation. After solving the integral equation, we can find the functions ϕ_i, then the field $F(\phi_i)$ in all the regions M_i in particular the coefficients of the plane wave expansions. It may be verified that, as seen in Sects.3.4 and 5, the integral equation includes all the conditions of the problem.

A little modification permits us to deal with the case where P_{Q+1} is perfectly conducting. In this case, it is sufficient to start with the function $U_Q(x,y)$. In the TE case, the value of U_{Q+} on P_Q is zero and the first unknown ϕ_1 is the value of the normal derivative dF/dn_+ on P_Q. On the other hand, in the TM case, the normal derivative on P_Q is zero and the first inknown function ϕ_1 is the value of the field F_+ on P_Q.

3.7 The Grating in Conical Diffraction Mounting

The term of conical diffraction is used as soon as the incident wave vector is not in the xOy plane. For a long time, interest in this type of mounting was restricted to a few rare experimenters [3.42,43]. It seems that the first rigorous theoretical studies on the subject were achieved in 1971 using an integral [3.44] or differential [3.45] method. Some time later, MAYSTRE and PETIT showed theoretically that, under particular conditions, the efficiency of a perfectly conducting ruled grating in conical diffraction could be constant and close to unity in a large domain of wavelength [3.46]. Finally, the same authors established a theorem of invariance *which gives the efficiency of a perfectly conducting grating in conical diffraction from the efficiencies of the same grating in two relevant fundamental cases of polarization, the wave vector lying this time in the xOy plane* [3.47]. The theorem of invariance has been the starting point of many theoretical and numerical studies. Since this theorem is not valid for metallic or dielectric gratings, an integral method was achieved in 1974 to investigate the relevance of conical diffraction mounts for metallic gratings in ultraviolet and visible regions [3.16,48]. This method needed the resolution of two coupled integral equations. More recently, a differential method [3.49], able to deal with coated gratings, confirmed the validity of the integral program. Since the numerical results for metallic gratings have shown that, with some modifications described in Chap.6, the invariance theorem is a good empiric rule for investigating the behavior of metallic gratings, we will be content with giving a demonstration of this theorem. This time, the direction of the incident wave is not only characterized by an angle of incidence θ (angle of the projection $\underline{k}_1^*$ of $\underline{k}_1$ on xOy plane with Oy axis) but also by the angle

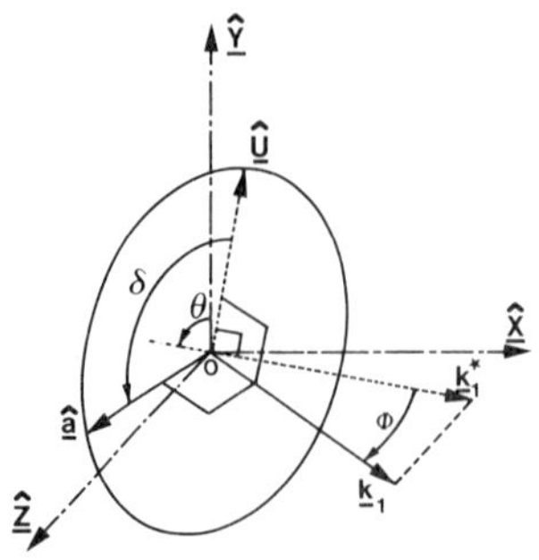

Fig. 3.7. Definition of the incidence angles θ, ϕ and the polarization angle δ, $\underline{k}_1^*$ is the projection of $\underline{k}_1$ on the xOy plane. The circle is constructed in the plane perpendicular to $\underline{k}_1$, and $\hat{\underline{U}}$ is the vector of the xOy plane which is orthogonal to $\underline{k}_1^*$ (and also to $\underline{k}_1$). All the vectors of xOy are drawn with a dashed line

ϕ between $\underline{k}_1$ and $\underline{k}_1^*$ (Fig.3.7), in such a way that $k_1^* = k_1 \cos\phi$. In these circumstances, the components α, $-\beta$ and γ of the incident wave vector $\underline{k}_1$ are given by

$$\alpha = k_1 \sin\theta \cos\phi \quad , \tag{3.69}$$

$$\beta = k_1 \cos\theta \cos\phi \quad , \tag{3.70}$$

$$\gamma = k_1 \sin\phi \quad . \tag{3.71}$$

The invariance of the problem after translation of the grating about Oz axis and the uniqueness of the solution, invoked here from the physical meaning, show that any component of the fields may be expressed in the form $u(x,y) \exp(i\gamma z)$. Thus, for simplicity's sake, the complex amplitude of such a component will be defined as the function $u(x,y)$. Using these notations, the incident electric field will be expressed in the following form:

$$\underline{E}^i(x,y) = \hat{\underline{a}} \exp[i(\alpha x - \beta y)] \quad . \tag{3.72}$$

The polarization vector $\hat{\underline{a}}$ (which is assumed to be real and of unit amplitude) which lies in the plane perpendicular to $\underline{k}_1$ (and oriented by $\underline{k}_1$) will be given by its angle δ with the vector $\underline{U}(\beta,\alpha,0)$ of this plane which is perpendicular to Oz (it is useful to notice that $\hat{\underline{U}}$ is the vector of xOy perpendicular to $\underline{k}_1^*$). It follows that

$$a_x = \cos\delta \cos\theta - \sin\delta \sin\theta \sin\phi \quad , \tag{3.73}$$

$$a_y = \cos\delta \sin\theta + \sin\delta \cos\theta \sin\phi \quad , \tag{3.74}$$

$$a_z = \sin\delta \cos\phi \quad . \tag{3.75}$$

From (3.72), Maxwell equations lead to the expression of the incident magnetic field

$$\underline{H}^i(x,y) = \underline{k}_1 \wedge \underline{E}^i(x,y)/\omega\mu_1 \quad .$$

Then, introducing $E_z^d \exp(i\gamma z)$ and $H_z^d \exp(i\gamma z)$ in Maxwell equations and taking into account the boundary conditions of the grating ($\hat{\underline{n}} \wedge \underline{E}_+ = 0$ and $\hat{\underline{n}} \cdot \underline{H}_+ = 0$), one shows that each of the two functions E_z^d and H_z^d satisfies conditions very close to those obtained for the two fundamental cases of polarization TE and TM (Sect.3.4). One finds, for example, that E_z^d which satisfies an OWC for $y \to +\infty$, also verifies the following equations:

$$E_z^d + k_1^{*2} E_z^d = 0 \quad \text{in} \quad M_1 \quad ,$$

$$E_z^d(M) = -E_z^i(M) = -a_z \exp[i(\alpha x - \beta y)] \quad \text{on} \quad P \quad .$$

By noting that $\alpha^2 + \beta^2 = k_1^{*2}$ and that $\alpha/\beta = \mathrm{tg}(\mathcal{C})$, we deduce that E_z^d/a_z is the solution of a classical TE fundamental case with the incidence θ and the wavelength $\lambda_1^* = 2\pi/k_1^* = \lambda_1/\cos\phi$. Because of the linearity of the Maxwell equations, we easily derive that when $y > \sup[f(x)]$

$$E_z^d = \sum_{m=-\infty}^{+\infty} a_z B_m \exp[i(\alpha_m x + i\beta_m y)] \quad ,$$

the coefficients B_m being naturally associated with the fundamental case TE (Sect. 3.4.1) treated with the incidence θ and the wavelength λ_1^*. In the same manner, one finds that when $y > \sup[f(x)]$

$$H_z^d = [(\alpha a_y + \beta a_x)/\omega\mu_1] \times \sum_{m=-\infty}^{+\infty} B_m' \exp(i\alpha_m x + i\beta_m y) \quad ,$$

the coefficients B_m' being now associated with the TM case of polarization (Sect. 3.4.2).

The components E_z and H_z being thus determined, the other components of the electromagnetic field are readily calculated by using formulas well known in waveguide theory [3.50] and resulting directly from the Maxwell equations.

After rather tedious calculations, we find that if $y > \sup[f(x)]$

$$E^d = \sum_n (P_n\hat{\underline{x}} + Q_n\hat{\underline{y}} + R_n\hat{\underline{z}} \exp[i(\alpha_n x + \beta_n y)] \quad , \tag{3.76}$$

with

$$\alpha_n = \alpha + nk \quad , \tag{3.77}$$

$$\beta_n{}^2 = \sqrt{k_1^{*2} - \alpha_n{}^2} = \sqrt{k_1^2 - \alpha_n^2 - \gamma^2} \text{ or } i\sqrt{\alpha_n{}^2 - k_1^{*2}} \quad , \tag{3.78}$$

$$P_n = -\frac{1}{k_1 \cos\phi}(\beta_n B_n' \cos\delta + \alpha_n B_n \sin\phi \sin\delta) \quad , \tag{3.79}$$

$$Q_n = \frac{1}{k_1 \cos\phi}(\alpha_n B_n' \cos\delta - \beta_n B_n \sin\phi \sin\delta) \quad , \tag{3.80}$$

$$R_n = B_n \sin\delta \cos\phi \tag{3.81}$$

If we remember that the component on z of the wave vector diffracted in the order n is γ, we can deduce from (3.76-78) the components of the unit vector $\hat{\underline{u}}_n$ of the direction of propagation of the order n

$$u_{nx} = \sin\theta \cos\phi + n\lambda_1/d \quad , \tag{3.82}$$

$$u_{ny} = \sqrt{1 - u_{nx}^2 - u_{nz}^2} \quad , \tag{3.83}$$

$$u_{nz} = \sin\phi \quad . \tag{3.84}$$

Denoting by θ_n the angle of Oy with projection $\underline{u}^*_n$ of $\hat{\underline{u}}_n$ on xOy plane, and by ϕ_n the angle between $\hat{\underline{u}}_n$ and $\underline{u}^*_n$, we can derive the generalized formula of gratings

$$\sin\theta_n = \sin\theta + n\,\lambda_1/d\,\cos\phi \quad , \tag{3.85}$$

$$\phi_n = \phi \quad . \tag{3.86}$$

The result is therefore particularly simple: the new grating formula may be derived from the classical one by replacing the actual wavelength λ_1 by a longer fictitious wavelength $\lambda_1^* = \lambda_1/\cos\phi$. Figure 3.8 permits a geometrical construction of the diffracted directions.

Finally, (3.76,79-81) allow us to calculate the efficiency e_n of the grating

$$e_n = \frac{\beta_n}{\beta}\,[\sin^2\delta\,|B_n|^2 + \cos^2\delta\,|B_n'|^2] \quad . \tag{3.87}$$

The efficiency reaches its extreme values for $\delta = 0$ and $\delta = \pi/2$ and, in addition, these two values correspond to the two fundamental states of polarization which we denoted TE and TM in classical diffraction. By generalization, we will call these two fundamental cases of polarization for conical diffraction TE and TM.

From (3.87), we can finally state the *"Theorem of invariance"*:

If the incident wave varies so that the projection of its wave vector on the principal section plane of the grating is a fixed vector (i.e., the incidence θ and $\lambda_1^* = \lambda_1/\cos\phi$ are fixed):

a) *The projection on the same plane of the wave vector diffracted in an arbitrary order m also remains a fixed vector,*

b) *If the grating is perfectly conducting, the efficiency in the order m does not vary either, provided the state of polarization of the incident wave* (angle δ of the incident electric field with $\underline{k}_1 \wedge \hat{\underline{z}} = k_1^* \wedge \hat{\underline{z}}$) *is fixed.*

We will see in Chap. 6 some interesting applications of this theorem.

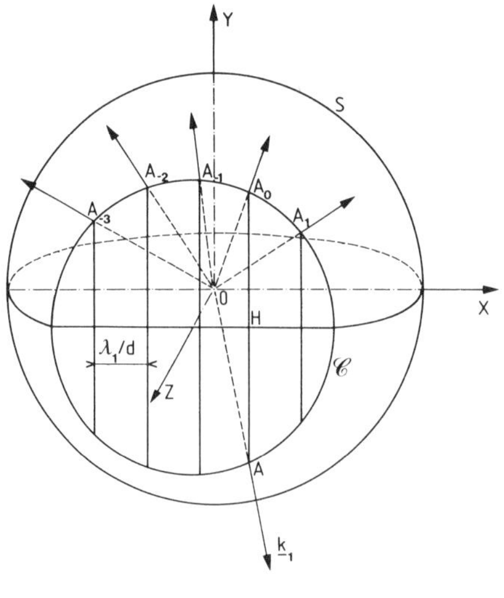

Fig. 3.8. Let S be a sphere of center O and unit radius and A the intersection of the hemisphere (y < 0) with its diameter parallel to the incident wave vector $\underline{k}_1$. The plane Π passing through A and perpendicular to Oz cuts S along a circle C. If $A_0, A_1, A_2, \ldots, A_j$ are the intersections with C of a family of lines parallel to Oy and equidistant from λ_1/d, we verify according to (3.82-86) that OA_0, OA_1, $OA_2, \ldots OA_j$ are, respectively, equal to the unit vectors $\hat{\underline{u}}_1, \hat{\underline{u}}_2, \ldots \hat{\underline{u}}_j$ of the directions of diffraction

3.8 Numerical Application

Using the computer, we now have to solve numerically the set of integral equations presented above. We shall restrict the study to the equations (3.33,37,40, 47,67) obtained for uncoated (perfectly conducting, dielectric or metallic) gratings but the methods proposed later can easily be extended to the other cases. An arbitrary integral equation of this chapter can be written in the following symbolic form:

$$\mu\, u(x) = v(x) + \int_0^d W(x,x')\, u(x')\, dx' \quad , \qquad (3.88)$$

where u, v and W are periodic functions, with W possibly singular but integrable, μ being equal to 1 or 0 according to whether the integral equation is of the second or of the first kind[9].

3.8.1 A Fundamental Preliminary Choice

Essentially two types of methods may be distinguished for solving (3.88).

The first type of method, called FSM in Chap. 1 consists in representing the unknown function u(x) by the coefficients of its exponential Fourier series

$$u(x) = \sum_{n=-\infty}^{+\infty} u_n \exp(inKx) \quad .$$

Then, denoting by v_n and W_{nm} the Fourier coefficients of v(x) and W(x,x') leads us to transform the integral equation (3.88) in an infinite set of linear equations with a right-hand member

$$\forall n \in (-\infty, +\infty) : \sum_{m=-\infty}^{+\infty} (W_{nm} - \mu\delta_{n,m})u_m = -v_n \quad . \qquad (3.89)$$

A numerical resolution of (3.89) needs a truncation, viz., the limitation of the Fourier series of u(x) and v(x) to 2N + 1 terms

$$\forall n \in (-N, +N) : \sum_{m=-N}^{N} (W_{nm} - \mu\delta_{n,m})u_m = -v_n \quad . \qquad (3.90)$$

So, we obtain a system of 2N+1 linear equations with a right-hand member, containing 2N+1 unknown coefficients u_m. The solving of such a set of equations can easily be performed by computer, using for example the Gauss method. Let us notice that this method requires essentially the computation of (2N+1) double integrals to determine the coefficients W_{nm}

$$W_{nm} = \frac{1}{d^2} \int_0^d \int_0^d W(x,x') \exp(-inKx - imKx')\, dx\, dx' \quad ,$$

9 PAVAGEAU proposed an iterative method for the solving of its integral equation of the second kind [3.10,51]. Unfortunately, after implementation of this method, we have been led to the conclusion that it does not always converge for any commercial grating [3.31]. This is the reason why this interesting method which does not require the inversion of a matrix will not be presented here.

$W(x,x')$ being a series. This explains why, in practice, the field of application of this method has been restricted to the cases where these integrals can be analytically calculated, without numerical integration. This occurs when the profile of the grating can be described by a small number of segments of lines: W_{nm} is then obtained by summing a series whose terms are known in closed form. The method has been applied to ruled gratings (see, e.g. [3.8,30]) or trapezoidal gratings [3.52]. Specially due to these practical limitations, this method, to our knowledge, has nowadays been practically abandoned.

The most commonly used method is called the point-matching method. Here the unknown function $u(x)$ is represented by its J values $u_j = u(x_j)$ at J points x_j of the interval $(0,d)$. Then, setting $v_j = v(x_j)$ allows us to write (3.88) in the form

$$\forall j \in (1,J), \quad \mu u_j = v_j + \int_0^d W(x_j,x')\, u(x')\, dx' \quad . \tag{3.91}$$

We have to compute the J integrals in the right-hand member of (3.91) in terms of a linear combination of the J unknown values u_j, in order to obtain an inhomogeneous system of J linear equations with J unknowns. So, we are led to set now

$$\forall j \in (1,J), \quad \int_0^d W(x_j,x')\, u(x')\, dx' \simeq \sum_{\ell=1}^{J} W_{j\ell} u_\ell \quad , \tag{3.92}$$

and finally, we find the following set of linear equations:

$$\forall j \in (1,J), \quad \sum_{\ell=1}^{J} (W_{j\ell} - \delta_{j\ell}) u_\ell = -v_j \quad . \tag{3.93}$$

The whole problem is the determination of the J^2 coefficients $W_{j\ell}$, in order to satisfy (3.92) as well as one can. Naturally, this requires some preliminary information on the functions $u(x)$ and $W(x,x')$. In the following, we assume that $u(x)$ is a continuous function. That is obviously the case when the grating profile has no edges. We will see (Sect.3.8.5) how the problem of edges has been treated but, for the moment, we assume the continuity of $u(x)$. On the other hand, $W(x,x')$ is equal to $G^{(j)}(x,x')$, $N^{(j)}(x,x')$, $N'^{(j)}(x,x')$ defined above and sometimes to a sum of convolutions of these functions. Therefore, we first have to study these three kernels.

3.8.2 Study of the Kernels

In order to simplify the expressions of the various kernels, it is convenient to set now

$$(x - x') = a \quad , \quad |y - y'| = b \quad ,$$

and to develop β_n in terms of α_n, using (3.8)

$$\beta_n^{(j)} = i|\alpha_n| - \frac{ik_j^2}{2|\alpha_n|} - \frac{i}{8}\,\frac{k_j^4}{|\alpha_n|^3} + \quad \dots \quad .$$

This allows us to give the asymptotic values ($n\to\infty$) of $\beta_n^{(j)}$ and $1/\beta_n^{(j)}$, used in the following:

$$\beta_n^{(j)} \simeq i|\alpha_n| = i|nK + k_1 \sin\theta| \quad , \tag{3.94}$$

$$\frac{1}{\beta_n^{(j)}} \simeq \frac{-i}{|\alpha_n|} \simeq \frac{-i}{|n|K} \quad . \tag{3.95}$$

First, let us consider the kernel $G^{(j)}$ given by (3.19)

$$G^{(j)}(x,x') = \sum_{n=-\infty}^{+\infty} \frac{1}{2id\beta_n^{(j)}} \exp(i\beta_n^{(j)}a + inKb) \quad . \tag{3.96}$$

It is worth noting that $a = b = 0$ when $x = x'$; thus taking into account (3.95) the series does not converge. Naturally a very slow convergence can be expected when x' is close to x. Different techniques have been used for accelerating the convergence of the series, For example, NEUREUTHER and ZAKI have employed a transformation technique based on the use of Mellin transforms [3.12]. As for us, in order to get over these difficulties, we have set

$$G^{(j)} = G_\infty + (G^{(j)} - G_\infty) \quad , \tag{3.97}$$

where G_∞ is obtained from $G^{(j)}$ by replacing $1/\beta_n^{(j)}$ and $\beta_n^{(j)}$ by their asymptotic expression stated in (3.94,95)

$$\begin{aligned} G_\infty(x,x') = &\sum_{n=1}^{\infty} -\frac{1}{4\pi n} \exp(-k_1 a \sin\theta) \exp[-nKa + inKb] \\ &+ \sum_{n=-1}^{-\infty} \frac{1}{4\pi n} \exp(k_1 a \sin\theta) \exp[nKa + inKb] \quad . \end{aligned} \tag{3.98}$$

The term in parenthesis in (3.97) is a series, each term of which is obtained from the two n^{th} terms in (3.96,98). Furthermore, it is easy to show that, by adding now the terms $+n$ and $-n$ of this new series, we finally obtain a rapidly convergent series from 1 to ∞, whose terms decrease like $1/n^3$ when $x' = x$. Moreover, this series is continuous when $x = x'$ and its value is simply given by

$$G^{(j)}(x,x) - G_\infty(x,x) = \frac{1}{2id\beta_0} + \sum_{n\neq 0} \left(\frac{1}{2id\beta_n} - \frac{1}{4\pi|n|}\right) \quad . \tag{3.99}$$

We deduce that a good approximation of the sum of this series can readily be computed by summing the terms for which $|n| < T$, we will see later how T must be chosen. Obviously, the singularity and the slow convergence of $G^{(j)}$ have been carried over to the series G_∞, represented by the first term in the right-hand member of (3.97). Fortunately, since G_∞ is the sum of two series having the form

$$\sum_1^\infty \frac{q^n}{n} \quad ,$$

it can be summed in closed form

$$G_\infty = \frac{1}{4\pi} \exp(-k_1 a \sin\theta) \log[1 - \exp(-Ka + iKb)] + \frac{1}{4\pi} \exp(+k_1 a \sin\theta) \log[1 - \exp(-Ka - iKb)] \quad . \tag{3.100}$$

Clearly, this function is singular when x = x'. After some calculations, it can be found from (3.100) that

$$\lim_{x' \to x} G_\infty(x,x') = \frac{1}{4\pi} \log[K^2(a^2 + b^2)] \quad .$$

Hence, noting that ~~b~~ a → [(x' - x)f'(x)] when x' → x yields

$$\lim_{x' \to x} G_\infty(x,x') = \frac{1}{2\pi} \left[\log(2\pi) + \frac{\log(1 + f'^2)}{2} + \log \frac{|x' - x|}{d} \right] \quad . \tag{3.101}$$

We conclude that G_∞ is the sum of a regular part and of a term which having a logarithmic singularity, is therefore integrable.

Let us consider now the kernel $N^{(j)}(x,x')$ given by (3.23)

$$N^{(j)}(x,x') = \frac{1}{2d} \sum_n \left\{ \operatorname{sgn}[f(x) - f(x')] - \frac{\alpha_n}{\beta_n^{(j)}} f'(x) \right\} \times \exp(i\beta_n^{(j)} a + inKb). \tag{3.102}$$

The term sgn[f(x) - f(x')] would seem to indicate that this kernel is not continuous when x' = x. In fact, proceeding in the same way as in the above section, we can define an asymptotic value $N_\infty(x,x')$ of $N^{(j)}(x,x')$ and we set now

$$N^{(j)}(x,x') = N_\infty(x,x') + [N^{(j)}(x,x') - N_\infty(x,x')] \quad , \tag{3.103}$$

with

$$N_\infty(x,x') = \frac{1}{2d} \operatorname{sgn}[f(x) - f(x')] + \frac{1}{2d} \{\operatorname{sgn}[f(x) - f(x')] - if'(x)\} \times \sum_{n=1}^{\infty} \exp(-k_1 a \sin\theta) \exp(-nKa + inKb) + \frac{1}{2d} \{\operatorname{sgn}[f(x) - f(x')] + if'(x)\} \times \sum_{n=-1}^{-\infty} \exp(k_1 a \sin\theta) \exp(nKa + inKb) \quad , \tag{3.104}$$

$$= \frac{1}{2d} \operatorname{sgn}[f(x) - f(x')] + \frac{1}{2d} \{\operatorname{sgn}[f(x) - f(x')] - if'(x)\} \exp(-k_1 a \sin\theta) \times \frac{1}{\exp(Ka - iKb) - 1} + \frac{1}{2d} \{\operatorname{sgn}[f(x) - f(x')] + if'(x)\} \times \exp(k_1 a \sin\theta) \times \frac{1}{\exp(Ka + iKb) - 1} \quad . \tag{3.105}$$

Now, the term in parenthesis in (3.103) must be considered as a series, each term of which is obtained from the corresponding terms of (3.102,104). It converges more rapidly than the series given by (3.102), is continuous and can be computed by summing the terms for which $|n| < T$. The first term in the right-hand side of (3.103) must be considered as the function given by (3.105). Calculating the limit of the two terms contained in the right-hand side of (3.103) when $x' \to x$, we may deduce, after tedious calculations, that $N^{(j)}(x,x')$ is continuous when $x = x'$ and that its limit is given by

$$N^{(j)}(x,x) = \lim_{x' \to x} [N^{(j)}(x,x')]$$
$$= -f'(x)\left[\frac{1}{2d}\sum_{n=-\infty}^{+\infty}\frac{\alpha_n}{\beta_n} + \frac{i}{2\pi}k_1 \sin\theta\right] - \frac{1}{4\pi}\frac{f''(x)}{1+f'^2(x)} , \qquad (3.106)$$

$f''(x)$ denoting the second derivative of the function $f(x)$. This curious result, established by PAVAGEAU [3.51], is very important for the numerical applications. The series contained in the right-hand member of (3.106) converges like $1/n^3$ (after addition of the nth and $-n^{th}$ terms). Let us note that the calculation of $f''(x)$ clearly requires the continuity of $f'(x)$, i.e., the absence of edges, as stated at the beginning of this section.

Obviously, the study of $N'^{(j)}(x,x')$ given by (3.20) can be derived from the study of $N^{(j)}(x,x')$ by replacing $f'(x)$ by $f'(x')$ in (3.102-105).

After calculations, this permits us to show the continuity of $N'^{(j)}(x,x')$:

$$N'^{(j)}(x,x) = \lim_{x' \to x} [N'^{(j)}(x,x')] \qquad (3.107)$$
$$= -f'(x)\left[\frac{1}{2d}\sum_{n=-\infty}^{+\infty}\frac{\alpha_n}{\beta_n} + \frac{i}{2\pi}k_1 \sin\theta\right] + \frac{1}{4\pi}\frac{f''(x)}{1+f'^2(x)} .$$

3.8.3 Integration of the Kernels

The above section shows that $N^{(j)}$ and $N'^{(j)}$ are continuous. On the other hand, $G^{(j)}$ is the sum of a continuous series and of a singular function. Thus, it appears very advisable to investigate the rules concerning the integration of a continuous periodic function over the whole period. Since such a function may have a discontinuous first derivative, the well-known formulae giving the precision of integrations performed using classical methods (trapezoidal integration, Simpson, Weddle, Gauss and so on) are inadequate. Indeed, these formulae require the knowledge of the maximum value of the nth derivative of the integrand ($n > 2$) for the whole period.

In order to find an accurate method for integrating such a function, one must notice that the choice of the two ends of the interval (respectively equal to 0 and d in our equations) may be arbitrarily chosen, provided that their distance is equal to d. Thus, no point of the interval (0,d) plays a particular role in the integration and it may seem reasonable to exclude any method in which the choice of the weights attributed to the points of integration, or the distance between two consecutive points, is not constant. Hence, we are led to the rectangular rule, with constant segment length. It is not our purpose to prove the superiority of this method from a mathematical point of view, but we want to give an intuitive explanation of its relevance.

Let us consider the following integral:

$$\rho = \int_0^d a(x)\, dx \quad ,$$

where $a(x)$ is a continuous function of period d.

By sampling the interval (0,d) with J points x_j, in such way that

$$0 = x_1 < x_2 < x_3 \ldots < x_J < d \quad ,$$

an arbitrary method of integration leads us to approximate the integral by a finite sum ρ'

$$\rho \simeq \rho' = \sum_{j=1}^{J} P_j\, a(x_j) \quad , \quad \text{with} \sum_{j=1}^{J} P_j = d \quad . \tag{3.108}$$

Since the function $a(x)$ is periodic, we can more generally define a function $\chi(\xi)$ by

$$\chi(\xi) = \sum_{j=1}^{J} P_j\, a(\xi + x_j) \quad .$$

Obviously, the function $\chi(\xi)$ has a period d and it is interesting to note that

$$\frac{1}{d}\int_0^d \chi(\xi)\, d\xi = \rho \quad . \tag{3.109}$$

So by setting

$$\chi(\xi) = \chi_0 + \chi_1(\xi) \quad , \quad \text{with} \int_0^d \chi_1(\xi)\, d\xi = 0 \quad ,$$

it follows easily from (3.107) that $\chi_0 = \rho$, and that $\chi_1(\xi)$, fluctuation of $\chi(\xi)$ over the interval (0,d) is nothing else than the error in the integration. In the general case, $\chi_1(\xi)$, whose average value is zero, has a period d. On the other hand, with the rectangular rule (which here is identical to the trapezoidal one), we get

$$P_j = d/J \quad \text{and} \quad x_j = (j - 1)d/J \quad , \tag{3.110}$$

and it can easily be deduced that the period of $\chi_1(\xi)$ is d/J, i.e., J times less than the period of χ_1 in the general case. For sure, this is not a mathematical proof of the superiority of the rectangular rule. Still, it is easy to demonstrate that this method is preferable to any other method using equal length segments of integration (Simpson, Weddle ...). Indeed, let us consider the J values $\chi(x_j)$ obtained by setting $\xi = x_j$ in an arbitrary method using J equal length segments. It can elementarily be shown that

$$\frac{1}{J}\sum_{j=1}^{J} \chi(x_j) = \chi_R(0) \quad ,$$

where $\chi_R(0)$ is the value of the integral performed by the rectangular method using J samples. So, $\chi_R(0)$ is the average value of the J integrals $\chi(x_j)$. Intuitively, this is a good indication of the superiority of the rectangular rule and mathematically speaking, this result implies that, at least, one of the integrals $\chi(x_j)$ is less good than $\chi_R(0)$.

By noting that the exact value ρ of the integral is nothing else than the product of d by the zeroth coefficient a_0 of the expansion of a(x) in Fourier series

$$a(x) = \sum_n a_n \exp(inKx) \quad ,$$

a simple calculation shows that the error ε_R in the rectangular method is given by

$$\varepsilon_R = \frac{1}{d}\sum_{k\neq 0} a_{k'} \qquad \text{with} \quad k' = k \times J \quad ,$$

where k denotes an integer different from zero. Since a(x) is continuous, the coefficients a_n at least decrease like $1/n^2$ and, in practice, the order of magnitude of the error is given by a_J, hence we conclude that the error decreases more rapidly than $1/J^2$.

The implementation of the rectangular rule is very easy and using (3.92,100), it follows elementarily that

$$W_{j\ell} = W[(j - 1)d/J, (\ell - 1)d/J] \times d/J \quad . \tag{3.111}$$

Using this relation allows us to integrate the expression containing $N^{(j)}(x,x')$ and $N'^{(j)}(x,x')$ by using (3.103-107).

The above relation also enables us to integrate the continuous part of G^j [viz., the second term of the right-hand member of (3.97), using (3.99)].

On the other hand, the first term of (3.97), which is a singular function in closed form, requires a more sophisticated treatment. The integration of $G_\infty(x,x')\tilde{\phi}(x')$ requires the subtraction from (3.100) of the singular logarithmic part expressed in (3.101). Unfortunately, $\log |x' - x|/d$ is not periodic and we know that the rectangular rule is very poor when applied to a nonperiodic function. In order to get a ride of this first difficulty, it is useful to note that the function G'_∞ given by

$$G'_\infty = \frac{1}{2\pi}\left[\log \frac{|x' - x|}{d} + \log\left(1 - \frac{|x' - x|}{d}\right)\right] \quad , \tag{3.112}$$

which has the same singularity as $(1/2\pi) \log |x' - x|/d$ in the interval (0,d)[10] can be considered as the restriction in the interval (0,d) of a periodic function, continuous everywhere except on the singularity (x' = x). Indeed, if x is fixed,

---

10 Indeed, d - |x' - x| never vanishes when x' and x remain in the interval (0,d).

all the derivatives of $G'_\infty(x,x')$ with respect to x' are the same in x' = 0 and x' = d (see [3.16]).

We then perform the integration by setting

$$\int_0^d G_\infty(x,x')\tilde{\phi}(x')dx' = \int_0^d [G_\infty(x,x')\tilde{\phi}(x') - G'_\infty(x,x')\tilde{\phi}(x)]dx' + \tilde{\phi}(x) \int_0^d G'_\infty(x,x')dx' \quad . \tag{3.113}$$

The first term of the right-hand side of (3.113) is continuous and can be integrated using the rectangular rule, if we know that

$$\lim_{x'\to x} [G_\infty(x,x')\tilde{\phi}(x') - G'_\infty(x,x')\tilde{\phi}(x)] = \frac{1}{2\pi}\left[\log 2\pi + \frac{1}{2}\log(1 + f'^2(x))\right]\tilde{\phi}(x) \quad .$$

Furthermore, the second term, which contains the singular part, can be analytically found

$$\forall x, \quad \int_0^d G'_\infty(x,x')dx' = -d/\pi \quad . \tag{3.114}$$

In conclusion, first the introduction of G_∞ enables us to define a series $G^{(j)} - G_\infty$ which is continuous, rapidly convergent and hence which can easily been integrated by the rectangular rule. Second, the integration of the term containing the function in closed form G_∞ is performed by defining a new function G'_∞ which is periodic, has the same singularity as G_∞ and can be analytically integrated.

3.8.4 Particular Difficulty Encountered with Materials of High Conductivity

When ν_j is the index of a highly conducting material (for example aluminum above 0,4 μm), the considerations described in the preceding section fall down. Indeed, as explained in Sect.3.5.6, the function $G^{(j)}$, $N^{(j)}$ and $N'^{(j)}$ tend towards delta distributions. This entails that the two terms of the integral of (3.92) are fundamentally different in nature: the term u(x') (the unknown function) is continuous and might be easily integrated using the rectangular rule, but since the kernel W(x,x') tends towards a delta distribution, the integration of W(x,x')u(x') becomes more and more difficult when $|\nu_j|$ increases, and we are forced to adopt more and more samples. Finally, the size J × J of the linear system becomes too great for the computer.

In order to get over this difficulty, it is interesting to use the integral form of the Hankel function which allows us to write [3.16]

$$G^{(j)}(x,y) = \frac{1}{4i}\sum_m \exp(-imk_1 d \sin\theta)\, H_0^+(k_j|\underline{r} + md\hat{\underline{x}}|) \quad , \tag{3.115}$$

where $\underline{r}$ denotes the radius vector associated with the point P(x,y) of the space. By "putting the pieces together again" in (3.115), we can show that in the integrals containing $G^{(j)}$, we can leave out the sign $\sum$ if we extend the integral to infinity

$$G^{(j)}\{\tilde{\phi}\} = \frac{-i}{4} \int_{-\infty}^{+\infty} H_0^+(k_j\ MM') \exp[ik_1(x' - x) \sin\theta]\tilde{\phi}(x')dx' \quad , \tag{3.116}$$

where MM' is the distance between M[x,f(x)] and M'[x',f(x')].
In the same way, it can be shown that

$$N'^{(j)}\{\tilde{\phi}\} = \frac{k_j i}{4} \int_{-\infty}^{+\infty} \frac{[f(x') - f(x)] - (x' - x)f'(x')}{MM'} H_1^+(k_j MM') \times \exp[ik_1(x' - x) \sin\theta]\tilde{\phi}(x')dx' \quad . \tag{3.117}$$

When $|\nu_j|$ is large, $H_0^+(k_j MM')$ and $H_1^+(k_j MM')$ tend towards delta distributions and we have used the "local summation method". At the first order (the interested reader may report to [3.16] for better approximation), this method consists in setting that $\tilde{\phi}(x')$ is constant and equal to $\tilde{\phi}(x)$ in the very small interval of x' where the Hankel functions are not very close to zero. For example, from (3.116), it follows

$$G^{(j)}\{\tilde{\phi}\} = \frac{-i}{4} \tilde{\phi}(x) \int_{-\infty}^{+\infty} H_0^+(k_j\ MM') \exp[ik_1(x' - x) \sin\theta]dx' \quad . \tag{3.118}$$

Noting that, when x' is close to x

$$MM' \simeq \sqrt{1 + f'^2(x)}\ |x' - x| \quad ,$$

$$f(x') - f(x) - (x' - x)f'(x') \simeq -f''(x)(x' - x)^2/2 \quad ,$$

$$\exp[ik_1(x' - x) \sin\theta] \simeq 1 \quad ,$$

we deduce that the integral in the right-hand member of (3.118) can be analytically performed and it yields

$$G^{(j)}\{\tilde{\phi}\} = \frac{\tilde{\phi}(x)}{2ik_j\sqrt{1 + f'^2}} \quad ,$$

and in the same way

$$N'^{(j)}\{\tilde{\phi}\} = \frac{i}{4k_j} \frac{f''(x)\ \tilde{\phi}(x)}{(1 + f'^2)^{3/2}} \quad .$$

It is surprising to note that in this approximation, the calculation of the integral requires neither summation of the kernel, nor integration: the operators $G^{(j)}$ and $N'^{(j)}$ tend towards multiplicative operators and the coefficients $W_{n\ell}$ tend towards zero, except when $n = \ell$ where they can be known in closed form. The numerical results have shown that this method not only can be applied in the domain when the classical integration fails but also remains valid with a relative accuracy of one percent in a large part of the domain where the classical integration works. For example, with aluminum, both methods work in the visible ($0.4 < \lambda < 0.8$ [μm]). In consequence, better approximation for the "local summation" may be avoided.

3.8.5 The Problem of Edges

When the grating surface has edges, fundamental difficulties appear. First, the uniqueness of the solution is not ensured. In fact, it has been shown in [3.16] that the hypothesis of integrability of the unknown u(x) of the integral equation (necessary for the solving of this equation) is equivalent to the Meixner condition of integrability and ensures the uniqueness. The second problem is the integration in the vicinity of the edges. The functions $N^{(j)}$ and $N'^{(j)}$ become discontinuous or, worse, meaningless, according to whether x is the abscissa of the edge or not [see (3.105-107)]. Moreover, the unknown function may be singular on the edge (see for example [3.51]). To overcome this difficulty, one can use appropriate transformations for the kernels, the unknown functions and the variables x and x' [3.16] in order to obtain nonsingular and well-defined functions. In our opinion, a simpler and very efficient solution consists in replacing the actual profile f(x) of the grating by the profile $\tilde{f}_F(x)$ corresponding to the summation of the F first coefficients of its Fourier series

$$\tilde{f}_F(x) = \sum_{p=0}^{F} c_p \cos(pKx) + \sum_{p=1}^{F} s_p \sin(pKx) \quad .$$

The new profile so obtained has no edges and by using various tests (for example by testing the convergence of the results when F is increased), we observed that accurate results are obtained as soon as the number of points J becomes greater than about $4 \times F$, provided that the number F exceeds about 8 [3.19][11]. It is very important to note that this very simple method allows us to deal with some types of profiles which are not described by a function f(x). This is the case for lamellar gratings [since the function $\tilde{f}_F(x)$ may be defined] provided that F exceeds about 10.

3.8.6 Precision on the Numerical Results

In order to check the numerical programs, several tests can be used. First, we can verify the convergence of the results when J and T are increased. Roughly, by using the numerical solutions described above and for commercial gratings, this convergence is obtained as soon as $J > 6\nu_1 d/\lambda_0$ and $T > 4\nu_1 d/\lambda_0$, provided that J and T do not fall below 20 and 15, respectively. In other words, the rule is to adopt about 6 samples per wavelength and to take $T = 2J/3$. Naturally, these numbers must be increased when the groove depth exceeds about 0.4d. On the other hand, these numbers can be notably reduced when the wavelength to period ratio becomes small and some computations have been made with two samples by wavelength, in a case where the wavelength to period ratio was of the order of 1/15.

---

11 Mathematical considerations on the kernels may also explain this result.

Numerous other tests have been employed. For perfectly conducting or dielectric gratings, we have implemented the energy balance criterion [see (1.63)]. For perfectly conducting ruled gratings, we have used the theorem of MARECHAL and STROKE [3.53] which gives analytically the efficiencies of this type of grating in S polarization, with some particular conditions of incidence and wavelength, or its generalization to the lamellar grating [3.16]. We have also verified (3.35,42). For dielectric and metallic gratings, we have checked that all the efficiencies vanish when the groove depth tends towards zero, except the efficiency in the zero order which takes the value given by the Fresnel formulae. In all the cases, we have verified that the theorem of reciprocity described in Chap 2 was satisfied [3.54].

All these criteria have been made in the resonance region with a relative accuracy better than 1%, for all types of gratings whatever the metal and the wavelenth, provided that J and T are chosen as explained above. Concerning the shape of the grating, the case where the groove depth exceeds 3d has been treated for sinusoidal profiles. Some computations have been made with $\lambda_0/d \simeq 1/20$. i.e., practically in the scalar domain. The reader will see in the following chapters that many other tests have been realized, in particular the comparison with experiments, with the results issued from the differential formalism and, for lamellar gratings, with the results issued from the modal theory. If one is content with the precision of 1%, the computation time on a CDC 7600 computer is of the order of a fraction of a second for an uncoated perfectly conducting grating, one second for an uncoated metallic or dielectric grating, and several seconds for a coated metallic or dielectric grating, except for the "buried grating" where only one interface is not plane and which requires the same computation time as the uncoated grating. Greater precision requires a longer computation time.

A last advantage of the integral theory should be noticed. For a given shape of the gratings, one generally wants the computation of efficiencies for both polarizations. Furthermore, it can be interesting to compare the efficiencies of a metallic grating with those obtained when the lower medium is assumed to be perfectly conducting. So, it frequently occurs that the solving of two or four mathematical problems is needed (TE and TM polarizations, metallic and perfectly conducting gratings). Consequently, we have to solve two or four integral equations. Now, we have seen that these integral equations are very close related, therefore, the major part of the numerical calculations of these problems can be conducted simultaneously. Of course, this is very interesting for the computer time and this fortunate feature must be taken into account for an actual comparison with the performances of other methods.

References

3.1 R. Courant, D. Hilbert: *Methods of Mathematical Physics*, Vol. 1 (Interscience, New York 1965) Chap. 3, pp. 112-163
3.2 P.M. Morse, H. Feshbach: *Method of Theoretical Physics*, Part 1 (Mc Graw-Hill, New York 1953) Chap. 8, pp. 896-997
3.3 L. Schwartz: *Methodes Mathématiques pour les Sciences Physiques* (Hermann, Paris 1965)
3.4 L. Schwartz: *Théorie des Distributions* (Hermann, Paris 1966)
3.5 R. Petit, M. Cadilhac: C.R. Acad. Sci. Paris 259, 2077 (1964)
3.6 A. Wirgin: Rev. Opt. 9, 449 (1964)
3.7 J.L. Uretski: Ann. Phys. 33, 400 (1965)
3.8 R. Petit: C.R. Acad. Sci. Paris 260, 4454 (1965)
3.9 R. Petit: Rev. Opt. 45, 249 (1966)
3.10 J. Pavageau, R. Eido, H. Kobeissé: C.R. Acad. Sci. Paris 264, 424 (1927)
3.11 A. Wirgin: Rev. Cethedec 5, 131 (1968)
3.12 A. Neureuther, K. Zaki: Alta Freq. 38, 282 (1969)
3.13 P.M. Van den Berg: Thesis, Delft, the Netherlands (1971)
3.14 D. Maystre: Opt. Commun. 6, 50 (1972)
3.15 D. Maystre: Opt. Commun. 8, 216 (1973)
3.16 D. Maystre: Thèse d'Etat, Marseille A0 9545 (1974)
3.17 R. Petit, D. Maystre, M. Nevière: Space Optics Proc. 9th Congr. I.C.O., 667 (1972)
3.18 M. Nevière, P. Vincent, R. Petit: Nouv. Rev. Opt. 5, 65 (1974)
3.19 D. Maystre: J. Opt. Soc. Am. 68, 490 (1978)
3.20 L.C. Botten: Opt. Acta 25, 481 (1978)
3.21 D. Maystre: Opt. Commun. 26, 127 (1978)
3.22 L.C. Botten: Ph. D. Thesis, Tasmania, Hobart (1978)
3.23 J. Meixner: IEEE Trans. AP-20, 442 (1972)
3.24 G. Dumery, P. Filippi: C.R. Acad. Sci. Paris 270, 137 (1970)
3.25 J. Bass: *Cours de Mathématiques*, Vol. 3 (Masson, Paris 1971)
3.26 H. Kalhor, A. Neureuther: J. Opt. Cos. Am. 61, 43 (1971)
3.27 P.M. van den Berg: Appl. Sci. Res. 24, 261 (1971)
3.28 A. Wirgin: Thesis, Paris (1967)
3.29 R. Green: IEEE Trans. MTT-18, 313 (1970)
3.30 D. Maystre, R. Petit: C.R. Acad. Sci. 271, 400 (1970)
3.31 D. Maystre, R. Petit: Opt. Commun. 2, 309 (1970)
3.32 R.C. McPhedran: Ph. D. Thesis, Tasmania, Hobart (1973)
3.33 R. Petit: Nouv. Rev. Opt. 6, 129 (1975)
3.34 R.C. McClellan, G.W. Stroke: J. Math. Phys. 45, 383 (1966)
3.35 A.W. Maue: Z. Phys. 126, 601 (1949)
3.36 R.C. McPhedran, D. Maystre: Opt. Acta 21, 413 (1974)
3.37 R.C. McPhedran, D. Maystre: Nouv. Rev. Opt. 5, 241 (1974)
3.38 E.G. Loewen, D. Maystre, R.C. McPhedran, I. Wilson: Jpn. J. Appl. Phys. 14, 143 (1975)
3.39 E.G. Loewen, M. Nevière, D. Maystre: Appl. Opt. 16, 2711 (1977)
3.40 A. Wirgin: Opt. Commun. 1, 65 (1973)
3.41 M. Nevière, M. Cadilhac, R. Petit: Opt. Commun. 6, 34 (1972)
3.42 G.H. Spencer, M.V. Murty: J. Opt. Soc. Am. 52, 672 (1962)
3.43 W. Werner: Thesis (1970)
3.44 D. Maystre, R. Petit: Opt. Commun. 4, 97 (1971)
3.45 M. Nevière, M. Cadilhac: Opt. Commun. 4, 13 (1971)
3.46 D. Maystre, R. Petit: Opt. Commun. 5, 35 (1972)
3.47 R. Petit, D. Maystre: Rev. Phys. Appl. 7, 427 (1972)
3.48 D. Maystre, R. Petit: J. Spectr. Soc. Jpn. 23 suppl. 61 (1974)
3.49 P. Vincent, M. Nevière, D. Maystre: Nucl. Instrum. Methods 152, 123 (1978)
3.50 A.C. Hewson: *An Introduction to the Theory of Electromagnetic Waves*, Mathematical Physics series (Longman Group, London 1970)
3.51 J. Pavageau, J. Bousquet: Opt. Acta 17, 469 (1970)
3.52 D. Maystre, R. Petit: Opt. Commun. 4, 25 (1971)
3.53 A. Marechal, G.W. Stroke: C.R. Acad. Sci. Paris 249, 2042 (1959)
3.54 D. Maystre, R.C. McPhedran: Opt. Commun. 12, 164 (1974)

4. Differential Methods

P. Vincent

With 11 Figures

From a mathematical point of view, the rigorous computation of the electromagnetic field diffracted by a grating reduces to the resolution of partial derivative equations with suitable boundary conditions. It is shown in Chap. 3 how this problem is equivalent to the resolution of Fredholm integral equations, but, as outlined in Chap. 1, it is possible to solve directly these differential equations. The method related in this chapter is equivalent to, but slightly different from, the scheme previously described in Chap. 1. It takes a place among the so-called differential methods that we shall classify into three categories:

I) The direct differential methods where Maxwell equations are projected onto Cartesian coordinates, giving a set of partial derivative coupled equations. These equations are solved by a finite difference scheme using a two-dimensional point matching. The formulation is very simple (see, e.g. [4.1]), but often requires long computing times and is limited by the numerical instabilities of the algorithm, if the grating depth is not very small compared to its period. These methods were not used in our laboratory and will not be described here.

II) This class includes the method we are concerned with. The projection of propagation equations onto a suitable basis of functions gives a set of *ordinary differential coupled equations*. For gratings the pseudo-periodicity of the field leads us to use an exponential basis. Efficient integration algorithms such as Runge-Kutta ones or others can be used because the unknowns are then functions of only one variable.

III) A third possibility exists: it is possible, using a conformal mapping technique to map the profile of the grating onto a plane. After such a coordinate transformation, we obtain of course a new propagation equation which can be projected on an exponential basis and solved using a differential algorithm [4.2,3].

The method II) mainly described in this chapter looks like an efficient compromise between simplicity of formulation, versatility and precision. Numerical experiment has shown that our programs are able to compute the electromagnetic field diffracted by a grating in most of the problems encountered by opticians, with special emphasis in the UV and XUV domains.

4.1 Introductory Remarks

4.1.1 Historical Survey

The numerical integration of the system of differential equations obtained by projection on the exponential basis was first suggested by PETIT [4.4] at least when the electric field $\underline{E}$ is parallel to the grooves ($E_{\parallel}$ case). The method was developped by CERUTTI-MAORI [4.5,6]. Good results were obtained for dielectrics with moderate permittivity, but serious difficulties arose for conducting materials. NEVIERE [4.7] has shown that the use of more sophisticated integration algorithms removes this restriction. In collaboration with the author, he has treated the $H_{\parallel}$ case, using the propagation equations proposed by CERUTTI-MAORI. The versatility of differential methods allows us to study more complex objects than simple gratings, such as gratings covered by a dielectric layer [4.9] or grating couplers [4.10,11]. Recently, the differential method has been extended to off-plane mountings [4.12,13] and to crossed gratings [4.14,15]. It is also used with other kinds of scatterers, such as cylinders [4.16-18] or nonperiodic obstacles [4.19,20].

4.1.2 Definition of Problem

The differential method is not restricted to the simple problem of a classical grating ruled on a plane homogeneous substrate (Fig.4.1). More complicated objects such as the array of cylinders of Fig.4.2 or the periodic waveguide of Fig.4.3 can be studied using the differential method. Let us define with more precision the diffraction problem we try to solve. The electromagnetic properties of the diffracting obstacle, at a given angular frequency ω, must be contained in two scalar functions: the magnetic permeability μ and the dielectric permittivity ε. For the sake of simplicity, we assume that μ is everywhere equal to the permeability μ_0 of the vacuum. In fact, this restriction can be removed easily at the price of a little extra complication of the propagation equations. Using a rectangular coordinate system, the formulation is based on the following properties for the permittivity $\varepsilon(x,y)$: it is a real or complex function, independent of the z coordinate;

I) constant and equal to ε_1 if $y > a$;
II) constant and equal to ε_2 if $y < 0$;
III) with period d and piecewise continuous with respect to x, if $0 < y < a$.

Furthermore, we suppose that the incident field is a plane, monochromatic wave [wave vector $\underline{k}^i$ with components $(\alpha_0, -\beta_0, \gamma)$]. The case $\gamma = 0$ is called "classical diffraction case", the other that of "conical diffraction".

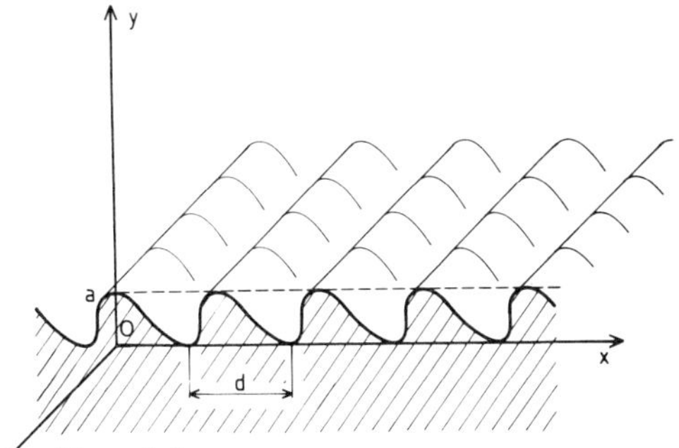

Fig. 4.1

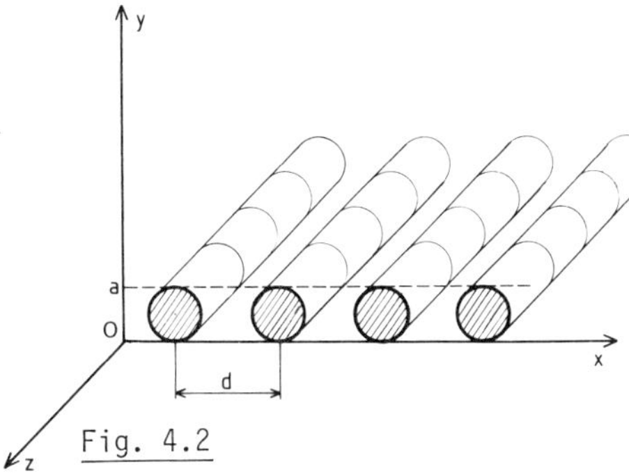

Fig. 4.2

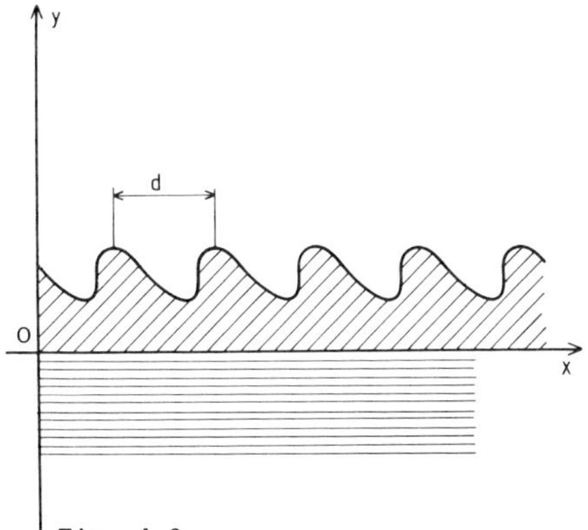

Fig. 4.3

Fig. 4.1. Schematic representation of a grating

Fig. 4.2. Grating of period d made by an infinite array of identical cylinders of finite conductivity and arbitrary profile

Fig. 4.3. A dielectric grating deposited on a plane substrate can be used has a periodic waveguide

4.2 The $E_{\parallel}$ Case

The idea of the differential method is given in Chap. 1 using this simple case. We shall describe here a slightly different version and detail the numerical implementation. Let us recall the propagation equation, valid in the sense of distributions

$$\Delta E + k^2 E = 0 \quad , \quad \text{with } k^2 = \omega^2 \varepsilon\mu \quad . \tag{4.1}$$

Let us define $K = 2\pi/d$ and $\alpha_n = \alpha_0 + nK$ (n is an integer). Expanding the pseudo-periodic functions (cf Appendix A, Sect.5) in terms of $\exp(i\alpha_n x)$, we write

$$E_z(x,y) = \sum_n E_n(y) \exp(i\alpha_n x) \quad . \tag{4.2}$$

We also define the Fourier expansion of the periodic function $k^2(x,y)$ by

$$k^2(x,y) = \sum_n (k^2)_n \exp(inKx) \quad . \tag{4.3}$$

It is worth noting that the product $k^2(x,y)\, E(x,y)$ is a function expandable on the same $\{\exp(i\alpha_n x)\}$ basis with coefficients $\sum_m (k^2)_{n-m} E_m$ which are functions of y. Consequently, the projected propagation equation leads to a set of ordinary coupled differential equations

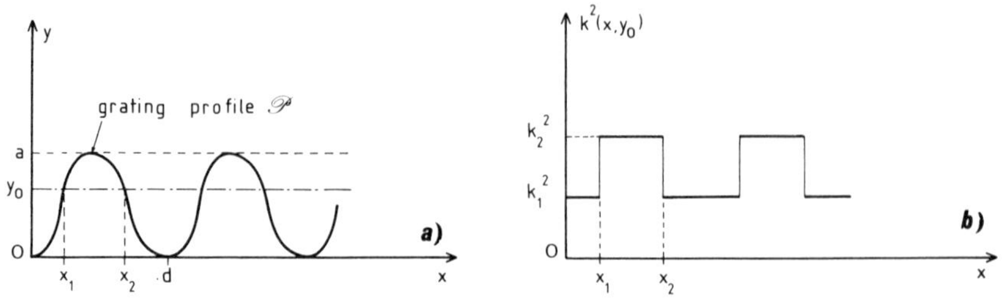

Fig. 4.4. For homogeneous materials, $k^2(x,y) = \omega^2\,\varepsilon(x,y)\,\mu_0$ is a step function of variable x at $y = y_0$ fixed. The abscissa x_1 and x_2 are determined by the crossing if the grating profile P with the $y = y_0$ line

$$\frac{d^2E_n}{dy^2} = \alpha_n{}^2E_n - \sum_m (k^2)_{n-m}E_m \quad . \tag{4.4}$$

For equally spaced values of y, the $(k^2)_n$ coefficients are computed using the usual formula for Fourier series with respect to the x variable. As an example, taking a sinusoidal grating (Fig.4.4a), for $y = y_0$, we have to find the Fourier coefficients of the function drawn in Fig.4.4b. When the value of y_0 is changed, only x_1 and x_2 changed as well.

4.2.1 The Reflection and Transmission Matrices

Outside the modulated zone $0 < y < a$, the dielectric permittivity ε is a constant, and the propagation equation (4.1) reduces to a Helmholtz equation with analytical solutions. When $y > a$, let us define

$$k_1{}^2 = \omega^2\mu_0\varepsilon_1 \quad , \quad (k_1{}^2)_n = k_1{}^2\delta_{n,0} \quad , \quad \beta_n{}^{(1)} = (k_1{}^2 - \alpha_n{}^2)^{1/2} \quad .$$

From (4.4), we get

$$y > a\colon\ E_n(y) = A_n{}^{(1)}\exp(-i\beta_n{}^{(1)}y) + B_n{}^{(1)}\exp(i\beta_n{}^{(1)}y) \quad . \tag{4.5}$$

When $y < 0$

$$k_2{}^2 = \omega^2\mu_0\varepsilon_2 \quad , \quad (k_2{}^2)_n = k_2{}^2\delta_{n,0} \quad , \quad \beta_n{}^{(2)} = (k_2{}^2 - \alpha_n{}^2)^{1/2} \quad ,$$

$$E_n{}^{(y)} = A_n{}^{(2)}\exp(-i\beta_n{}^{(2)}y) + B_n{}^{(2)}\exp(i\beta_n{}^{(2)}y) \quad . \tag{4.6}$$

The determination of the square roots which give β_n constants is given by: $\mathrm{Re}\,\{\beta_n\} + \mathrm{Im}\{\beta_n\} > 0$ (see Chap.5). As a consequence, for real permittivity and real orders the A_n coefficients describe plane waves propagating towards the $y < 0$ direction and the B_n waves propagating in the opposite direction (Fig.4.5).

We suppose that the grating is illuminated by a wave propagating downwards, that is to say $B_n{}^{(2)} = 0$, $\forall n$. We call $\psi_A{}^{(1)}$, $\psi_B{}^{(1)}$ and $\psi_A{}^{(2)}$ the column matrices built, respectively, with coefficients $A_n{}^{(1)}$, $B_n{}^{(1)}$ and $A_n{}^{(2)}$. Our hypothesis im-

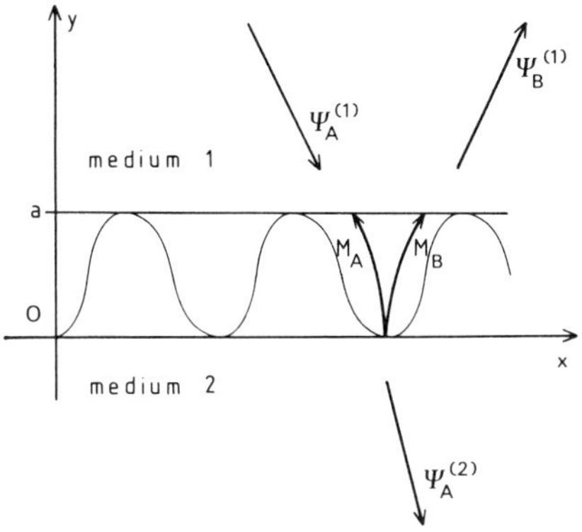

Fig. 4.5. Outside the modulated zone, the waves propagating towards the y < 0 direction are described by matrix ψ_A build with A_n coefficients, the others by ψ_B or B_n

plies the linearity of the diffraction problem, i.e., the matrices are related linearly. Let us introduce square matrices M_A, M_B and R by

$$\psi_A^{(1)} = M_A \psi_A^{(2)} \quad , \tag{4.7}$$

$$\psi_B^{(1)} = M_B \psi_A^{(2)} \quad , \tag{4.8}$$

$$\psi_B^{(1)} = R \psi_A^{(1)} \quad . \tag{4.9}$$

From these definitions, it follows immediately that

$$R = M_B (M_A)^{-1} \quad . \tag{4.10}$$

The reflection matrix R is an obvious generalization of the usual reflection coefficient well known in thin film theory. In the same way, the matrix $T = (M_A)^{-1}$ is the generalization, in the grating case, of the transmission coefficient. Thus, the resolution of the diffraction problem of gratings reduces to the determination of R and T matrices, which are deduced from M_A and M_B. The following section is devoted to the numerical determination of these matrices.

4.2.2 The Computation of Transmission and Reflection Matrices

The computation of matrices M_A and M_B requires the numerical integration of the differential system (4.4) from y = 0 to y = a. Standard algorithms start with known values of the function and its derivative. These values are obtained by matching the exponential expansion (4.2) [valid in the modulated zone with the Rayleigh expansion (4.6) with $B_n^{(2)} = 0$ because of the outgoing wave condition] onto the y = 0 plane: the continuity of the tangential electric component implies

$$E_n^{(0)} = A_n^{(2)} \quad , \tag{4.11}$$

and the continuity of its derivative

$$\frac{dE_n}{dy}(0) = -i \beta_n^{(2)} A_n^{(2)} \quad . \tag{4.12}$$

Thus, giving a matrix $\psi_a^{(2)}$, relations (4.11,12) determine $E_n(0)$ and $dE_n(0)/dy$ so that we are ready to perform a numerical integration of (4.4). Of course, we must suppose that the field is accurately represented by N components on the $\{\exp(i\alpha_n x)\}$ basis, and truncate the infinite set of differential equations to order N. Classical algorithms (Runge-Kutta, Numerov, ...) allow us to compute $E_n(a)$ and $dE_n(a)/dy$. Matching these values with the Rayleigh expansion (4.5) onto the $y = a$ plane will give us the $A_n^{(1)}$ and $B_n^{(1)}$ coefficients. The continuity of the normal derivative dE_n/dy on the $y = a$ plane implies

$$\frac{dE_n}{dy}(a) = i\beta_n^{(1)} \, [B_n^{(1)} \exp(i\beta_n^{(1)}a) - A_n^{(1)} \exp(-i\beta_n^{(1)}a)] \quad . \tag{4.13}$$

From (4.5,13), we get

$$B_n^{(1)} = \left[E_n(a) + \frac{1}{i\beta_n^{(1)}} \frac{dE_n}{dy}(a)\right] \exp(-i\beta_n^{(1)}a) \quad , \tag{4.14}$$

$$A_n^{(1)} = \left[E_n(a) - \frac{1}{i\beta_n^{(1)}} \frac{dE_n}{dy}(a)\right] \exp(i\beta_n^{(1)}a) \quad . \tag{4.15}$$

Finally, the numerical integration through the modulated zone gives the matrices $\psi_A^{(1)}$ and $\psi_B^{(1)}$ from $\psi_A^{(2)}$. In fact we don't know the real field in the $y = 0$ plane but only the incident field in medium 1. Nevertheless, it is possible to compute the M_A and M_B matrices: it is well known that a linear application acting on a finite dimensional space is completely determined when the image of a basis of this space is given. We can use this property because we have supposed the field to be accurately represented by N components. Consequently, we have only to take arbitrary independent vectors $\psi_A^{(2)}$ and compute their images $\psi_A^{(1)}$ and $\psi_B^{(1)}$ by N numerical integrations. The matrix obtained by juxtaposing vectors $\psi_A^{(1)}$ gives M_A and vectors $\psi_B^{(1)}$ gives M_B. In order to have these matrices directly in the $[\exp(i\alpha_n x)]$ basis, we shall take vectors $\psi_A^{(2)}$ with all their components equal to zero but one equal to one. From M_A and M_B, the transmission matrix is obtained by performing a matrix inversion with the usual Gauss-Jordan method, and then the reflection matrix by a simple matrix product.

For a single incident plane wave, the matrix representation $\psi_A^{(1)}$ has only one element different from zero, and the reflected field is represented by $\psi_B^{(1)}$ proportional to a column of the reflection matrix R.

4.2.3 Numerical Algorithms

The most difficult part of the computation is the numerical integration of (4.4) in the modulated zone. CERUTTI-MAORI, using a simple, but not very accurate algorithm, did not get good results for metallic gratings. Thus, NEVIERE selected the very powerful Numerov algorithm [4.8], even if this choice complicates the programming.

Let us define the column matrix ψ with elements $E_n(y)$. Thus, the differential system (4.4) is equivalent to the matrix equation

$$\frac{d^2\psi}{dy^2} = V\psi \quad , \tag{4.16}$$

with matrix elements for V : $V_{n,m} = \alpha_n^2 \delta_{n,m} - (k^2)_{n-m}$.

As required by the Numerov algorithm, this equation does not include derivatives of the first order. The algorithm, usually introduced for scalar functions, is used here with matrices; h stands for the integration step, constant in the whole modulated zone. Defining matrix ξ

$$\xi = \psi - \frac{h^2}{12}\frac{d^2\psi}{dy^2} = \left(I - \frac{h^2}{12} V\right)\psi \quad , \tag{4.17}$$

we have [4.8] the Numerov formula:

$$\xi(y+h) = [2I + h^2 V(y) + h^4 V^2(y)/12]\, \xi(y) - \xi(y-h) + O(h^6) \quad . \tag{4.18}$$

This formula is a very efficient tool for computing the matrix ξ, from the knowledge of this matrix at the two preceding points. Note that this algorithm is not self-starting, and that it is necessary to know $\xi(0)$ and also $\xi(h)$ to use it. Thus, we must compute with another formula $\psi(h)$. This can be done, using a classical Runge-Kutta algorithm [4.21]. We find

$$\psi(h) = \left[I + \frac{h^2}{6} V(0) + \frac{h^2}{3} V(h) + \frac{h^4}{24} V(h)V(0)\right] \psi(0)$$
$$+ h\left[I + \frac{h^2}{6} V(h)\right] \frac{d\psi}{dy}(0) + O(h^5) \quad . \tag{4.19}$$

Then, formula (4.16) gives $\xi(h)$ immediately.

Another problem arises at the end of the integration. The Numerov algorithm gives $\xi(a)$ but the matrices needed are ψ and $d\psi/dy$ for $y = a$. In fact, $\psi(a)$ is computed by inverting formula (4.17)

$$\psi(a) = I + \frac{h^2}{12} V(a) + \frac{h^4}{144} V^2(a) \quad \xi(a) + O(h^6) \quad . \tag{4.20}$$

This derivative can be found by combining a classical formula giving the derivative of a function from its values in seven points, and a formula cited by RAYNAL [4.28] giving $\psi(y)$ from $\xi(y + h)$, $\xi(y)$ and $\xi(y + h)$. The result is

$$\frac{d\psi}{dy}(a) = [10\xi(a - 7h) + 28\xi(a - 6h) - 485\xi(a - 5h) + 1778\xi(a - 4h)$$
$$- 3325\xi(a - 3h) + 3740\xi(a - 2h) - 3150\xi(a - h) - 360\xi(a)] \tag{4.21}$$
$$/(720h) + 147\psi(a)/(60h) + O(h^6) \quad .$$

All these formulae look rather complicated, but numerical experiments show that they allow accurate results with short computation times, even with good conductors or gratings with deep grooves.

Let us summarize the successive operations performed by the computer program:

a) Giving N arbitrary values for the vector $\psi_A^{(2)}$.
b) Deducing the corresponding values for $\psi(0)$ and $d\psi(0)/dy$ using (4.11,12).
c) Computing $\xi(0)$ using (4.17).
d) Computing $\psi(h)$ using Runge-Kutta formula (4.19), then $\xi(h)$ by (4.17).
e) Performing N simultaneous integrations with the Numerov formula (4.18).
f) Returning from $\xi(a)$ to $\psi(a)$ by (4.20).
g) Computing $d\psi/dy$ for $y = a$ by (4.21).
h) Computing M_A and M_B using (4.14,15).
i) Computing the transmission matrix T by inverting M_A.
j) Computing the reflection matrix R using (4.10).
k) Deducing the efficiencies from the usual formula given in the preceding chapters.

4.2.4 Alternative Matching Procedures for Some Grating Profiles

Contrary to the case for lamellar gratings, the V matrix for sinusoidal or echelette gratings has continuous derivatives with respect to the y coordinate. In this latter case, it is possible to simplify the procedure by increasing the zone where a numerical integration is performed, in order to use simultaneously Rayleigh expansions and the Numerov formula.

Let us start the numerical integration with values of $\xi(-h)$ and $\xi(0)$ deduced directly from the Rayleigh expansion (4.6). We get for the nth component of ξ

$$\xi_n(0) = \left[1 - \frac{h^2}{12}\left(\beta_n^{(2)}\right)^2\right] A_n^{(2)} \quad ,$$

$$\xi_n(-h) = \xi_n(0)\,\exp\left(i\beta_n^{(2)}h\right) \quad .$$

The same technique can be used (with the same restrictions) at the top of the grating: we perform the numerical integration up to $y = a + h$, return from ξ to ψ, then match it with the Rayleigh expansion (4.5) for $y = a$ and $y = a + h$ planes. $A_n^{(1)}$ and $B_n^{(1)}$ coefficients are deduced from the linear system obtained.

4.2.5 Field of Application

As with any numerical procedure, the differential method is not universal, and the user must pay attention to the range of the data. The results can be checked by the use of several criteria: the reciprocity theorems and the stability of the results when the number of terms used in the expansions or the length of integration steps are varied. For dielectrics without losses, the energy balance can also be used. Of course, comparison with results given by other formalisms or by experiments is also useful.

The limits are strongly dependent on three parameters: the number of diffracted orders, the depth of the grating (in wavelength), the modulus of the refractive index. For example, a precision greater than 10^{-2} can easily be obtained for an aluminum grating having a depth of 0.4 wavelength used in the visible region with five propagating diffracted orders. The high number of parameters involved makes the complete description of the field of the differential method difficult. As a rule of thumb, we can say that for the $E_{\parallel}$ case the differential method is convenient for usual gratings in the UV and visible regions, but is not well adapted to infrared problems. In some case it has shown surprising capability. NEVIERE has obtained acceptable efficiencies in UV with the energy concentrated around the zero order, though with several hundred theoretically diffracted orders[1]. At the present time, the improvement of the precision and stability of the numerical integration means that this part of the method is not critical; the main limitation seems to arise from the inversion of matrix M_A. We think that it is possible to improve this point as well.

4.3 The $H_{\parallel}$ Case

4.3.1 The Propagation Equation

Let us now suppose that the magnetic field vector $\underline{H}$ is parallel to the grating grooves. Again, this problem is a scalar one, but this time the unknown function is H_z. The normal derivative of this function is not continuous through a discontinuity surface for the permittivity. Consequently, the propagation equation $\Delta H_z + k^2 H_z = 0$ is usually not valid in the sense of distributions. It is not surprising, because, obviously, this case of polarization is different from the preceding one, and this remark implies that the propagation equation introduced in the numerical integration must be different. The differential method needs a propagation equation where all the derivations with respect to the y coordinate are performed on continuous quantities. In order to find this equation, let us write the Maxwell equations

$$\mathrm{curl}(H_z\hat{\underline{z}}) = -i\omega\varepsilon\underline{E} \quad , \tag{4.22}$$

$$\mathrm{curl}(\underline{E}) = i\omega\mu H_z\hat{\underline{z}} \quad . \tag{4.23}$$

Using the well-known identity: curl $(H_z\hat{\underline{z}}) = \mathrm{grad}(H_z)\wedge\hat{\underline{z}}$, we get

$$\mathrm{curl}(\underline{E}) = \mathrm{curl}\left[\frac{1}{-i\omega\varepsilon}\,\mathrm{grad}(H_z)\wedge\hat{\underline{z}}\right] = i\omega\mu_0 H_z\hat{\underline{z}} \quad . \tag{4.24}$$

1 Of course, the numerical integration is performed with N < 30 terms for the Fourier expansions, thus some real propagating orders are neglected during the computation. The result is accurate only if the energy carried by these orders is negligible.

With $k^2 = \omega^2 \varepsilon\mu$, (4.24) gives

$$\operatorname{div}\left[\frac{1}{k^2}\operatorname{grad}(H_z)\right] + H_z = 0 \quad . \tag{4.25}$$

We can check that the divergence operator acts on a continuous quantity.

Outside the modulated zone, k^2 is a constant and the propagation equation reduces to a Helmholtz equation. Thus we can use the Rayleigh expansions for H_z. Hence the scheme of the differential method remains the same for the $E_{\shortparallel}$ and the $H_{\shortparallel}$ case. But now, the matrices are linked to H_z instead of E_z.

Thus, we have to

- give N arbitrary linearly independent values for $\psi_A^{(2)}$,
- integrate the propagation equation (4.25) from $y = 0$ to $y = a$,
- perform linear combinations to get M_A and M_B,
- invert M_A to compute the reflection matrix R and the transmission matrix T.

The details of the numerical treatment differ from the $E_{\shortparallel}$ case because (4.25) exhibits the first derivatives which makes the Numerov algorithm no longer convenient.

4.3.2 Numerical Treatment

The propagation equation (4.25) can be written

$$\frac{\partial}{\partial x}\left(\frac{1}{k^2}\frac{\partial H_z}{\partial x}\right) + \frac{\partial}{\partial y}\left(\frac{1}{k^2}\frac{\partial H_z}{\partial y}\right) + H_z = 0 \quad . \tag{4.26}$$

Even if k^2 is discontinuous, k^{-2} *grad* H_z exhibits a continuous normal derivative. This property makes the numerical integration possible but implies that we never split the product $k^{-2}\,\partial H_z/\partial y$. For that reason, we rewrite (4.26) as a system of two equations of the first order. Let us define

$$\tilde{E} = \frac{1}{k^2}\frac{\partial H_z}{\partial y} = \frac{E_x}{i\omega\mu_0} \quad . \tag{4.27}$$

We get, from (4.26)

$$\frac{\partial H_z}{\partial y} = k^2\,\tilde{E} \quad , \tag{4.28}$$

$$\frac{\partial \tilde{E}}{\partial y} = -\frac{\partial}{\partial x}\left(\frac{1}{k^2}\frac{\partial H_z}{\partial x}\right) - H_z \quad . \tag{4.29}$$

Projecting on the $\{\exp(i\alpha_n x)\}$ basis, with $\tilde{E}_n$ and H_n the components for $\tilde{E}$ and H_z, we obtain an infinite set of ordinary coupled differential equations

$$\frac{dH_n}{dy} = \sum_m (k^2)_{n-m}\,\tilde{E}_n \quad , \tag{4.30}$$

$$\frac{d\tilde{E}_n}{dy} = \alpha_n \sum_m \alpha_m \left[\frac{1}{k^2}\right]_{n-m} H_m - H_n \quad . \tag{4.31}$$

This time ψ is a column matrix built with the H_n and $\tilde{E}_n$ functions. The differential system (4.30,31) is written, in matrix form

$$\frac{d\psi}{dy} = V\psi \quad ,$$

the elements of matrix V being defined by (4.30,31). If the function H_z is accurately represented by N components on the $[\exp(i\alpha_n)]$ basis, the matrix ψ has 2N coefficients and the square matrix V has size 2N.

The differential system is a classical first order one and can be integrated using usual methods. We use a Runge-Kutta algorithm with four approximations [4.21] for the first step, and continue with an Adams-Moulton predictor-corrector formula [4.22] to decrease the computation time. All these algorithms are very common and can be found in many numerical analysis books.

The matching formula on the y = 0 plane differs form the $E_{\parallel}$ case because we need $\tilde{E}(0)$ to start the numerical integration instead of the derivative. The connection between these two quantities is straightforward: E_x being a tangential component for the y = 0 plane, $\tilde{E}$ is continuous and (4.28) immediately gives $\tilde{E}(0)$ from $\partial H_z/\partial y$.

The matching in the y = a plane uses the same property: $\partial H_z/\partial y$ is computed from E(a) and $\tilde{E}(a)$, and is used in formulae analogous to (4.14,15).

4.3.3 Field of Application

The $H_{\parallel}$ case is numerically less attractive than the $E_{\parallel}$ one. Numerical experiments show that for good conductors such as aluminum or silver, the integral formalism must be prefered in the visible region , but the differential method is competitive for shorter wavelengths, where the refractive indices are smaller. Nevertheless, good results are obtained for a silver grating at λ = 0.5 μm, with three to five diffracted orders and a groove depth about 0.07 μm. The user must be careful, because the reciprocity relation is verified by the differential formalism even if N is too small. Here the main criterion seems to be the stability of the results.

4.4 The General Case (Conical Diffraction Case)

Up to now, we have assumed that the incident wave vector is perpendicular to the grating grooves. In this section, we remove this restriction and consider incident monochromatic waves with any possible direction of propagation. Let us write γ the z component of the incident wave vector $\underline{k}^i$. Thus, the incident electromagnetic field has an $\exp(i\gamma z)$ dependence. The important point is the invariance of the scatterer with respect to any translation parallel to the Oz axis. The same reasons that allow the use of a complex amplitude technique with respect to the time dependence, work here for the Z dependence: γ and z are conjugated quantities like

ω and t. Consequently, we shall always use the complex vectors $\underline{E}(x,y)$ and $\underline{H}(x,y)$, but now the field is obtained by taking the real parts of $\underline{E}(x,y)\exp(i\gamma z - i\omega t)$ and $\underline{H}(x,y)\exp(i\gamma z - i\omega t)$.

As explained in the preceding chapter, it is impossible, when γ is not zero, to split the equations in TE and TM and the problem of conical diffraction is not a scalar one. Nevertheless calculation based on Maxwell equations, well known in waveguide theory [4.23] show that, for cylindrical problems, all the components of the field can be expressed in terms of any two of them. Here, we choose the axial components E_z and H_z as main components.

4.4.1 The Reflection and Transmission Matrices

Outside the modulated zone $0 < y < a$, the Maxwell equations reduce to Helmholtz ones

$$\Delta E_z + (k^2 - \gamma^2)E_z = 0 \quad , \tag{4.32}$$

$$\Delta H_z + (k^2 - \gamma^2)H_z = 0 \quad . \tag{4.33}$$

In these regions, we use Rayleigh expansions for E_z and H_z, analogous to (4.5,6), substituting $k^2 - \gamma^2$ for k^2 into the definitions of β_n. The column matrices $\psi_A^{(1)}$, $\psi_B^{(1)}$, $\psi_A^{(2)}$ are built with the Rayleigh coefficients. Note that one matrix contains the coefficients describing E_z and the coefficients for H_z. If one component of the field is accurately represented by N coefficients, $\psi_A^{(1)}$, $\psi_A^{(2)}$ and $\psi_B^{(1)}$ matrices have 2N terms. The definition of matrices M_A , M_B, R and T remains the same, but now the size of these square matrices is 2N. This matrix formulation allows us to use in the conical diffraction case, exactly the same method as in the preceding ones. The only differences are in numerical details, depending on the particular form of the equations.

4.4.2 The Differential System

In the modulated zone, Rayleigh expansions may not be convergent, and it is necessary to integrate numerically the propagation equations. The ideas used to find the differential system are the same as previously worked in the $H_{\parallel}$ case. We need a set of coupled equations where differential operators act on continuous quantities. By projection onto the $\{\exp(i\alpha_n x)\}$ basis, these equations give an infinite set of ordinary coupled differential equations, where the unknowns are functions of the y coordinate.

In [4.12,17] we first established second-order propagation equations including only E_z and H_z and then, split them in a system of first-order coupled equations. It is simpler to project the Maxwell equations directly on the axes and then put them into convenient form. We shall use here this second method.

The reader should keep in mind that for our purpose, the partial derivative $\partial/\partial y$ plays a role different from the $\partial/\partial x$ and $\partial/\partial z$ ones. The complex amplitude technique shows that $\partial/\partial z$ is equivalent to a multiplication by $i\gamma$. The role of $\partial/\partial x$ using the $\{\exp(i\alpha_n x)\}$ basis is similar, because this operator is equivalent to the multiplication of each nth component by $i\alpha_n$.

Let us start with the Maxwell harmonic equations

$$\text{curl}[\underline{E}\ \exp(i\gamma z)] = i\omega\mu_0\ \underline{H}\ \exp(i\gamma z) \quad , \qquad (4.34)$$

$$\text{curl}[\underline{H}\ \exp(i\gamma z)] = -i\omega\varepsilon\ \underline{E}\ \exp(i\gamma z) \quad . \qquad (4.35)$$

These equations are projected on the cartesian coordinates Oxyz, and the $\exp(i\gamma z)$ dependence omitted

$$\partial E_z/\partial y - i\gamma E_y = i\omega\mu_0 H_x \quad , \qquad (4.36)$$

$$i\gamma E_x - \partial E_z/\partial x = i\omega\mu_0 H_y \quad , \qquad (4.37)$$

$$\partial E_y/\partial x - \partial E_x/\partial y = i\omega\mu_0 H_z \quad , \qquad (4.38)$$

$$\partial H_z/\partial y - i\gamma H_z = -i\omega\varepsilon E_x \quad , \qquad (4.39)$$

$$i\gamma H_x - \partial H_z/\partial x = -i\omega\varepsilon E_y \quad , \qquad (4.40)$$

$$\partial H_y/\partial x - \partial H_x/\partial y = -i\omega\varepsilon E_z \quad . \qquad (4.41)$$

From (4.40,37) , we get

$$E_y = (i\gamma H_x - \partial H_z/\partial x)/(-i\omega\varepsilon) \quad , \qquad (4.42)$$

$$H_y = (i\gamma E_x - \partial E_z/\partial x)/(i\omega\mu_0) \quad . \qquad (4.43)$$

The four remaining equations can be rewritten as a set of first-order coupled equations

$$\partial E_z/\partial y = i\gamma E_y + i\omega\mu_0 H_x \quad , \qquad (4.44)$$

$$\partial H_z/\partial y = i\gamma H_y - i\omega\varepsilon E_x \quad , \qquad (4.45)$$

$$\partial E_x/\partial y = \partial E_y/\partial x - i\omega\mu_0 H_z \quad , \qquad (4.46)$$

$$\partial H_x/\partial y = \partial H_y/\partial x + i\omega\varepsilon E_z \quad . \qquad (4.47)$$

Let us define E_n^z by

$$E_z(x,y) = \sum_n E_n^z(y)\ \exp(i\alpha_n x) \quad . \qquad (4.48)$$

H_n^z, E_n^y and H_n^y are introduced in the same way. If we project (4.42-47) on the $\{\exp(i\alpha_n x)\}$ basis, we get[2]

---

2 As defined in (4.4), $(f)_n$ is the nth Fourier coefficient of f.

$$E_n^y = \sum_m \left(\frac{1}{-i\omega\varepsilon}\right)_{n-m} (i\gamma H_m^x - i\alpha_m H_m^z) \quad , \tag{4.49}$$

$$H_n^y = \frac{1}{i\omega\mu_0} (i\gamma E_n^x - i\alpha_n E_n^z) \quad , \tag{4.50}$$

and

$$\begin{aligned} dE_n^z/dy &= i\gamma E_n^y + i\omega\mu_0 H_n^x \quad , \\ dH_n^z/dy &= i\gamma H_n^y - \sum_m (i\omega\varepsilon)_{n-m} E_m^x \quad , \\ dE_n^x/dy &= i\alpha_n E_n^y - i\omega\mu_0 H_n^z \quad , \\ dH_n^x/dy &= i\alpha_n H_n^y + \sum_m (i\omega\varepsilon)_{n-m} E_m^z \quad . \end{aligned} \tag{4.51}$$

Let us define ψ by juxtaposing the components E_n^x, H_n^x, E_n^z, H_n^z. Thus, the differential system obtained by substituting the expressions of E_n^y and H_n^y given by (4.49, 50) into (4.51) can be written in matrix form

$$d\psi/dy = V\psi \quad . \tag{4.52}$$

As in the $H_{\parallel}$ case, this differential system is integrated using classical Runge-Kutta algorithms from $y = 0$ to $y = a$.

4.4.3 Matching with Rayleigh Expansions

The x and z components are tangential components with respect to the $y = 0$ and $y = a$ planes on which the matching with Rayleigh expansions is made; thus, they are continuous. The numerical integration of (4.52) requires the determination of $E_n^x(0)$ and $H_n^x(0)$. The classical expressions of transverse components as functions of axial ones [4.23] gives these values from $E_n^z(0)$ and $H_n^z(0)$. We get

$$E_n^x(0) = \left[\omega\mu_0\beta_n^{(2)} H_n^z(0) - \gamma\alpha_n E_n^z(0)\right] / (k^2 - \gamma^2) \quad , \tag{4.53}$$

$$H_n^x(0) = \left[-\omega\varepsilon_1\beta_n^{(2)} E_n^z(0) - \gamma\alpha_n H_n^z(0)\right] / (k^2 - \gamma^2) \quad . \tag{4.54}$$

The numerical integration gives $E_n^z(a)$, $H_n^z(a)$, $E_n^x(a)$ and $H_n^x(a)$. The Rayleigh coefficients can be deduced from these quantities by solving a linear system of order four.

4.4.4 Field of Application

The field of application for the differential method in the general case ranges between the field given for the $E_{\parallel}$ case and the $H_{\parallel}$ one. We used it essentially in UV and XUV regions where the ITCD mounting (cf Sect.6.3.5) seems very interesting. The low conductivity of materials at these wavelengths makes the results accurate without any numerical problem. The versatility of the method is interesting for studying the propagation in periodic waveguides used in integrated optics. For example, the computation of a diffracted efficiency in the general case takes only a few seconds and 30.000 words of memory storage on a CDC 7600 computer.

4.5 Stratified Media

4.5.1 Stack of Gratings

The differential methods can be easily extended to a stack of gratings or to gratings built on a layered substrate, provided all the gratings have the same period. Let us consider the two gratings shown in Fig.4.6. The space is now divided into five parts: two modulated zones ($0 < y < y_1$ and $y_2 < y < y_3$) and three homogeneous regions where Rayleigh expansions are valid (regions 1 , 2 and 3). Column matrices ψ_A and ψ_B are defined in the same way as for a single grating.

Having chosen N arbitrary linearly independent values for $\psi_A^{(3)}$, we perform a numerical integration in the modulated zone $0 < y < y_1$, then obtain N values for $\psi_A^{(2)}$ and $\psi_B^{(2)}$. We deduce the corresponding values for the field in the $y = y_2$ plane, from which we integrate in the modulated zone $y_2 < y < y_3$ and obtain $\psi_A^{(1)}$ and $\psi_B^{(1)}$. In other words, the whole of the system between $y = 0$ and y_3 can be considered as a single grating, but the validity of Rayleigh expansion between $y = y_1$ and $y = y_2$ allows us to avoid a numerical integration between those two planes.

If $y_2 < y_1$, which is the usual case for gratings covered with a dielectric layer protecting them, we perform rhe numerical integration from $y = 0$ up to y_3. The only difference from a single bare grating is that the function $k^2(x,y)$ that we must develop in Fourier series is a little more complicated (Fig.4.7).

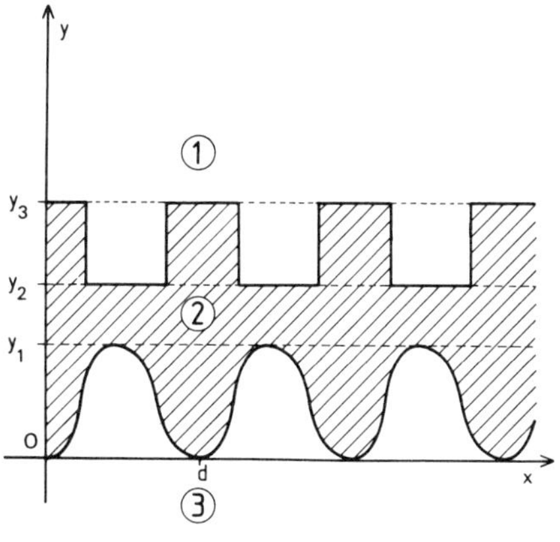

Fig. 4.6. An example of a stack of two gratings having the same period

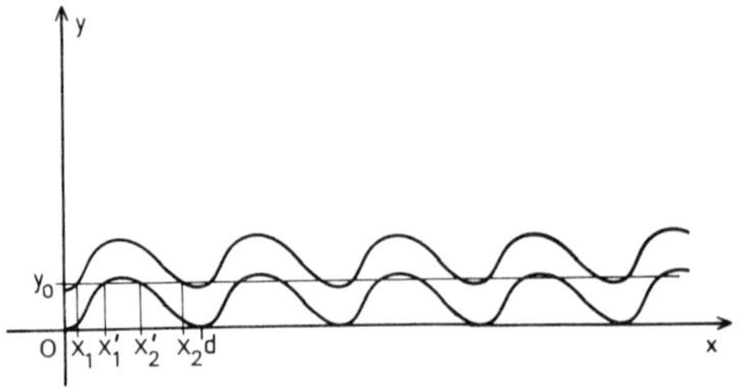

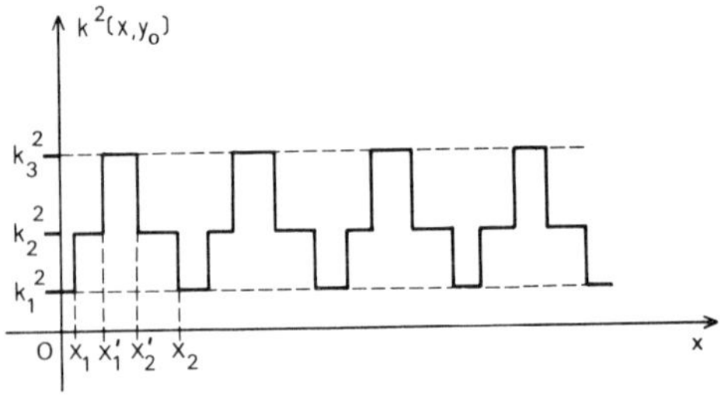

Fig. 4.7. Aspect of the $k^2(x,y_0)$ function for a dielectric coated grating

4.5.2 Plane Interfaces Between Homogeneous Media

When the stack of gratings includes a plane interface between two homogeneous media (Fig.4.8), the relation between matrices in the media 1 and 2 can be expressed analytically. In TE and TM cases, the expressions are classical and used in thin film theory. In the conical diffraction case, the relations between the z components of the field are found using the continuity of the tangential components of the field. If $A_n^{(1)}$, $A_n^{(2)}$, $B_n^{(1)}$, $B_n^{(2)}$ are the Rayleigh components of E_z, we can show that, if

$$u_e = \frac{\varepsilon_2}{\varepsilon_1} \frac{k_1^2 - \gamma^2}{k_2^2 - \gamma^2} \frac{\beta_n^{(2)}}{\beta_n^{(1)}} \quad ,$$

$$b^- = \exp[i(\beta_n^{(2)} - \beta_n^{(1)})y_1] \quad , \quad b^+ = \exp\left[i(\beta_n^{(2)} + \beta_n^{(1)})y_1\right] \quad ,$$

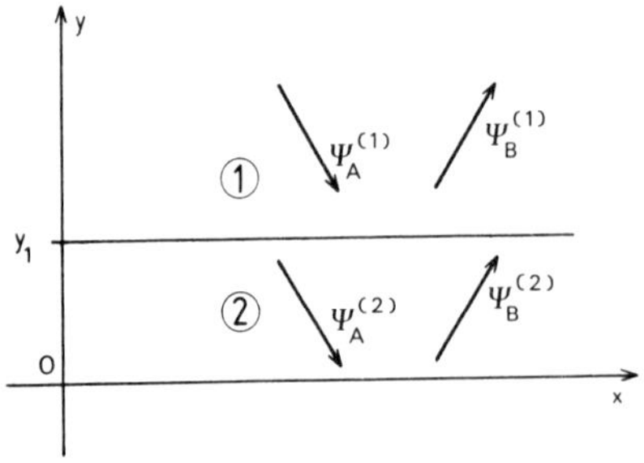

Fig. 4.8. The various matrices describing the field on each side of the plane interface at $y = y_1$ between medium 1 and 2

we get

$$\begin{pmatrix} B_n^{(1)} \\ A_n^{(1)} \end{pmatrix} = \frac{1}{2} \begin{pmatrix} (1 + u_e)b^- & (1 - u_e)/b^+ \\ (1 - u_e)b^+ & (1 + u_e)/b^- \end{pmatrix} \begin{pmatrix} B_n^{(2)} \\ A_n^{(2)} \end{pmatrix} \quad . \tag{4.55}$$

The same relation applies for H_z component if we change u_e into u_h

$$u_h = \frac{k_1^2 - \gamma^2}{k_2^2 - \gamma^2} \frac{\beta_n^{(2)}}{\beta_n^{(1)}} \quad .$$

In conclusion, the existence of a plane interface between two modulated zones is very easy to take into account. The general scheme of the method remains the same as for a single grating.

4.6 Infinitely Conducting Gratings: the Conformal Mapping Method

4.6.1 Method

The differential method described in the previous sections does not work for infinitely conducting gratings. Indeed, the permittivity ε, as a function of x for given values of y, tends to infinity inside the metal; thus, it is impossible to describe it by a Fourier expansion. The solution to this kind of problem is a method of the third type as defined in Sect. 4.1.1. If the boundary of the grating is transformed into a plane of coordinates, the difficulty of infinite permittivity is avoided. The following method has been developed by NEVIERE et al. [4.3,24]. It is given for in-plane (or classical) diffraction but, as said in Chap.3, for infinitely conducting gratings, the conical diffraction case reduces to two simultaneous classical cases. Thus the method can be used for all cases.

Let us consider the grating of period 2π, illuminated by a $E_\parallel$ wave in Fig.4.9. We define the complex variable $u = x + iy$, and suppose that we are able to find a conformal mapping relating u with $U = X + iY$ in such a way that the grating profile $\mathscr{P}$ is the transformed of the line Y=0. Outside the grating, the medium is homogeneous and the propagation equation for the E_z component of the electromagnetic field is a Helmholtz equation

$$(\Delta + k^2)\, E_z(x,y) = 0 \quad . \tag{4.56}$$

We can verify that, if $E_z(x,y)$ is transformed into $\hat{E}(X,Y)$, the Helmholtz equation is replaced by

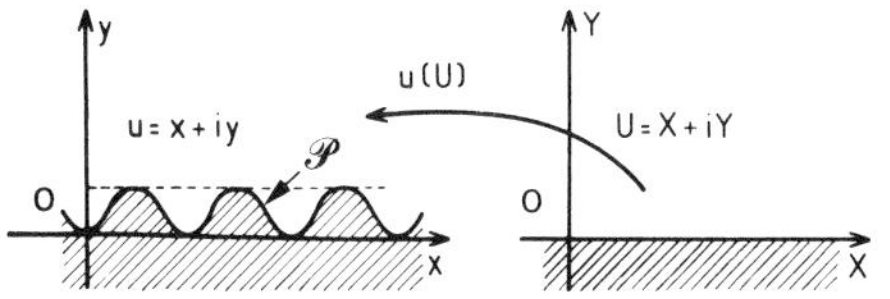

Fig. 4.9. The grating profile $\mathscr{P}$ is the image of the real axis OX through the conformal mapping function u(U)

$$\left(\Delta + k^2 \left|\frac{du}{dU}\right|^2\right) \hat{E}(X,Y) = 0 \quad . \tag{4.57}$$

As it is shown in [4.3], the conformal mapping function can be written

$$u = U + \sum_{n=0}^{\infty} b_n \exp(inU) \quad .$$

Thus, the coefficient $k^2|du/dU|^2$ is periodic with periodicity 2π, and tends exponentially towards k^2 if $Y \to +\infty$. The function $\hat{E}(X,Y)$ is also pseudo-periodic with respect to X and the boundary condition becomes

$$\hat{E}(X,0) = 0 \quad . \tag{4.58}$$

Furthermore, the outgoing wave condition for $Y \to \infty$ remains the same. The grating is equivalent to a dielectric layer deposited on a perfectly conducting plane, with a periodic variation of the optical index along OX having the same period as the grating. Theoretically, the width of this layer is infinite, but it is possible to show [4.3] that beyond the ordinate $Y = 2\pi$ (with grating period 2π) the function $k^2|du/dU|^2$ is practically equal to k^2 and the Rayleigh expansion is valid. Thus, this mathematical problem is very similar to the previous ones, and can be solved using the same method.

In the $0 < Y < 2\pi$ region, we write

$$\hat{E}(X,Y) = \sum_n \hat{E}_n(Y) \exp(i\alpha_n X) \quad .$$

In the same manner as for finite conductivity, N integrations of differential equation (4.57) give two matrices M_A and M_B. The initial values are N arbitrary independent functions for the normal derivative $\partial\hat{E}/\partial Y(X,0)$ while $\hat{E}(X,0)$ is null. The matrices M_A and M_B relate, respectively, the incoming and outgoing waves for $Y > 2\pi$ to the derivative of the field in the $Y = 0$ plane, i.e., on the grating profile. Thus the diffracted field ψ_B is obtained from the incident field ψ_A by

$$\psi_B = M_B M_A^{-1} \psi_A \quad . \tag{4.59}$$

The $H_{\shortparallel}$ case can be solved in the same way, using N arbitrary values for $\hat{E}(X,0)$, with $\partial E(X,0)/\partial Y = 0$, but, it is possible to solve simultaneously the two cases of polarization (with only N+1 numerical integrations performed in the opposite direction from $Y = 2\pi$ to $Y = 0$) because the only difference between the $E_{\shortparallel}$ and $H_{\shortparallel}$ cases is in the boundary condition in the $Y = 0$ plane. N linearly independent arbitrary values for ψ_B (with $\psi_A = 0$) give M_B^{-1} and the equivalent matrix M' of the $H_{\shortparallel}$ case. Another integration with the incident field as initial value for ψ_A gives a corresponding field $\psi(0)$ and its derivative $\psi'(0)$. Thus, the $E_{\shortparallel}$ case is the solution of $M_B^{-1} \psi_B + \psi(0) = 0$. The $H_{\shortparallel}$ case is solved by the resolution of a similar equation, with M' and $\psi'(0)$ instead of M_B^{-1} and $\psi(0)$. This technique allows us to compute simultaneously the TE and TM efficiencies at nearly the same cost as for one polarization only.

4.6.2 Determination of the Conformal Mapping

The method described in Sect.4.6.1 requires the determination of the complex function du/dU. Two cases must be considered:

I) for echelette, trapezoidal or lamellar gratings, the classical Schwarz-Christoffel transform is extended to periodic profiles [4.25] by

$$\frac{du}{dU} = \prod_j \left[1 - \exp\left|i\ (U - a_j)\right|\right]^{\mu_j} \quad , \tag{4.60}$$

where a_j and μ_j are found using analytical relations.

II) for other gratings, a numerical method is used. Note that the usual mapping problem is to find the function U(u), but that, on the contrary, we need the inverse u(U), i.e., to map the Y = 0 line into the grating profile y = f(x). This determination can be made by an original iterative method proposed by CADILHAC [4.25] and used by NEVIERE in his computer programs. As shown by numerical experiments, this method is convergent for usual profiles, but this convergence is not yet proved. It can be improved by the use of Padé Approximants or ε-algorithm techniques. The interested reader can find a comprehensive discussion of these questions in a recent book by GILEWICZ [4.26].

Let us define the complex variables

$$w = \exp(iu) = \exp(ix - y) \quad , \tag{4.61}$$

$$W = \exp(iU) = \exp(iX - Y) \quad . \tag{4.62}$$

In the complex plane W, the line Y = 0 is the unit circle ∂P, and the line ∂p defined by $w = \exp[ix - f(x)]$ represents the grating in the w complex plane (Fig. 4.10). Let P be the disc of radius unity, with its center on the origin of W. We assume that ∂p is a sufficiently regular curve and that its interior p is a simply connected area so that, according to the general theorems on conformal mapping, P and p are linked by a unique function w(W), which maps ∂P into ∂p, is analytic for $|W| < 1$, and is continuous for $|W| \leqq 1$, with $w(W) = 0$ if and only if $W = 0$ is a simple zero for the function w(W). Thus the function W/w(W) is analytic in P and we get

$$w(W) = W/\sum_{n=0}^{\infty} a_n W^n \quad . \tag{4.63}$$

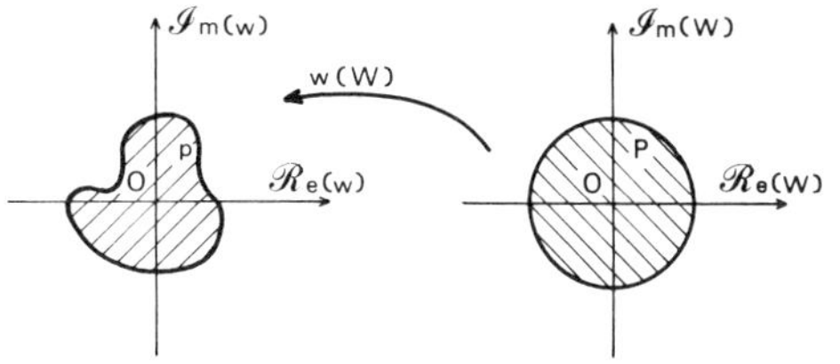

Fig. 4.10. In the complex w plane the grating profile is represented by a closed curve which is the image, through the mapping function w(W), of the circle of radius 1

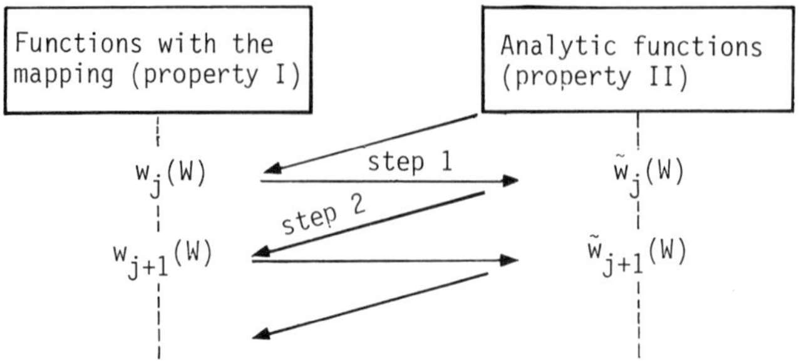

Fig. 4.11. Scheme of the iterative method used to found the conformal mapping function

An iterative procedure is used to determine the coefficients a_n. The idea is the following: we look for a complex function w(W) having two properties: I) it maps ∂P into ∂p; II) it is analytic in P. The difficulty is to enforce both these properties, but it is very simple to satisfy only one. Consequently each iteration is divided in two steps (Fig.4.11):

1) Starting with a function $w_j(W)$ having the mapping property I), we define a new one $\tilde{w}_j(W)$, analytic in P, but no longer possessing the mapping property I).
2) From $\tilde{w}_j(W)$, we deduce a new function w_{j+1} with the mapping property I) but no longer analytic.

The function W/w_j is a periodic function of U and its Fourier expansion is

$$\frac{W}{w_j} = \sum_{n=-\infty}^{+\infty} a_{n,j} \exp(inU) = \sum_{n=-\infty}^{+\infty} a_{n,j} W^n \quad . \tag{4.64}$$

Consequently, step 1) is achieved by computing

$$\tilde{w}_j = W/\sum_{n=0}^{\infty} a_{n,j} W^n \quad , \tag{4.65}$$

with the index running only to positive values in the summation. In fact, we need only the argument of $\tilde{w}_j$ for $W \in \partial P$

$$\tilde{x}_j = \arg(\tilde{w}_j) = X - \arg\left[\sum_{n=0}^{\infty} a_{n,j} \exp(inX)\right] \quad . \tag{4.66}$$

Using (4.61), the new function w_{j+1} is then defined by

$$w_{j+1} = \exp\left[i\,\tilde{x}_j - f(\tilde{x}_j)\right] \quad . \tag{4.67}$$

The numerical computation of $\tilde{w}_{j+1}$ for a given number of $\tilde{x}_j$ allows us to compute its Fourier expansion, i.e., the coefficients $a_{n,j+1}$ and to start a new iteration. At the end of the procedure, u(U) is easily deduced from w(W).

4.6.3 Field of Application

The conformal mapping technique for infinitely conducting gratings is competitive with the integral method. Usual values of the parameters give very short computation times (less than one second on a Univac 1110 computer). The determination of the conformal mapping is sometimes difficult for very high profiles, but some sophisticated methods (Padé approximant, ...) can improve the range of the method. Infinitely conducting gratings covered with dielectric an be studied by this method [4.27] if the boundary of the layer has the shape of an equipotential curve for the electrostatic related problem. It seems possible to use the same technique to study finite conductivity gratings, but two conformal mappings must be used: one for the air region and the other for the metal one.

References

4.1 M.K. Moaveni, H.A. Kalhor, A. Afrashteh: Comput. Elect. Eng. 2, 265-271 (1975)
4.2 M. Nevière, G. Cerutti-Maori, M. Cadilhac: Opt. Commun. 3, 48 (1971)
4.3 M. Nevière, M. Cadilhac, R. Petit: IEEE Trans. AP-21, 37 (1973)
4.4 R. Petit: Rev. Opt. 8, 353-370 (1966)
4.5 G. Cerutti-Maori, R. Petit, M. Cadilhac: C. R. Acad. Sci. (Paris) B8, 1060 (1969)
4.6 G. Cerutti-Maori: Thesis, Orsay (1970)
4.7 M. Nevière, P. Vincent, R. Petit: Nouv. Rev. Opt. 5, 65-77 (1974)
4.8 M.A. Melkanoff, J. Raynal, T. Sawada: *Methods in Computational Physics* (Academic Press, New York 1966)
4.9 M. Nevière, D. Maystre, P. Vincent: J. Opt. 8, 231-242 (1977)
4.10 M. Nevière, P. Vincent, R. Petit, M. Cadilhac: Opt. Commun. 9, 48-53 (1973)
4.11 M. Nevière, R. Petit, M. Cadilhac: Opt. Commun. 8, 113 (1973)
4.12 P. Vincent, M. Nevière, D. Maystre: Nucl. Instrum. Methods 152, 123-126 (1978)
4.13 M. Nevière, P. Vincent, D. Maystre: Appl. Opt. 17, 843-845 (1978)
4.14 P. Vincent: Opt. Commun. 26, 293-296 (1978)
4.15 D. Maystre, M. Nevière: J. Opt. (Paris) 9, 301-306 (1978)
4.16 P. Vincent, R. Petit: Opt. Commun. 5, 261-266 (1972)
4.17 P. Vincent: Appl. Phys. 17, 239-248 (1978)
4.18 G. Mur: Thesis, Delft (1978)
4.19 J.P. Hugonin, R. Petit: Opt. Commun. 20, 360 (1977)
4.20 J.P. Hugonin, R. Petit: Opt. Commun. 22, 221 (1977)
4.21 M. Abramowitz, I.A. Stegun: *Handbook of Mathematical Functions*, (Dover, New York 1972) pp. 896-897
4.22 P. Henrici: *Discrete Variable Methods in Ordinary Differential Equations* (John Wiley, New York 1962)
4.23 A.C. Henson: *An Introduction to the Theory of Electromagnetic Waves* (Longman Group, London 1970)
4.24 M. Nevière, G. Cerutti-Maori, M. Cadilhac: Opt. Commun. 3, 48-52 (1971)
4.25 M. Nevière, M. Cadilhac: Opt. Commun. 3, 379-383 (1971)
4.26 J. Gilewicz: *Approximants de Padé*, Lecture Notes in Mathematics, Vol. 667 (Springer, Berlin, Heidelberg, New York 1978)
4.27 M. Nevière, M. Cadilhac, R. Petit: Opt. Commun. 6, 34-37 (1972)
4.28 J. Raynal: Seminar Course on Computing as a Language of Physics, ICTP, Trieste (2-20 Aug. 1971)

5. The Homogeneous Problem

M. Nevière

With 25 Figures

The title of this chapter may seem obscure to many physicists. What we intend to deal with consists of several curious phenomena, well known to experimentalists, or recently discovered. They are grating anomalies, usually referred to as Wood anomalies, total absorption of a plane wave by a metallic grating (bare or coated with a thin dielectric layer) as well as the coupling of a laser beam into an optical waveguide by means of a holographic thin film coupler. Despite their different appearances, all these phenomena have the same origin in common. They all are connected with the excitation of surface waves along the periodic structure. Such a surface wave carries energy parallel to the mean plane of the surface, but is also slightly attenuated in that direction. Thus, it is referred to by many authors as a "leaky wave". From a mathematical point of view, it is a solution of Maxwell equations and the associated boundary conditions on the grating surface, without any wave impinging on the structure. It is the study of such a solution that we call the "homogeneous problem" and we wish to show how the resolution of the homogeneous problem can enlighten the study of the response of the structure to a given excitation.

Such a process is not new in physics. At an elementary level, let us consider a resonant dipole made by a resistance R, an inductance L and a capacitance C. Let us apply at the ends of the dipole a sinusoidal tension $V = V_0 \cos\omega t$ and let ω vary from zero to infinity. If the resistance R is small enough, and if we measure the intensity of the electric current which goes through the network we observe a strong maximum, called a resonance phenomenon, for a particular value ω_r of ω.

To study the response to the excitation V, it is useful first to determine the natural frequency $2\pi\omega_p$ of the network, i.e., to study the oscillations of the network without any exterior excitation, which is exactly what we call the homogeneous problem. Such natural frequency is found to be the complex pole of the dipole admittance (ω_p is a complex number since, in presence of energy losses, any oscillation is expected to vanish with time). Thus if it were possible to excite the system with a pulsatance $\omega = \omega_p$, the response would be infinite. Of course, since ω is real, such a phenomenon never occurs. But for small values of energy dissipation,

ω_p lies not far from the real axis in the complex plane. So, when ω passes near the real part of ω_p, which is closed to the natural pulsatance ω_0 of the lossless network, the admittance takes very high values and gives rise to the aforementioned resonance phenomenon.

For periodic structures such as metallic gratings, the mathematical analysis is much more complicated than this elementary exposition. But the basic idea still holds.

5.1 Historical Summary

In 1902, WOOD published the first paper [5.1] on pronounced variations of the light diffracted by a grating into its spectral orders. WOOD illuminated a grating with a continuous light source. Instead of observing a smooth spectrum, under convenient polarization he saw several regions containing very high or low quantities of light. Under certain conditions, the drop from maximum illumination to minimum, which was about 10 to 1, occurred within a range of wavelengths not greater than the distance between the sodium lines. He was so astounded to find such sudden variations that he called them "anomalies". At this time indeed, no theory was able to explain these rapid changes in efficiency, which were only found for TM polarization (vector $\underline{H}$ parallel to the grating grooves) and never when $\underline{H}$ is in the plane of incidence.

The first attempt of explanation was given by RAYLEIGH [5.2], who suggested that these anomalous changes of grating efficiency in a particular order could be linked with the passing-off of a higher order. He was able to show that the location of the anomalies observed by WOOD agrees fairly well with the positions that he derived from the passing off of spectral orders. Moreover, his theory predicted singularities only for TM polarization, as experimentally observed at that time. But his perturbation treatment [5.3] was unable to investigate the form taken by anomalies. Later experimental work was done by several authors [5.4-9], which concerns the discovery of anomalies for the other fundamental case of polarization (TE polarization), the study of the shape of the anomalies, the reluctance of two anomalies to coincide, the influence of grating grooves, surface conductivity, wavelength and incidence, as well as the modifications due to the introduction of a dielectric layer.

In recent years, COWAN and ARAKAWA [5.10] have shown that for TM polarization anomalous features are markedly affected by quite thin dielectric layers. Measurements made on holographic gratings [5.11,12] have shown the importance of the absorption effects associated with the passing off of orders in TM polarization, and drastic effects due to the introduction of a new kind of anomalies in TE polarization [5.13].

On the theoretical side, too many contributions have been published to be reported here. Many studies were based on a plane wave expansion of the diffracted field above the grating surface. Thus they were only valid for low modulated gratings. Two early papers have contributed to our understanding of the physical processes causing anomalous behavior. FANO [5.14] was the first to suggest that anomalies could be associated with the excitation of surface waves along the grating. The same idea was also developed by HESSEL and OLINER [5.15]. By using a new approach in which the physical grating was replaced by an equivalent surface-impedance structure, they were able to predict the existence of both a maximum and a minimum, the occurrence of P anomalies under appropriate conditions as well as a new effect described in [5.8]. Also they were able to distinguish two different types of anomalous behavior: the Rayleigh anomalies, and the resonance ones. However the problem which remained was to be able to find the equivalent surface impedance corresponding to a given grating.

More recently, the development of the electromagnetic theory of gratings has made possible accurate computations which agree very well with anomalies observed in infrared or microwave domains (see, e.g. [5.16]). Moreover, after including the effects of a dielectric overcoating [5.17], a new kind of anomaly was predicted for TE polarization. But this formalism, based on infinite surface conductivity, failed to explain in quantitative manner the anomalies observed in the visible spectrum.

On the other hand, a separate approach was developed by people working on surface plasmons and solid state physics [5.18,19]. Starting from the plasmon resonances of a plane surface, they derived a perturbation treatment in which the groove profile is taken into account in the calculation of the probability of surface plasmon excitation. But, for high modulated gratings, they were not able to obtain quantitative agreement between theory and experimental observations, except by tuning the optical constants of the grating surface to fit the measured data.

The difficulty was overcome by MAYSTRE [5.20-22] who developed an efficient single integral equation formalism which takes into account the finite conductivity of the metal surface. A completely different approach elaborated by NEVIERE et al. [5.23] was able to include both finite conductivity and dielectric overcoating. Thus it gave a spectacular agreement [5.24] with the previous measurements published in [5.13].

Despite the high degree of accuracy given by these theories, such formalisms tend to be rather involved mathematically, and do not give much insight into the actual physical processes associated with anomalies. Thus, recent studies have been developed, which conciliate both aspects. The study of the anomalies of a metallic grating used in TM polarization was developed in a leaky wave (guided wave) viewpoint and studied with the electromagnetic theory as a tool [5.25,26]. The theory predicted an unexpected spectacular phenomenon. Under particular circumstances, a high reflecting grating was able to absorb an incident plane wave in totality. The fact was thoroughly confirmed by experiments [5.27]. A similar study for dielectric coated gratings used in TE polarization [5.28] was able to account for the new kind of anomalies observed in [5.13,24]. It also predicted a phenomenon of total absorption, again observed experimentally [5.29]. A recent extension of the theory allows us to predict near an anomaly, the repartition of the energy between several spectral orders [5.30].

While the above papers deal with optics, the total absorption of an incident plane wave due to the excitation of a leaky (surface) wave has been known for some time in acoustics [5.31]. It also should be noted that such leaky waves were studied from an electromagnetic point of view [5.32-34] to predict the coupling phenomenon between a laser beam and an optical waveguide by the use of a grating coupler. They also were extensively used by different authors [5.35-39a] to study not only holographic thin film couplers, but prism couplers, or laser beam displacement. Thus the strong absorption which occurs when the incident wave is phase-matched to the leaky wave field is not restricted to gratings, but may be supported by many other structures. However, we will limit the present study to the case of gratings and present a rigorous electromagnetic theory based on the existence and location in the complex plane of the poles and zeros of the scattering operator. Such poles appear to be equal to the propagation constants of the aforementioned leaky waves supported by the periodic structure.

The effect associated to the passing off of a spectral order, usually called "Rayleigh anomaly" is well understood and requires no further discussion. Thus we will concentrate on effects which are related to leaky waves, i.e., to solutions of the homogeneous problem.

5.2 Plasmon Anomalies of a Metallic Grating

5.2.1 Reflection of a Plane Wave on a Plane Interface

Interface Between Vacuum and a Lossless Dielectric

Figure 5.1 describes some notations which will be used hereafter. The incident plane wave falls on the interface xOz under the incidence θ, with the vector $\underline{H}$ parallel to axis Oz whose unit vector is $\hat{\underline{z}}$ [$\underline{H} = H(x,y)\hat{\underline{z}}$]. In the plane of incidence xOy, it gives rise to a reflected and a transmitted wave. Setting $\alpha = k_0 \sin\theta$ where k_0 is the wave number $2\pi/\lambda_0$, the reflexion and transmission factor γ and t are dependent on α. Defining the refractive index ν_2 of the dielectric and assuming the $\exp(-i\omega t)$ time dependence, the electromagnetic field in both regions can be written in the form region ① $(y > 0)$

$$H(x,y) = \exp[i(\alpha x - \beta^{(1)}y)] + r(\alpha)\exp[i(\alpha x + \beta^{(1)}y)] \tag{5.1}$$

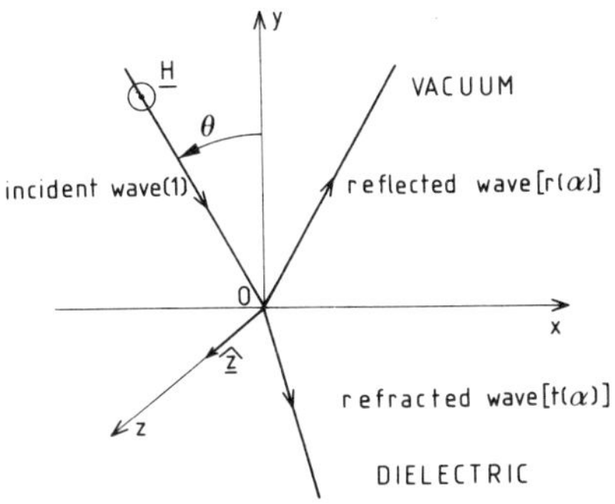

Fig. 5.1. Reflection of a plane wave on a plane interface

$$\text{where } \beta^{(1)} = k_0 \cos\theta = \sqrt{k_0^2 - \alpha^2} \quad , \tag{5.2}$$

region ② $(y < 0)$

$$H(x,y) = t(\alpha) \exp[i(\alpha x - \beta^{(2)} y)] \tag{5.3}$$

$$\text{with } \beta^{(2)} = \sqrt{k_0^2 \nu_2^2 - \alpha^2} \quad . \tag{5.4}$$

the continuity of the tangential components of the electromagnetic field at $y = 0$ leads to the Fresnel formulae

$$r(\alpha) = [\nu_2^2 \beta^{(1)} - \beta^{(2)}] / [\nu_2^2 \beta^{(1)} + \beta^{(2)}] \quad , \tag{5.5}$$

$$t(\alpha) = 2\nu_2^2 \beta^{(1)} / [\nu_2^2 \beta^{(1)} + \beta^{(2)}] \quad . \tag{5.6}$$

It is well known that the zero of $r(\alpha)$, obtained when $\nu_2^2 \beta^{(1)} = \beta^{(2)}$, gives, by the use of (5.2,4), the value $\hat{\alpha}$ of α corresponding to the Brewster incidence θ_B

$$\hat{\alpha} = k_0 \nu_2 / \sqrt{1+\nu_2^2} \quad , \tag{5.7}$$

which is equivalent to: $tg\theta_B = \nu_2$.

As will be seen later, it is often convenient in this chapter to get rid of the wave number k_0. To this end, let us introduce the normalized parameter $\delta = \alpha/k_0$. The value $\hat{\delta}$ of δ corresponding to the Brewster phenomenon is thus

$$\hat{\delta} = \frac{\hat{\alpha}}{k_0} = \nu_2 / \sqrt{1+\nu_2^2} \quad . \tag{5.7'}$$

Both reflection and transmission coefficients become functions of δ, and from now, will be noted $r(\delta)$ and $t(\delta)$.

A Remark on the Double Nature of the Root $\hat{\delta}$

An important remark needs to be made concerning the zero $\hat{\delta}$. Of course, provided ν_2 is real, $\hat{\delta}$ is a real number. However, it may be useful hereafter to consider the extension of the preceding problem to the complex plane. As soon as δ is allowed to be complex, the definition of $\beta^{(1)}$ and $\beta^{(2)}$ given by (5.2,4) is ambiguous, and the function $r(\delta)$ given by (5.5) in nonuniform. In order to get a uniform function, we have to introduce cuts in the complex plane to uniformize $\sqrt{k_0^2-\alpha^2}$ and $\sqrt{k_0^2\nu_2^2-\alpha^2}$.

Recalling that the refractive index of vacuum ν_1 is equal to unity, $\beta^{(1)}$ and $\beta^{(2)}$ can be written in a common form

$$\beta^{(j)} = \sqrt{k_0^2\nu_j^2-\alpha^2} = k_0\sqrt{\nu_j^2-\delta^2} \quad , \quad \text{with } j = 1,2 \quad . \tag{5.8}$$

Thus in the complex δ plane, the cuts start at the branch points $\pm\nu_j$ on the real axis and go to infinity. When δ moves between the two branch points on the real axis, $\beta^{(j)}$ is real and positive. Outside the interval, $\beta^{(j)}$ is chosen purely imaginary, with imaginary part positive. For the sake of continuity, we place the cuts as indicated in Fig.5.2. One can easily verify, for example, that the argument of $\beta^{(j)}$ is increased by $\pi/2$ when one goes from A to A', following the small half-circle in the lower half plane.

What is the image of the two cuts in the complex $\beta^{(j)}$ plane? When $\delta = \nu_j + i\eta$, with $\eta > 0$, $[\beta^{(j)}]^2 = k_0^2(\eta^2 - 2i\nu_j\eta)$, and its locus is a half parabola with axis Ox and tangent to the imaginary axis. Thus the locus of $\beta^{(j)}$ is the curve represented in Fig.5.3 which is tangent to the second bisector at the origin.

Provided that δ moves not too far from the real axis, a condition which will always be satisfied in practice, $\beta^{(j)}$ stays near the positive part of the two axis of Fig.5.3. Then, the image of the cut can be easily replaced by the second bisector without any consequence, a thing that we will do hereafter. In conclusion, a unique value of $\beta^{(j)}$ corresponding to a given complex value of δ is given by (5.8) and the condition

$$\mathrm{Im}\{\beta^{(j)}\} + \mathrm{Re}\{\beta^{(j)}\} > 0 \quad . \tag{5.9}$$

The corresponding part of the complex $\beta^{(j)}$ plane is the one which has a physical interest since when δ is real, $\beta^{(j)}$ has to be real and positive.

Now, what happens to the multivalued function $r(\delta)$? One can verify that whatever the chosen determinations of the two square roots may be, two and only two values of r correspond to a given value of δ, and that these two values are inverse to each other. Thus $r(\delta)$ may be considered as a single valued function on a two sheet Riemann surface. Forsaking these sophisticated considerations to come back to the

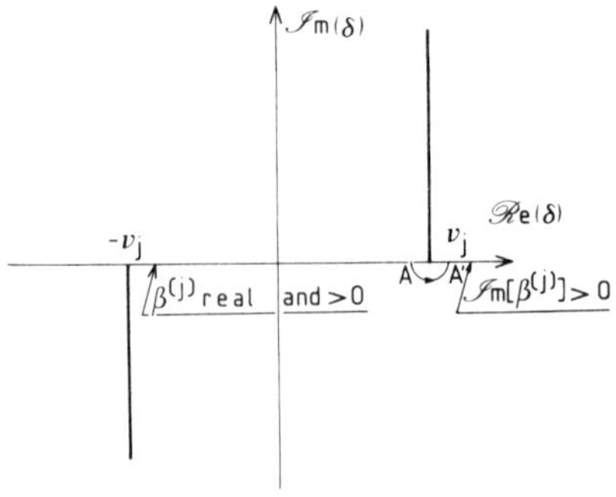

Fig. 5.2. Location of the cuts in the complex δ plane

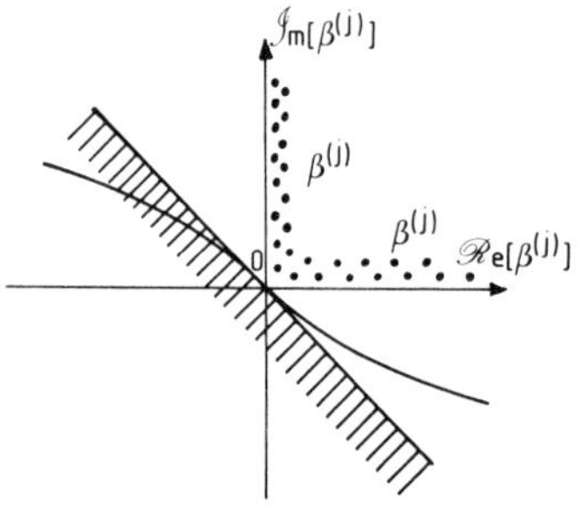

Fig. 5.3. Image of the cuts in the complex $\beta(J)$ plane and location of the values of $\beta(j)$ when δ moves near the real axis

more familiar language of multivalued functions, the function $r(\delta)$ presents two determinations. The first determination, r_I defined by (5.5,9) is the one with physical interest. The second determination r_{II} is deduced from r_I by replacing $\beta^{(1)}$ by $-\beta^{(1)}$ [1] and is equal to $1/r_I$. Thus a complex value $\hat{\delta}$ of δ which is a zero of r_I, is also a pole of r_{II}. Since replacing $\beta^{(1)}$ by $-\beta^{(1)}$ in (5.1) does not change the physical problem (only the incident and reflected wave exchange their nature), a given phenomenon such as Brewster absorption may be associated to a zero or a pole, depending on the determination that we chose. Hereafter δ^z and δ^p designate the eventual zeros and poles of r_I. From a numerical point of view, their determinations can be conducted with the same code.

Interface Between Vacuum and a Lossy Material

This paragraph is concerned with any lossy material whatever the origin of the losses may be (lossy dielectric, or metal with conduction losses). The Fresnel formula (5.5) still holds, but since the refractive index ν_2 is complex, $r(\delta)$ is complex too. So is the value $\hat{\delta}$ given by (5.7'). In the case of a good metallic reflector like silver for example, at wavelength 0.5 µm for which $\nu_2 = 0.05 + i\ 2.87$, $\hat{\delta} = 1.0668 + 0.00257\ i$. It is to be noted that, since the conduction losses are small, $\hat{\delta}$ is located near the real axis. The fact that $\mathrm{Re}\{\hat{\delta}\}$ is greater than unity requires some explanations.

To this end, Fig.5.4 shows the trajectory of $\hat{\delta}$ when, starting from the refractive index of glass for which $\hat{\delta}$ is real and lower than unity, one increases the imaginary part of ν_2, and then decreases its real part to arrive at the refractive

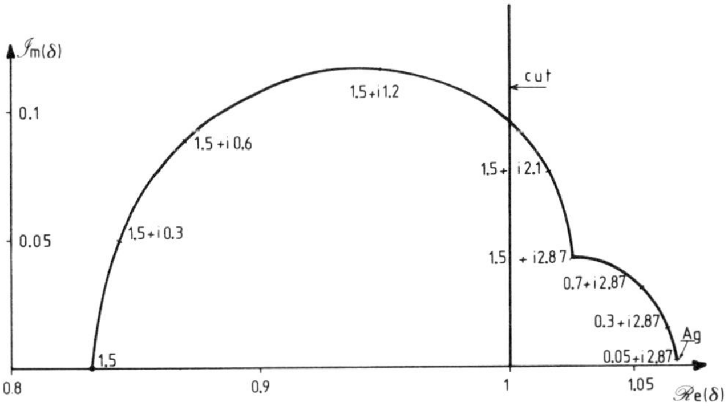

Fig. 5.4. Locus of $\hat{\delta}$ in the complex δ plane when ν_2 is varied from the value of glass (dielectric) to the value of silver (good reflector). On the left side of the cut, $\hat{\delta}$ is a zero of r_I. It is a pole on the right side

1 The same result is obtained by replacing $\beta^{(2)}$ by $-\beta^{(2)}$.

index of silver. When the trajectory of $\hat{\delta}$ crosses the cut, $\hat{\delta}$ which was a zero of r_I becomes a zero of r_{II}, i.e., a pole of r_I. The result is that for a metallic interface, the reflection coefficient presents a *pole* on the sheet of physical interest. Thus no zero can be detected by experiments, which means that no Brewster incidence can be observed.

On the other hand, the presence of the pole of $r(\delta)$ on the physical sheet which is also a pole of $t(\delta)$ given by (5.6) implies the existence of a reflected and a transmitted wave without an incident one. Of course, since $\hat{\delta}$ is complex, these waves are leaky waves which propagate nearly parallel to the interface, but rapidly decrease in the perpendicular direction. Such a solution of Maxwell equations is generally called a resonance and, in this case, is attributed to collective oscillations of electrons near the surface (plasmon resonance [5.39b]). When an incident plane wave strikes the interface under incidence θ, $\delta = \alpha/k = \sin\theta$ is always less than unity and can never excite such a resonance. Thus both resonance and Brewster phenomenon are never seen when one studies the reflection of a plane wave on a metallic interface. The question which arises is whether these phenomena could be observed by introducing a suitable modulation.

5.2.2 Reflection of a Plane Wave on a Grating

Definitions and Notations

Figure 5.5 gives a schematic representation of a metallic grating illuminated by a plane wave under the incidence θ. The groove spacing is d, the groove depth h, the wavelength in vacuum (region ①) is λ_0, and TM polarization is assumed. Space is now divided into 3 regions. In any homogeneous region, the field can be described as a superposition of plane waves. In region ①, it is the sum of an incident and a diffracted part

$$H(x,y) = \exp[i(\alpha_0 x - \beta_0^{(1)} y)] + \sum_{n=-\infty}^{+\infty} B_n \exp[i(\alpha_n x + \beta_n^{(1)} y)] \quad , \tag{5.10}$$

where $\alpha_0 = k_0 \sin\theta = \frac{2\pi}{\lambda_0} \sin\theta \quad ,$ (5.11)

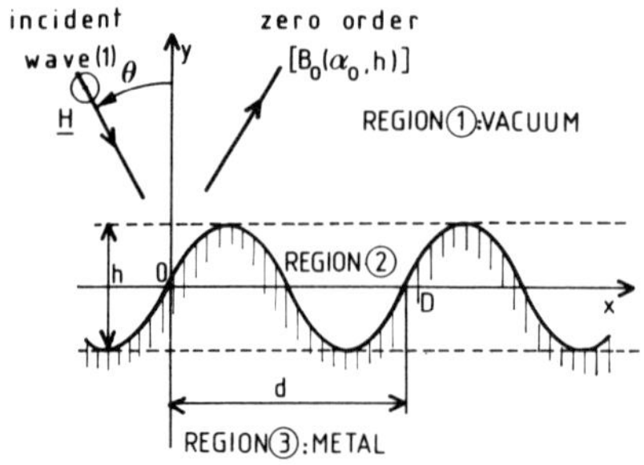

Fig. 5.5. Schematic representation of a grating

$$\alpha_n = \alpha_0 + n\,\frac{2\pi}{d} \quad , \tag{5.12}$$

$$\alpha_n^2 + [\beta_n^{(1)}]^2 = k_0^2 \quad . \tag{5.13}$$

In the region with the number ③ , the field is reduced to a diffracted one

$$H(x,y) = \sum_{n=-\infty}^{+\infty} A_n \exp[i(\alpha_n x - \beta_n^{(2)} y)] \quad , \tag{5.14}$$

$$\alpha_n^2 + [\beta_n^{(2)}]^2 = k_0 \nu_2^2 \quad . \tag{5.15}$$

Here too, (5.13,15) do not define $\beta_n^{(1)}$ and $\beta_n^{(2)}$ in a unique manner. Thus, we choose them in order to verify condition (5.9).

It is worth noting that in the second member of (5.10,14), only a finite number of terms correspond to propagating plane waves diffracted by the grating. The others are of evanescent nature. For the sake of simplicity, let us suppose that the incidence θ and the wavelength to groove spacing ratio are chosen in such a way that in the diffracted field, only the zero order propagates, with amplitude B_0. The grating then acts like a mirror, since the incident and diffracted orders are symmetrical with respect to the grating normal. Thus the only quantity of interest for the opticist is B_0 which is dependent on α_0 and on the groove shape. If we assume that the profile is sinusoidal, the groove shape is solely determined by the groove depth h, since the groove spacing has already been chosen. By introducing again the same normalized notations $\delta_n = \alpha_n/k_0$ as in Sect.5.2.1, we then find that B_0 is only dependent on δ_0 and h, and that, when the modulation tends toward zero

$$\lim B_0(\delta_0, h) = B_0(\delta_0, 0) = r(\delta_0) \quad . \tag{5.16}$$

Existence of a Complex Pole for the Function $B_0(\delta_0,h)$

As we previously did for the plane interface, it may enlighten our understanding of the diffraction phenomenon to consider the extension of the above problem to the complex plane. It means that we now consider an electromagnetic field given by (5.10,14) in which the real constants α_n have been replaced by complex numbers $\hat{\alpha}_n$. The research of poles for $B_0(\delta_0,h)$ implies that we try to see if it possible to find such solutions of Maxwell equations without any incident wave (homogeneous problem). Thus we search solutions of the form

region ①

$$H(x,y) = \sum_n \hat{B}_n \exp[i(k_0\hat{\delta}_n x + \hat{\beta}_n^{(1)} y)] \quad , \tag{5.17}$$

$$\text{with } \hat{\delta}_n = \hat{\delta}_0 + n\,\frac{\lambda_0}{d} \text{ and } -1 < \mathrm{Re}\{\delta_0\} < +1 \quad . \tag{5.18}$$

region ②

$$H(x,y) = \sum_n \hat{A}_n \exp[k_0\hat{\delta}_n x - \hat{\beta}_n^{(2)} y)] \quad , \tag{5.19}$$

where $\hat{B}_n$, $\hat{A}_n$ and $\hat{\delta}_n$ are dependent on the groove depth h.

Of course, if the modulation of the grating tends towards zero, the present problem tends towards the problem of the plane interface. Thus, among the infinity of values of $\hat{\delta}_n$, there exists a value n_0 of n for which

$$\lim_{h\to 0} \hat{\delta}_{n_0} = \hat{\delta} \quad . \tag{5.20}$$

Since Re{$\hat{\delta}$}is slightly greater than unity and since we have defined $\hat{\delta}_0$ so that $-1 < \mathrm{Re}\{\hat{\delta}_0\} < +1$, the index n_0 is found to be equal to 1. The result is that the terms corresponding to n = 1 in (5.17,19) tend towards the plasmon resonance of the metallic plane when $h \to 0$.

Special attention must be devoted to the term corresponding to n = 0. From (5.20), we get

$$\lim_{h\to 0} \hat{\delta}_0 = \hat{\delta} - \frac{\lambda_0}{d} \quad , \tag{5.21}$$

and Fig.5.6 shows the location of $\hat{\delta}_0$ in the complex plane. From its small imaginary part we can deduce that, when $h \to 0$, the term corresponding to n = 0 in (5.17) has a structure very close to a propagating wave going away from the very low modulated grating. Thus, on top of the plasmon resonance, we find a (slightly) leaky wave which provokes a loss of energy, and an infinite set of creeping waves near the grating surface.

If we find a value $\hat{\delta}_0$ for which (5.17,19) are the solution of the boundary value problem, it means that there exist finite coefficients $\hat{B}_n(\hat{\delta}_0,h)$ without any incident wave, which has the consequence that, for a given incident wave with a complex propagation constant $\hat{\delta}_0$, $\hat{B}_n(\hat{\delta}_0,h)$ are infinite. Thus such value $\hat{\delta}_0$ is found to be a pole $\delta^p(h)$ for the functions $B_n(\delta_0,h)$, and in paritcular, for the function $B_0(\delta_0,h)$.

The determination of the pole corresponding to a given grating problem has been conducted by an extension to the complex plane of the formalisms described in Chaps. 3 and 4. The cuts explained in Sect.5.2.1 have been introduced, and the problem reduced to the research of a complex zero of the complex function $1/B_0(\delta_0,h)$ for a given h. To this end, an iterative method has been developed [5.33], which requires the knowledge of an approximate value of the zero. For low modulations, (5.21) gives a good approximation to begin the determination. For high modulations, one must start from the value δ_0 numerically determined for a more shallow grating.

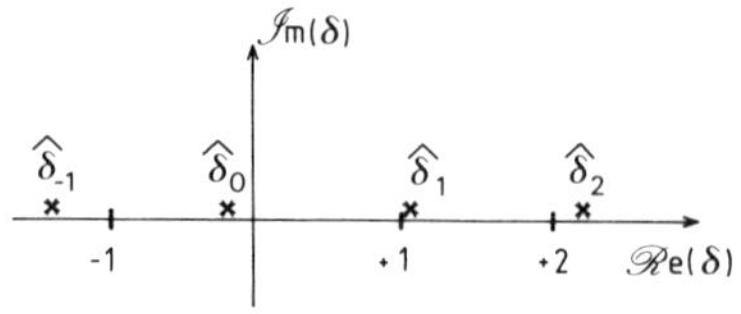

Fig. 5.6. Location of $\hat{\delta}_n$ in the complex δ plane

On the Existence of a Complex Zero for the Function $B_0(\delta_0,h)$

From the preceding paragraph, it follows that the function $B_0(\delta_0,h)$ may be written as

$$B_0(\delta_0,h) = \frac{u(\delta_0,h)}{\delta_0 - \delta^p(h)} \quad \text{where} \quad u[\delta^p(h),h] \neq 0 \;^2 \quad .$$

In the limit case of a plane structure, (5.16) implies that

$$B_0(\delta_0,0) = r(\delta_0) = \frac{u(\delta_0,0)}{\delta_0 - \delta^p(0)} \quad , \quad \text{or}$$

$$u(\delta_0,0) = [\delta_0 - \delta^p(0)]r(\delta_0) \quad .$$

Thus $\delta^p(0)$ is not only a pole for $B_0(\delta_0,0)$, but it is also a zero.

In the case of a modulated surface, continuity reasons imply that, at least for low modulated gratings, a complex zero associated with the complex pole exists and will be designated by $\delta^z(h)$.

Such existence can easily be established in the particular case of a perfectly conducting grating supporting only the zero diffracted order. There, the unitarity implies that if a pole $\delta^p(h)$ exists, an associated zero $\delta^z(h) = \overline{\delta^p(h)}$ must exist in order to find a diffracted efficiency equal to unity whatever the incidence may be. In the general case of a conducting grating supporting several spectral orders, no simple demonstration can be given. But numerical computations show the existence of zeros $\delta^z(h)$ which nearly have the same real part as $\delta^p(h)$, and have different imaginary parts. Such zeros are responsible for strong energy absorption phenomena, when a plane wave falls on the grating under suitable conditions.

Total Absorption of a Plane Wave by a Grating

In order to confirm the preceding theoretical considerations, many numerical experiments have been performed concerning the research of both poles and zeros, and to study their evolution in the complex plane when an arbitrary parameter (groove spacing, groove depth, refractive index, wavelength ...) is varied. Figure 5.7 gives an example of location and evolution when the groove depth is varied. Starting from a common value near the real axis equal to $\delta^p(0)$, the pole and zero separate and go towards opposite parts of the complex plane when h increases. The locus of the pole of course never crosses the real axis since a real value of $\delta^p(h)$ would correspond to an infinite diffracted field corresponding to a bounded incident plane wave. But the locus of the zero $\delta^z(h)$ crosses the real axis for a convenient groove depth $h_c = 0.021$ μm. It implies that the corresponding grating absorbs in totality an incident plane wave falling under the incidence θ given by $\sin\theta = \delta^z(h_c)$.

---

2 We assume that $\delta^p(h)$ is a single pole.

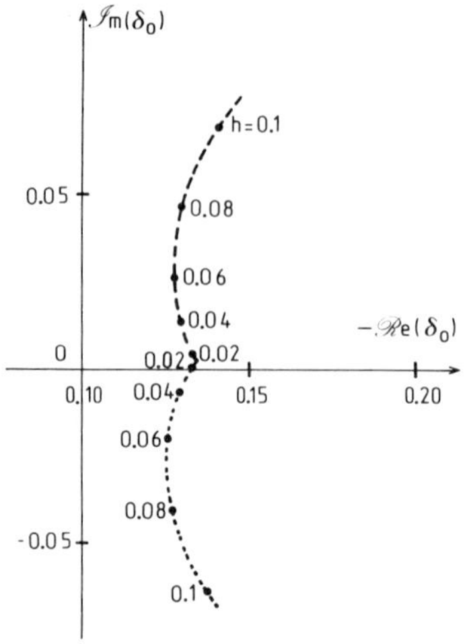

Fig. 5.7. Loci of the pole δ^p and the zero δ^z of $B_0(\delta_0,h)$ when the groove depth h is varied, for a 2400 gr/mm sinusoidal silver grating. The wavelength is 0.5 μm. (---): δ^p, (...): δ^z

It is worth noting that the critical groove depth h_c is generally very small. Thus it may appear unbelievable that, while a plane interface reflects more than 90% of the incident energy, a very shallow grating may absorb it in totality. But the phenomenon of total absorption predicted by the theory has been thoroughly confirmed by experiments [5.27]. Under the conditions determined by computer programs based on the preceding theory, no significant reflected energy was seen, and the incident energy was totally dissipated in heating the grating surface.

Variation of the Efficiency Near an Absorption Peak

The existence of a pole $\delta^p(h)$ and a zero $\delta^z(h)$ for the function $B_0(\delta_0,h)$ allows us to represent this function in the vicinity of an absorption peak by the expression

$$B_0(\delta_0,h) = w(\delta_0,h)\ [\delta_0 - \delta^z(h)]/[\delta_0 - \delta^p(h)] \quad , \tag{5.22}$$

where $w(\delta_0,h)$ is a complex regular function near δ^z and δ^p, and does not present any zero in their vicinity.

As previously mentioned, when $h \to 0$, $B_0(\delta_0,h) \to r(\delta_0)$. Since $\delta^z(h)$ and $\delta^p(h)$ tend towards the same value $\hat{\delta}$, the ratio in (5.22) tends towards unity. Thus

$$\lim_{h\to 0} w(\delta_0,h) = r(\delta_0) \quad .$$

Since in the vicinity of the absorption peak $w(\delta_0,h)$ is a slowly varying function, (5.22) can be approximated to a good degree of accuracy by

$$B_0(\delta_0,h) \simeq r(\delta_0)\ [\delta_0 - \delta^z(h)]/[\delta_0 - \delta^p(h)] \quad .$$

The zero-order efficiency $\mathscr{E}_0$ immediately follows[3]

3 In this chapter, the nth order absolute efficiency is denoted $\mathscr{E}_n$, instead of e_n.

$$\mathcal{E}_0(\delta_0,h) = |B_0|^2 \simeq R(\delta_0)\ |\delta_0 - \delta^z(h)|^2/|\delta_0 - \delta^p(h)|^2 \quad , \tag{5.23}$$

where $R(\delta_0)$ is the reflection factor for the energy.
Thus, as soon as the numerical values of $\delta^z(h)$ and $\delta^p(h)$ are known, $\mathcal{E}_0(\delta_0,h)$ can be obtained from (5.23) by means of a pocket calculator.

In order to illustrate these considerations, let us consider a 2400 groove/mm sinusoidal grating, coated with silver for which $R(\delta_0) \simeq 0.98$, $\forall\delta_0$. Let the wavelength λ_0 be equal to 0.5 µm and the groove depth to 0.05 µm. Under these conditions, only the zero order propagates provided $|\delta_0|$ is inferior to 0.2. Using the corresponding values of δ^z and δ^p plotted in Fig.5.7

$$\delta^z(0.05) = -0.1270 - i\ 0.01318$$

$$\delta^p(0.05) = -0.1278 + i\ 0.01853 \quad .$$

Figure 5.8 shows the variation of $\mathcal{E}_0(\delta_0,h)$ obtained through (5.23). For comparison, the results obtained from a direct computation based on the rigorous solution of the diffraction of a plane wave by a grating are shown. It is clear that (5.23) gives a very accurate approximation of $\mathcal{E}_0(\delta_0,h)$. No significant discrepancy occurs, as far as only the zero order propagates. When the -1 order arises, the sum of the two efficiencies $\mathcal{E}_0$ and $\mathcal{E}_{-1}$ again meets the approximate curve very well.

It can be seen on the curve that the minimum of $\mathcal{E}_0$ takes place at $\delta_0 = -0.0127$, which correspondes to $\mathrm{Re}\{\delta^z(0.05)\}$ and $\mathrm{Re}\{\delta^p(0.05)\}$. The corresponding efficiency $\mathcal{E}_{0m}$ can be easily deduced from (5.23).

$$\mathcal{E}_{0m} \simeq 0.98\ [\mathrm{Im}\{\delta^z(0.05)\}/\mathrm{Im}\{\delta^p(0.05)\}]^2 \quad ,$$

and its width at half depth is found to be close to 2 $\mathrm{Im}\{\delta^p(0.05)\}$. These simple rules allow us to predict the shift and evolution of the absorption peak, using the numerical values plotted in Fig.5.7. Figure 5.9 gives some typical examples. Starting from a uniform reflectance obtained when h = 0, a sharp peak is seen as soon as

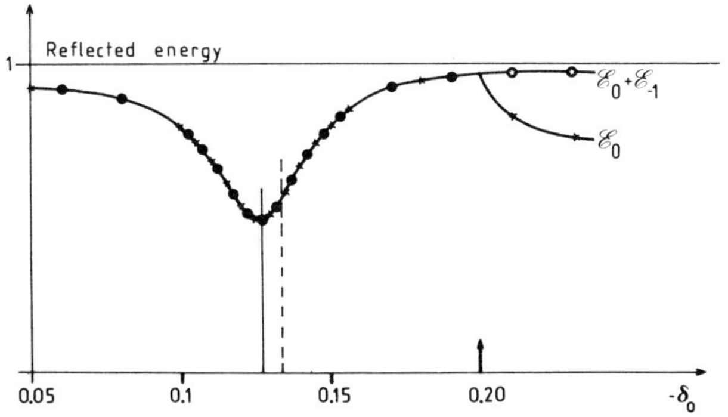

Fig. 5.8. Zero-order efficiency as a function of δ_0. The arrow indicates the apparition of the -1 order. (*) rigorous computation; (•) values given by (5.23)

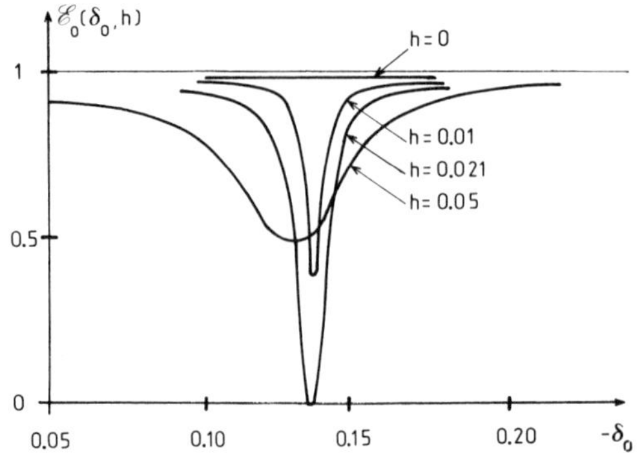

Fig. 5.9. Zero-order efficiency as a function of δ_0 for several groove depths

$h = 0.01$. The minimum value falls down to zero when $h = h_c = 0.021$ µm since, for this groove depth, $\mathrm{Im}\{\delta^z\} = 0$ as we previously remarked. When $h = 0.05$, a weaker absorption peak is found. Its width at half depth is increased in the ratio of $\mathrm{Im}\{\delta^p\}$, while the location of the minima follows the evolution of $\mathrm{Re}\{\delta^z\}$ and $\mathrm{Re}\{\delta^p\}$. For higher modulations, only weak absorption peaks are expected to be found, since $|\mathrm{Im}\{\delta^p\}|$ and $|\mathrm{Im}\{\delta^p\}|$ become of the same order of magnitude. The conclusion is that the absorption phenomenon is strongly dependent on the modulation of the grating. But high modulated gratings do not necessarily present a stronger absorption than low modulated ones.

5.3 Anomalies of Dielectric Coated Reflection Gratings Used in TE Polarization

Efficiency curves of bare reflection gratings are generally very smooth and free of anomalies. Although, after the work of PALMER, it is well established that anomalies may exist for TE polarization, they are never found in the resonance domain, i.e., in the blaze region. These anomalies are only seen with deeply modulated gratings, working with values of the ratio $\lambda/h \ll 1$. The consequence is that a commercial grating will exhibit such anomalies only for low values of the wavelength to groove spacing ratio λ/d. Thus, near the blaze peak where gratings are generally used, no anomaly occurs. Anyway, when they occur, such anomalies coincide with Rayleigh wavelengths and, as previously explained, are out of the scope of this chapter.

The situation becomes quite different when one superimposes a dielectric layer on top of a metallic grating. Then the groove depth or groove spacing no longer matters (nor does the conductivity of the surface). Grating anomalies only require a minimum thickness of the dielectric coating to appear.

Such coatings are customarily deposited on aluminum gratings to be used in VUV, in order to prevent the serious losses of efficiency which result from absorption by the thin oxide layer which forms on the uncoated metal. They are also sometimes used on top on silver gratings working in near infrared region. With the development of highly powerful lasers, stacks of layers are deposited on top of very low modulated gratings working in infrared region to realize beam samplers free of polarization effects [5.40]. Also they were recently used in the visible domain on top of commercial gratings used as laser end mirror tuning elements, in order to increase their efficiency [5.41].

Despite their widespread use, little work appears to have been done on their effects on diffraction efficiency. For some time it has been shown for TE [5.7] and TM polarizations [5.42] that anomalous features of a bare grating are markedly affected by an overcoating, even a very thin one. But the introduction of a new kind of anomaly linked with the existence of the layer seems to have been established more recently.

The first theoretical study [5.17], based on infinite conductivity of the grating surface, predicted new anomalies for both TE and TM polarizations related to the influence of the layer. Measurements made on holographic gratings [5.13] pointed out the drastic effects due to the use of thick layers on gratings, and reproduced by a theory including both the finite conductivity of the metal and the presence of a dielectric layer [5.24]. The link with leaky waves guided in the corrugated waveguide constituted by the dielectric layer was recently established [5.28] by a rigorous electromagnetic study that we will now present briefly.

5.3.1 Determination of the Leaky Modes of a Dielectric Slab Bounded by Metal on One of Its Sides

Although this problem has been solved for a while [5.43], we think it interesting to present another approach. Contrary to [5.43] which neglects the imaginary part of the propagation constant of the leaky mode, the following study determines it without any approximation.

Space is now divided into three parts by the planes y = 0 and y = -e (Fig.5.10). Each part is filled with a medium characterized by its refractive index ν_i. Region ① is vauum ($\nu_1 = 1$). Region with the number ② is a lossless dielectric (ν_2 is a real number) and region ③ is a metal (ν_3 is a complex number).

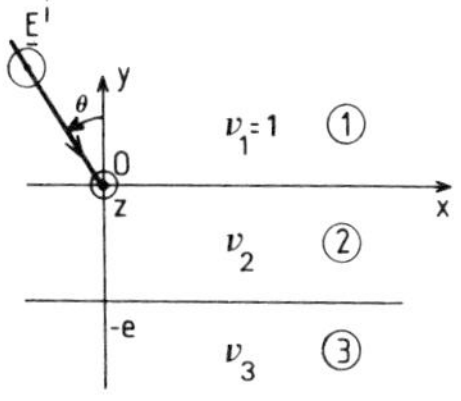

Fig. 5.10. Schematic representation of a dielectric slab

An incident plane wave with wavelength λ impinges on the structure under incidence θ, with the electric vector $\underline{E}^i$ parallel to axis Oz

$$\underline{E}^i = E^i(x,y)\,\hat{\underline{z}} \quad .$$

Determination of the Reflection Coefficient

Using our previous notations: $\alpha = k_0 \sin\theta$, $\delta = \alpha/k_0 = \sin\theta$, and $\beta^{(j)} = k_0\sqrt{\nu_j^2 - \delta^2}$, the electric field can be described in the different regions by

region ① $(y > 0)$

$$E(x,y) = \exp[i(\alpha x - \beta^{(1)}y)] + r(\alpha)\exp[i(\alpha x + \beta^{(1)}y)] \quad ,$$

region ② $(-e < y < 0)$

$$E(x,y) = t(\alpha)\exp[i(\alpha x - \beta^{(2)}y)] + t'(\alpha)\exp[i(\alpha x + \beta^{(2)}y)] \quad ,$$

region ③ $(y < -e)$

$$E(x,y) = p(\alpha)\exp[i(\alpha x - \beta^{(3)}y)] \quad .$$

In these expressions r, t, t' and p are four unknown functions of α or δ. The former is the reflection coefficient of the structure for TE polarization. The continuity of the electric field $E(x,y)$ and its normal derivative at each interface allows the derivation of four equations from which the four preceding functions can be found. Let us introduce

$$\zeta_1^+ = \beta^{(2)} + \beta^{(1)}$$

$$\zeta_1^- = \beta^{(2)} - \beta^{(1)}$$

$$\zeta_2^+ = \beta^{(2)} + \beta^{(3)}$$

$$\zeta_2^- = \beta^{(2)} - \beta^{(3)} \quad .$$

Then the reflection coefficient is given by

$$r(\delta) = \frac{\zeta_1^+ \zeta_2^- \exp(i\beta^{(2)}e) - \zeta_1^- \zeta_2^+ \exp(-i\beta^{(2)}e)}{\zeta_1^+ \zeta_2^+ \exp(-i\beta^{(2)}e) - \zeta_1^- \zeta_2^- \exp(i\beta^{(2)}e)} \quad . \tag{5.24}$$

The study of $r(\delta)$ when δ varies from -1 to +1 (i.e., when θ varies from $-\pi/2$ to $+\pi/2$) shows that the reflection coefficient is a slowly varying function. Except under very special circumstances, no zero is found, which is natural since no classical Brewster phenomenon exists for TE polarization. The study of $1/r(\delta)$ also shows the absence of pole on the real axis. But, as we did in Sect.5.2, it may be interesting to study the extension of $r(\delta)$ into the complex δ plane. It requires the use of the cuts shown in Fig.5.2, whose image in the complex $\beta^{(j)}$ plane can be approximated by the second bisector (See Fig.5.3). The determination of $\beta^{(j)}$ which verifies $\mathrm{Im}\{\beta^{(j)}\} + \mathrm{Re}\{\beta^{(j)}\} > 0$ will be called $\beta_I^{(j)}$ and the opposite $\beta_{II}^{(j)}$. From (5.24), it immediately results that the change of $\beta_I^{(1)}$ into $\beta_{II}^{(1)}$

turns $r(\delta)$ into its inverse, which means that $r_{II}(\delta) = 1/r_I(\delta)$. This remark allows a simultaneous research of both poles and zeros of the reflection coefficient. Both problems are reduced to the research of the complex zeros of the complex functions given by (5.24), with the two possible determinations for $\beta^{(1)}$. Hereafger, we will denote "pole" (resp. "zero") of $r(\delta)$ a pole (resp. zero) of $r_I(\delta)$.

Determination of the Propagation Constants of the Leaky Modes of the Structure

Since the refractive index ν_3 is a complex number, a wave guided by this structure is necessarily leaky, which means that its propagation constant is a complex number. Such a wave again corresponds to a solution of Maxwell equations without any incident wave. It implies that the associated value of δ, called $\hat{\delta}$, is a pole of the reflection coefficient $r(\delta)$.

In order to find the complex pole $\hat{\delta}$, it is convenient to begin with the research of the real pole δ_r corresponding to a lossless structure which can support lossless guided waves. This lossless structure is derived from the one in Fig.5.10 by taking a real value for ν_3 less than ν_2. We then get a dielectric slab whose solution is well known [5.44]. Usually, δ_r is determined by resolving (5.25)

$$\beta^{(2)}e = \phi_3 + \phi_1 + m\pi \quad , \tag{5.25}$$

where $\phi_3 = \tan^{-1}\left(\dfrac{\beta^{(3)}}{i\beta^{(2)}}\right)$,

and $\phi_1 = \tan^{-1}\left(\dfrac{\beta^{(1)}}{i\beta^{(2)}}\right)$.

Of course this determination implies finding the root of a transcendental equation on the real axis. Since from now on, an accurate determination of the propagation constant is not necessary, a plot of the graph of $1/r_I(\delta)$ on the [-1,+1] interval is sufficient to give an approximate value of δ_r. The study may be restricted to the interval $[\max(\nu_1,\nu_3),\nu_2]$ since $\beta^{(2)}$ has to be real in order to get a guided wave, but $\beta^{(1)}$ and $\beta^{(3)}$ must be imaginary to avoid propagating waves outside the layer. We then introduce a small imaginary part for the refractive index ν_3, and compute the corresponding complex pole of $r(\delta)$ in the vicinity of δ_r, using the iterative method described in [5.33]. Then the imaginary part of ν_3 is slightly increased, and a new value of the pole is determined in the vicinity of the preceding one. The process continues till ν_3 has reached the value of the refractive index of the metal filling region ③. The whole calculation requires about 1 second on a Univac 1110 computer, and gives the complex propagation constant $\hat{\delta}$ of a leaky mode of the structure.

Figure 5.11 illustrates the determination in the case of a 0.1 μm thickness, SiO dielectric layer ($\nu_2 = 1.54$) deposited on an aluminum plane ($\nu_3 = 0.47 + i\ 4.84$)

for the wavelength 0.436 μm. Starting from the value $\nu_3 = 0.47$ which, of course, does not correspond to a physical situation, we find a real value for δ_r in the interval $[\nu_1,\nu_2]$. When the imaginary part of ν_3 increases, the imaginary part of the pole increases, presents a maximum, then decreases and reaches the complex value $\hat{\delta}$. The existence of the maximum is not surprising since, if $\mathrm{Im}\{\nu_3\} \to \infty$, $\hat{\delta}$ must tend towards a real value again, corresponding to lossless Zenneck guided waves [5.45]. Of course Fig.5.11 recalls Fig.5.4, except that we have taken a starting point which does not correspond to a real situation ($\nu_3 = 0.47$).

For certain values of the thickness e, the locus of $\hat{\delta}$ may meet the parallel to the imaginary axis: $\mathrm{Re}\{\hat{\delta}\} = 1$ when $\mathrm{Im}\{\nu_3\}$ varies. Then the guided mode disappears. On the other hand, if the refractive index ν_3 is kept constant, but the thickness e is varied, the guiding phenomenon vanishes under a cut-off thickness which depends on wavelength, indices and polarization. Figure 5.12 gives an example for the two polarizations.

It may be noted that the cut-off thickness for TE polarization is about twice as small as for TM polarization. In both cases, $\mathrm{Re}\{\hat{\delta}\}$ is an increasing function of e and tends towards ν_2 when $e \to \infty$. The imaginary part of $(\hat{\delta})$ presents a maximum not far from the cut-off thickness, and then decreases when e increases. This point may be explained by the fact that for a given length of guide, the number of reflections of the wave on the metal decreases. So does the dissipation of energy.

5.3.2 Reflection of a Plane Wave on a Dielectric Coated Reflection Grating Used in TE Polarization

Definitions and Notations

The problem is similar to the one illustrated in Fig.5.5 except that region ② now includes a dielectric overcoating and we deal with TE polarization $[\underline{E} = E(x,y)\hat{\underline{z}}]$. All the other parameters are the same. In any homogeneous region, the electric field may be represented by a superposition of plane waves

Region ①

$$E(x,y) = \exp[i(\alpha_0 x - \beta_0^{(1)} y)] + \sum_{n=-\infty}^{+\infty} B_n \exp[i(\alpha_n x + \beta_n^{(1)} y)] \quad ,$$

Region ③

$$E(x,y) = \sum_{n=-\infty}^{+\infty} A_n \exp[i(\alpha_n x - \beta_n^{(2)} y)] \quad ,$$

where α_n and $\beta_n(j)$ have been previously defined, and B_n and A_n are new and unknown coefficients. Assuming that the ratio λ_0/d is well chosen in order to get only a propagating order, the problem can be solved exactly in the same way as in Sect. 5.2. Introducing again the reduced variables $\delta_n = \alpha_n/k_0$, the continuity of the phenomenon when $h \to 0$ shows that $B_0(\delta_0,h)$ presents complex poles $\hat{\delta}_n$ and that $\lim \hat{\delta}_0 = \hat{\delta} - \lambda_0/d$. The finding of these poles may be conducted in the same way as in Sect.

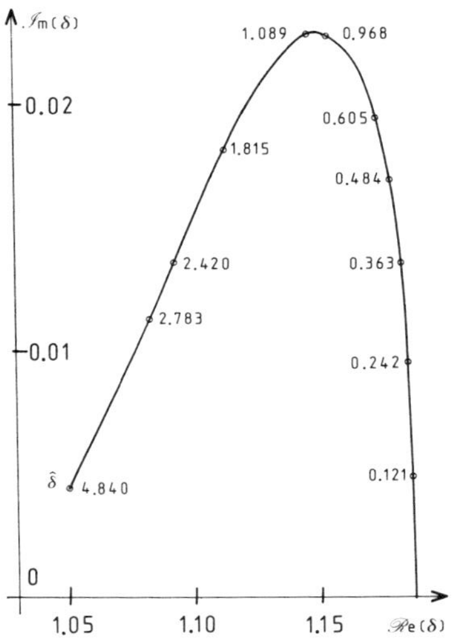

Fig.5.11. Locus of the pole of $r(\delta)$ in the complex δ plane when $Im\{\nu_3\}$ grows from 0 to the value of aluminum, 4.84

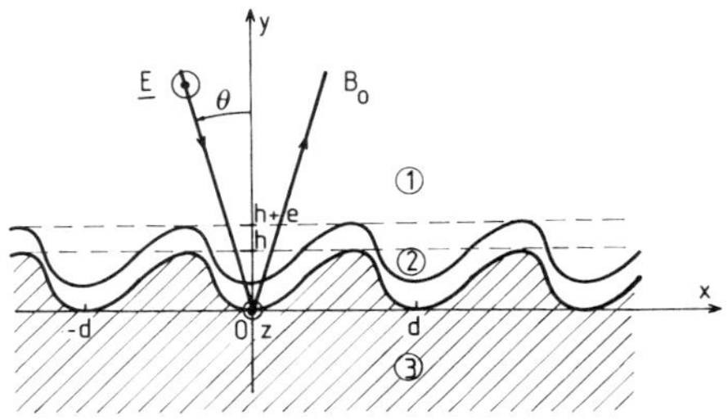

Fig.5.13. Schematic representation of a dielectric coated grating

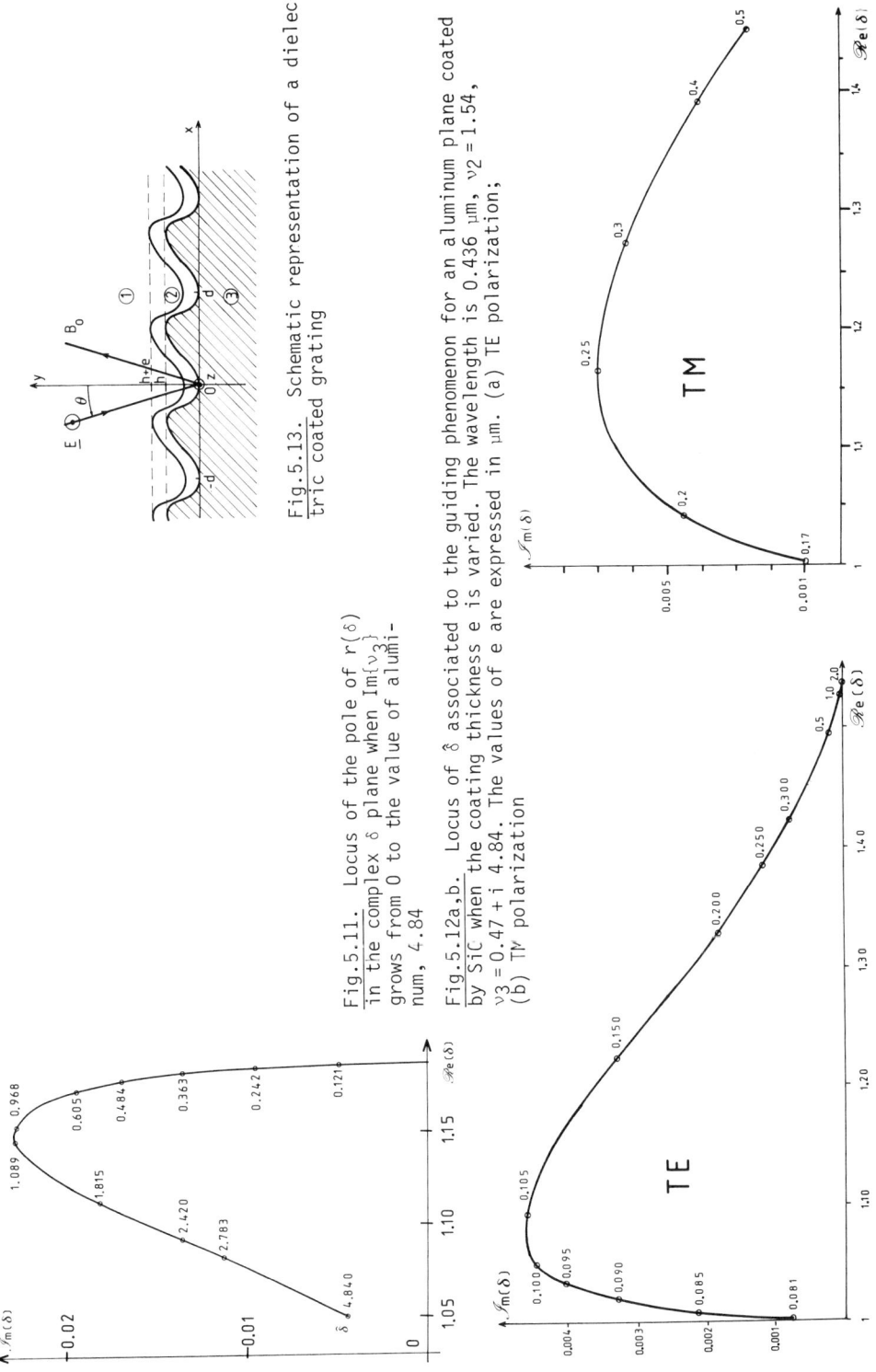

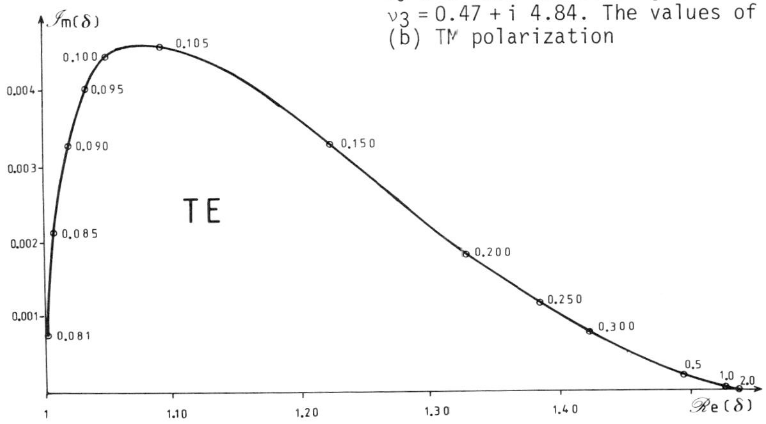

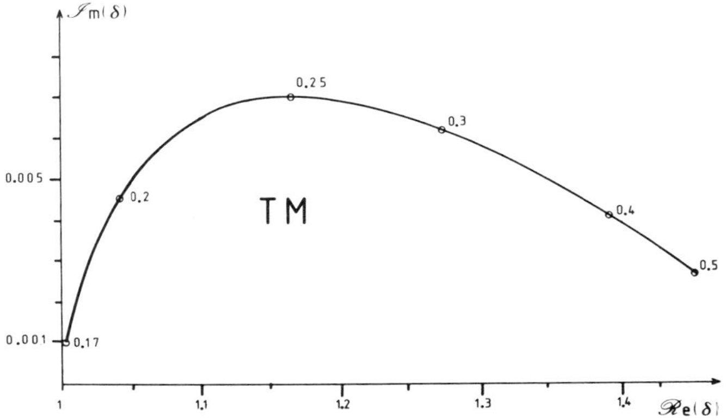

Fig.5.12a,b. Locus of $\hat{\delta}$ associated to the guiding phenomenon for an aluminum plane coated by SiC when the coating thickness e is varied. The wavelength is 0.436 μm, $\nu_2 = 1.54$, $\nu_3 = 0.47 + i\ 4.84$. The values of e are expressed in μm. (a) TE polarization; (b) TM polarization

5.2.2 starting from the propagation constant $\hat{\delta}$ of the leaky modes of the plane structure. The determination is again conducted by an iterative method, and requires the extension to complex variables of formalisms including a dielectric overcoating. As previously explained, the introduction of the grating modulation implies the existence of a pole and a zero, which have different imaginary parts and real parts very close to each other. Their determination and evolution in the complex plane can be conducted, with the aid of sophisticated computer programs, as in Sect. 5.2.2, for any varying parameter (groove depth, dielectric thickness, refractive index, wavelength ...). Figure 5.14 gives an example of their loci when the dielectric thickness e is varied. Here too, the locus of the zero $\delta^z(e)$ crosses the real axis for e = 0.165 μm. Then a phenomenon of total absorption is found when the incidence is given by

$$\sin\theta = \mathrm{Re}\{\delta^z(0.165)\} - \frac{\lambda_0}{d} \quad .$$

This time, the phenomenon is not linked with plasmon oscillations of electrons under the surface of the metal, since there is no plasmon oscillations for TE

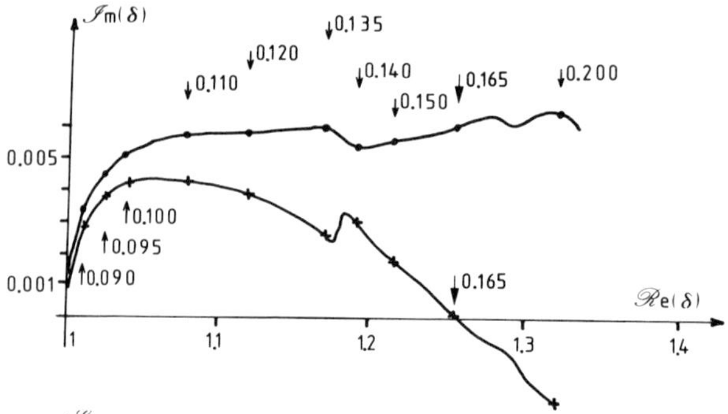

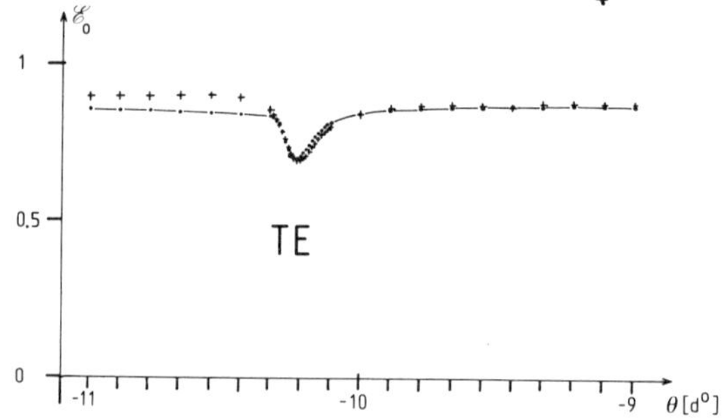

Fig. 5.14. Loci of the poles and the zero of B_0 in the complex δ plane when the dielectric thickness e is varied, for a 0.37 μm groove spacing, 0.02 μm groove depth aluminum grating coated with SiO. The wavelength is 0.436 μm and the thicknesses are expressed in μm. (•) pole; (+) zero

Fig. 5.15. Zero-order efficiency as a function of θ for an aluminum grating coated with SiO. (•) exact values given by the electromagnetic theory. (+) approximate values given by (5.23). d = 0.37 μm, h = 0.02 μm, e = 0.082 μm, λ = 0.436 μm, polarization; but it has been completely confirmed by experiments [5.29]. As soon as the pole and zero are known, the form of the absorption peak can again be predicted by (5.23), where $R(\delta_0) = |r(\delta_0)|^2$ and $r(\delta_0)$ is given by (5.24). Figure 5.15 shows a comparison with direct calculations and gives an idea of the accuracy of the approximation (5.23).

The introduction of strong absorption peaks on efficiency curves of a metallic grating shows that the use of dielectric overcoatings to enhance the efficiency of a grating must be made carefully. Whenever possible, the use of thickness under the cut-off seems to be preferable. If one increases the overcoating thickness, the number of leaky modes increases, and then the number of absorption peaks. The result is complete destruction of the efficiency curve of a grating [5.23].

5.4 Extension of the Theory

5.4.1 Anomalies of a Dielectric Coated Grating Used in TM Polarization

Plane Structure

For TM polarization, two kinds of poles corresponding to two kinds of different surface waves are found. The guided leaky waves appear for a dielectric thickness greater than the cut-off thickness, and $\mathrm{Re}\{\delta\}$ is located on the interval $[1, \nu_2]$, where ν_2 still designates the refractive index of the layer. The plasmon leaky waves appear whatever the dielectric thickness may be, and are not restricted to the preceding interval. Starting from the value $\hat{\delta}$ of the metallic plane, the locus of the propagation constant when the layer thickness varies is shown in Fig.5.16. When $e \to \infty$, $\delta \to \nu_2 \cdot \nu_3/\sqrt{\nu_2^2 + \nu_3^2}$ which, in the case of an aluminum plane coated by SiO is equal to 1.621604 + i 0.017316. As soon as e = 0.69 μm, our programs give a propagation constant $\hat{\delta}$ = 1.62160 + i 0.01736. Thus the asymptotic value is reached with a discrepancy less than 10^{-5} as soon as e/λ_0 is about 1.5.

For a thickness above the cut-off, the two phenomena coexist, but the propagation constants have very different values, as one can see by comparing Fig.5.12b and Fig.5.16.

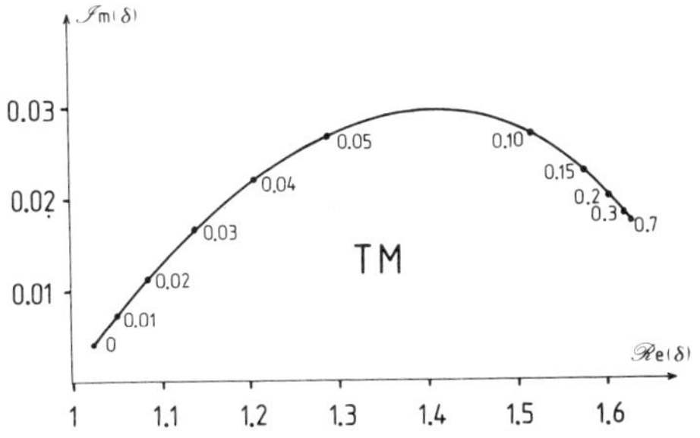

Fig. 5.16. Locus of the pole $\hat{\delta}$ associated to a plasmon resonance for an aluminum plane coated by SiO when the coating thickness e (expressed in μm) is varied. λ = 0.436 μm, TM polarization

Modulated Structure

Starting from the preceding two kinds of poles, two kinds of poles and zeros can be found when one introduces a modulation. Figure 5.17, relative to the plasmon leaky wave, shows an example of such a splitting. For e ≃ 0.37 μm, the zero locus crosses the real axis, which produces again a phenomenon of total absorption. The effect of the dielectric overcoating on both zeros and poles is generally to increase the imaginary parts (Fig.5.18). The result is a reduction of the absorption peak, which explains that TM anomalies are reduced by a dielectric overcoating, a fact pointed out by experiments [5.13].

Concerning the guided leaky waves, an increase of the overcoating has the same effects as for TE polarization, i.e., as the number of leaky modes increases, it tends to deteriorate the efficiency curve of a grating. But since the cut-off thickness is about twice as great as for TE polarization, the deterioration due to the guided modes only appears for higher dielectric thickness.

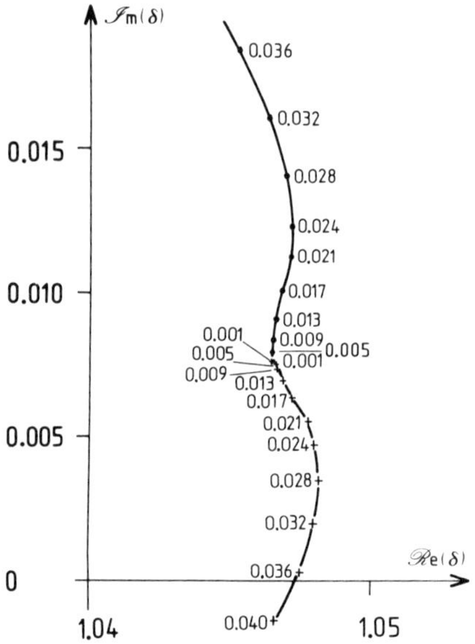

Fig. 5.17. Loci of the pole and zero corresponding to a plasmon leaky wave for an aluminum sinusoidal grating coated by SiO, when the groove depth (expressed in μm) is varied. d = 0.37 μm, e = 0.01 μm, λ = 0.436 μm. (•) pole; (+) zero. TM polarization

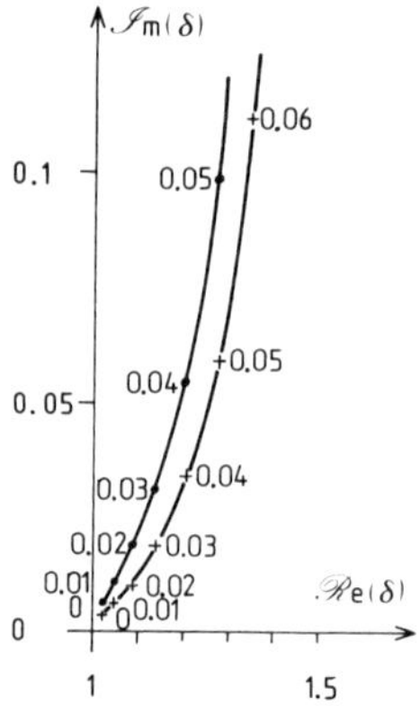

Fig. 5.18. Same as Fig. 5.17, but h is constant and equal to 0.02 μm and e (in μm) is varied.

5.4.2 Plasmon Anomalies of a Bare Grating Supporting Several Spectral Orders

Efficiency Behavior of the Various Spectral Orders

In order to generalize the theory, let us consider the more general case where two incident plane waves, with amplitudes A_0 and A_{-1}, linearly polarized, fall on the grating under incidences θ and θ_{-1} linked by the grating formula. For the sake of simplicity, let us suppose that the wavelength to groove spacing ratio is chosen in such a way that only two propagating orders ($n = 0$ and $n = -1$) are diffracted by the grating under the directions θ_0 ($\theta_0 = \theta$) and θ_{-1}. Above the modulated region, the total field is given by

$$\begin{aligned} H(x,y) &= A_0 \exp[i(\alpha_0 x - \beta_0^{(1)} y)] + A_{-1} \exp[i(\alpha_{-1} x - \beta_{-1}^{(1)} y)] \\ &+ B_0 \exp[i(\alpha_0 x + \beta_0^{(1)} y)] + B_{-1} \exp[i(\alpha_{-1} x + \beta_{-1}^{(1)} y)] \\ &+ \sum_{n \neq 0, n \neq -1} B_n \exp[i(\alpha_n x + \beta_n^{(1)} y)] \quad . \end{aligned}$$

As explained in the preceding chapter, the existence of the scattering matrix S with elements $S_{n,m}(\delta)$ allows us to write

$$B_n \sqrt{\beta_n^{(1)}} = \sum_m S_{n,m}(\delta)\, A_m \sqrt{\beta_m^{(1)}} \quad , \tag{5.26}$$

$$n \in [-1,0], \quad m \in [-1,0], \quad \delta = \alpha/k_0 \quad .$$

Introducing vectors $A = (A_{-1}\sqrt{\beta_{-1}^{(1)}}, A_0\sqrt{\beta_0^{(1)}})$ and $B = (B_{-1}\sqrt{\beta_{-1}^{(1)}}, B_0\sqrt{\beta_0^{(1)}})$ which represent the normalized incident and diffracted amplitudes, this expression can be written in matrix form

$$B = S(\delta) A \quad . \tag{5.27}$$

We will find a homogeneous solution of the problem if it is possible to find a non-null vector B with a null vector A (no incident wave), i.e., if the determinant Det $S(\delta)$ presents poles in the complex plane. Such poles δ^p depending on all the parameters of the problem, including grating geometry, their determination again requires the use of computer programs based on the electromagnetic theory. The associated zeros δ^z are determined in the same way.

If we remember that, when the mth order is incident on the grating, the absolute efficiency in the nth order $\mathcal{E}_n = |B_n|^2 \beta_n^{(1)}/\beta_m^{(1)}$ is equal to $|S_{n,m}|^2$, it is clear that the knowledge of δ^z and δ^p is not sufficient to predict the efficiencies in the various orders. The poles and zeros of each elements $S_{nm}(\delta)$ of the S matrix have to be determined.

A detailed numerical study made in our laboratory has led to the following conclusions:

the poles of all elements $S_{n,m}(\delta)$ are the same and equal to the pole δ^p of Det S. The fact that δ^p appears as the pole of *all* elements $S_{n,m}(\delta)$ is not clearly understood

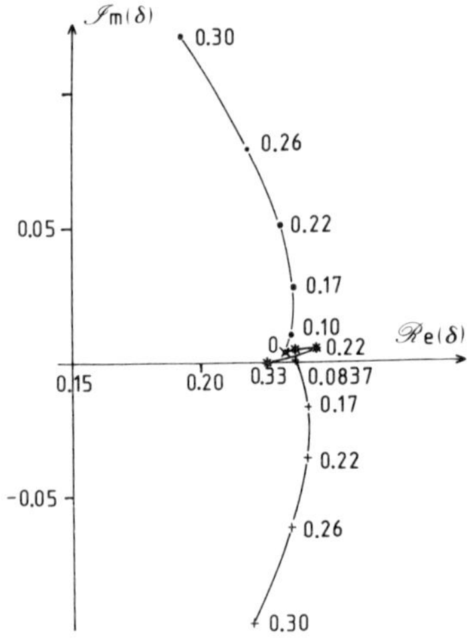

Fig. 5.19. Trajectories of the pole and the zeros for an aluminum, 1200 g/mm sinusoidal grating when the groove depth is varied. The wavelength is $\lambda = 0.647$ µm; (•) δ^p; (+) δ^z; (*) $\delta^z_{-1,0}$

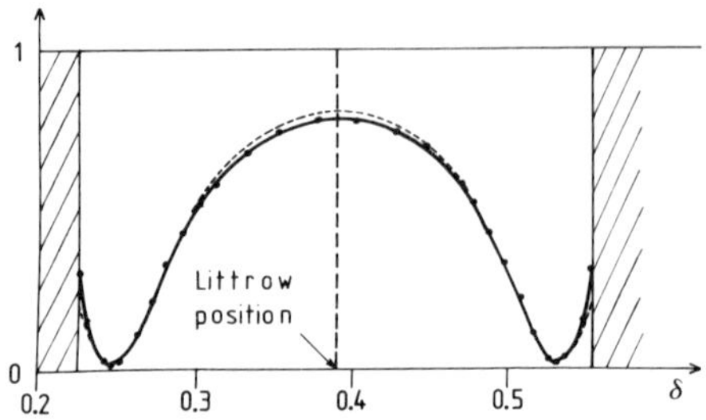

Fig. 5.20. Comparison between computed efficiencies (full line) and reconstituted efficiencies (dashed line) for the grating of Fig.5.19 and h = 0.22 µm. The hatched regions correspond to the values of δ which produce more than two diffracted orders

the zeros of elements $S_{n,m}(\delta)$ are different and distinct from the zero of Det $S(\delta)$. Thus, we designate by $\delta^z_{n,m}$ the zero of the element $S_{n,m}(\delta)$.

Figure 5.19 shows a determination of δ^p, δ^z and $\delta^z_{-1,0}$ for a 1200 g/mm sinusoidal aluminum grating supporting two spectral orders when the groove depth is varied. Again, when $h \to 0$, δ^z and δ^p, as well as $\delta^z_{-1,0}$ tend towards the same value. For $\delta \simeq \mathrm{Re}\{\delta^z\} \simeq \mathrm{Re}\{\delta^z_{n,m}\}$, $S_{n,m}(\delta)$ can be written in the following form:

$$S_{n,m}(\delta) = C_{n,m}\frac{\delta - \delta^z_{n,m}}{\delta - \delta^p} \quad , \tag{5.28}$$

where $C_{n,m}$ is a factor independent of δ, and for a given geometry, is only dependent on the groove depth. If $C_{n,m}$ has been determined, (5.28) allows us to reconstitute the efficiency $\mathcal{E}_n(\delta)$. Figure 5.20 shows an example for $\mathcal{E}_{-1}(\delta) = |S_{-1,0}(\delta)|^2$ which is in good agreement with values given by direct computation based on Maxwell equations. Since the reconstitution is done not only near $\delta = \mathrm{Re}\{\delta^z\}$, but in the whole interval of δ which produces two diffracted orders, it is necessary to use, on top of $\delta^z_{-1,0}$ and δ^p, the zero and pole $\delta'^z_{-1,0}$ and δ'^p corresponding to surface waves propagating in the opposite direction. Thus (5.28) has to be turned into

$$\mathcal{E}_{-1}(\delta) = |S_{-1,0}(\delta)|^2 = |C_{-1,0}|^2 \left|\frac{\delta - \delta^z_{-1,0}}{\delta - \delta^p}\right|^2 \left|\frac{\delta - (\frac{\lambda}{d} - \delta^z_{-1,0})}{\delta - (\frac{\lambda}{d} - \delta^p)}\right|^2 \quad .$$

The use of this formula implies that $\mathcal{E}_{-1}(\delta)$ is symmetrical with respect to $\alpha = \lambda/2d$, i.e., to Littrow position. The symmetry is still found on computed values and is due to reciprocity [5.46].

Though the method presented in this section allows the prediction of the shape of the efficiency curve in a given order, it cannot predict the total energy curve. To this aim, we have to derive a more general formalism.

Total Energy Curve

From the definition of vectors A and B, it immediately follows that the total incident energy and the total diffracted energy are, respectively, equal to the scalar products $\langle A,A\rangle$ and $\langle B,B\rangle$. Thus, we get the total efficiency $\mathcal{E}^d(\delta)$

$$\mathcal{E}^d(\delta) = \frac{\langle B,B\rangle}{\langle A,A\rangle} \quad ,$$

and (5.27) implies

$$\mathcal{E}^d(\delta) = \frac{\langle SA,\ SA\rangle}{\langle A\ ,\ A\rangle} = \frac{\langle S^*\ SA,\ A\rangle}{\langle A,\ A\rangle} \quad , \tag{5.29}$$

where S* is the adjoint of matrix S.

Contrary to what happened for the S matrix, matrix S*S is self-adjoint. Thus, it has two orthogonal eigenvectors V_1 and V_2 and two real and positive eigenvalues u_1 and u_2 (a fact implied by the particular form of this matrix). On top of that, the preceding constants δ^z and δ^p are also zero and pole for S*S.

Numerical studies have shown that this matrix is very suitable for efficiency studies. The eigenvectors V_1 and V_2, as well as eigenvalue u_2, are practically independent of δ near $\delta \simeq \mathrm{Re}\{\delta^p\}$. This implies that the second eigenvalue u_1, which is equal to Det $(S^*S)/u_2$ is given to a good accuracy by a simple formula similar to (5.28)

$$u_1(\delta) = |c|^2 \left|\frac{\delta - \delta^z}{\delta - \delta^p}\right| \quad , \quad \text{c: constant with respect to } \delta.$$

Rigorous computations for a particular value of δ allow the determination of the value of c. Thus, it is possible to check this approximate formula over a wide range of δ. Here too its validity has been thoroughly confirmed.

As soon as the eigenvalues and eigenvectors of S*S are known, (5.29) gives a simple expression of the total reflected energy. Expressing vector A in terms of its components on the orthogonal normalized eigenvectors V_1 and V_2

$$A = \langle A,V_1\rangle \cdot V_1 + \langle A,V_2\rangle \cdot V_2 \quad ,$$

we get after elementary calculations

$$\mathcal{E}^d(\delta) = u_1\ |\langle A,V_1\rangle|^2 + u_2\ |\langle A,V_2\rangle|^2 \quad . \tag{5.30}$$

Thus, the total diffracted energy can be easily predicted, and is found to be in good agreement with the value given by computer programs.

Of course if $u_1 = 0$, (5.30) reduces to

$$\mathcal{E}^d(\delta) = u_2|<A,V_2>|^2 \quad .$$

This happens if $\delta = \delta^z$, which implies that δ^z is real, a situation found for a particular groove depth h = 0.0837 µm given by Fig.5.19. So, if the two incident waves are chosen in such a way that $A = V_1$, $E^d(\delta) = 0$ and the grating absorbs in totality the two incident waves. It may seem surprising that, starting from this situation, a simple change in the relative phases or amplitudes of the incident waves may produce a total reflected energy close, to u_2, i.e., close to the reflectance of the metal. Under the preceding conditions, no total absorption phenomenon is found if only an incident plane wave strikes the grating.

5.4.3 General Considerations on Anomalies of a Grating Supporting Several Spectral Orders

Though the theory presented in the preceding paragraph is applicable for both polarizations to any kind of modulated structure, it may be troublesome to compute the eigenvalues and eigenvectors of S*S. In many cases, it is sufficient to be able to predict the location of anomalies and not their detailed shape. Thus the knowledge of the zeros of S is generally enough.

Figure 5.21 gives for example the total energy curve of a 7° blaze angle, 1264 grooves/mm aluminum grating, coated with a 0.104 µm thick SiO layer when the angle

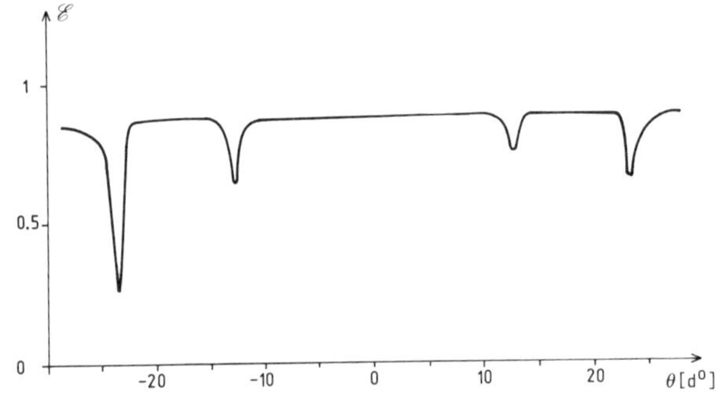

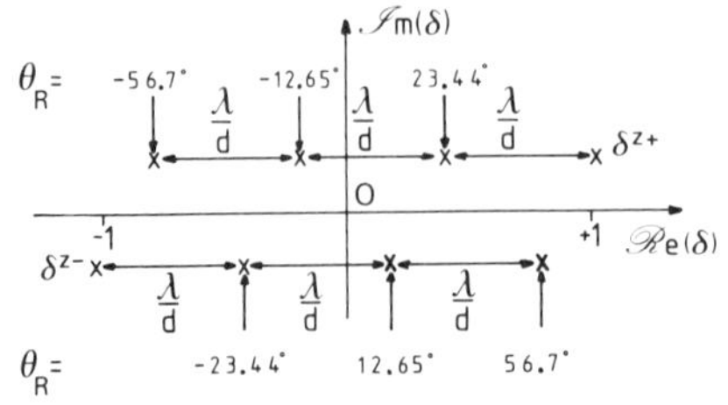

Fig. 5.21. Total energy curve of a 7° blaze angle, 1264 grooves/mm aluminum grating, coated with a 0.104 µm thick SiO layer, as a function of the angle of incidence. The apex angle of the grating is 110° and the wavelength 0.488 µm

Fig. 5.22. Positions of the absorption peaks in the total energy curve of the grating of Fig. 5.21, deduced from (5.31)

of incidence θ is varied. The apex angle of the grating is 110°, and the wavelength 0.488 μm. The curve given by direct computation shows four absorption peaks near -23.5°, -12.6°, $+12{,}7^{\circ}$ and 23.5°. Two other absorption peaks not plotted in the figure were also found near $\theta = \pm 57^{\circ}$.

The determination of the zeros of the S matrix gives two symmetrical values

$$\delta^{Z+} = 1.01546 + i\,0.00391 \quad \text{and} \quad \delta^{Z-} = -1.01546 - i\,0.00391 \quad .$$

When the incidence θ is chosen in such a way that there exists an integer n so that

$$\sin\theta + n\frac{\lambda}{d} = \mathrm{Re}\{\delta^{Z+}\} \quad \text{or} \quad \mathrm{Re}\{\delta^{Z-}\} \quad , \tag{5.31}$$

a surface wave is excited and an anomaly is expected to appear. Figure 5.22 sums up the values of θ obtained from (5.31). It is clear that the number and location of anomalies coincide with the results from direct calculations.

5.5 Theory of the Grating Coupler

Holographic thin film couplers are used in integrated optics to couple a laser beam into and out of optical waveguides. Figure 5.23 shows a microscopic description of a typical structure as well as the associated macroscopic arrangement. A few millimeter wide photoresist grating is superimposed on top of a dielectric slab and is struck by a limited beam with width L. For certain values θ_R of the macroscopic incidence θ_0, an important fraction of the incident energy is coupled into the leaky waveguide formed by the thin dielectric slab and its modulated coating. Though the modulation ends at a certain distance from the edge of the laser beam, in our analysis, it will be assumed unlimited in the x direction in order to keep the periodicity of the problem.

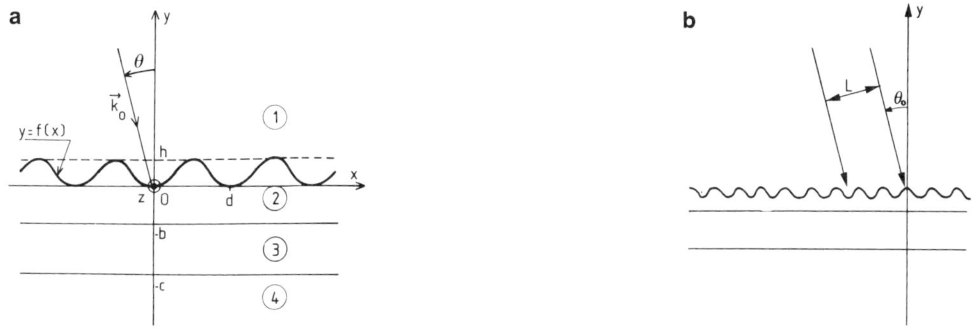

Fig. 5.23. (a) Microscopic desription of a typical grating coupler system; (b) Associated macroscopic arrangement

Here too, the coupling phenomenon is related to the excitation of surface waves. Thus, the angle of resonance θ_R will still be given by (5.31). But the problem differs from the preceding ones in that a limited beam impinges on the structure. Moreover we not only want to determine the coupling angles θ_R, but we also hope to determine the coupling coefficient and to optimize the width of the laser beam. To this end we adopt the following approach:

1) Describe the incident beam as a superposition of plane waves by means of the Fourier transform;
2) Determine the response of the system to a plane wave;
3) Add up the responses corresponding to all the plane waves and study the properties of the resultant field.

5.5.1 Description of the Incident Beam

Let $F^i(x,y)$ be the z component of the incident field $\underline{E}$ or $\underline{H}$, depending on the polarization and let $p(\alpha)$ be a density function with a width at half height $\Delta\alpha \ll k_0$. A limited beam which impinges on the coupler under the incidence θ_0 can be represented by

$$F^i(x,y) = \int_{-k_0}^{+k_0} p(\alpha-\alpha_0)\exp(i\alpha x)\exp(-i\beta^{(1)}y)\,d\alpha \quad , \tag{5.32}$$

where $\alpha = k_0 \sin\theta$, $\alpha_0 = k_0 \sin\theta_0$ and $\beta^{(1)} = \sqrt{k_0^2 - \alpha^2}$.

In the vicinity of α_0, the function $\beta^{(1)}(\alpha)$ can be approximated by the first two terms of its Taylor series

$$\beta^{(1)}(\alpha) = \beta^{(1)}(\alpha_0) - \frac{\alpha_0}{\beta^{(1)}(\alpha_0)}(\alpha - \alpha_0) \quad .$$

If we remember that $\Delta\alpha \ll k_0$, (5.32) becomes

$$F^i(x,y) \simeq \exp[i(\alpha_0)x - \beta^{(1)}(\alpha_0)y]\int_{-\infty}^{+\infty} p(\alpha-\alpha_0)$$
$$\times \exp\left[i(\alpha-\alpha_0)x + \frac{i\alpha_0}{\beta^{(1)}(\alpha_0)}(\alpha-\alpha_0)y \quad d\alpha\right] \quad .$$

Introducing the Fourier transform $q(x)$ of the density $p(\alpha)$

$$q(x) = \int_{-\infty}^{+\infty} p(\xi)\exp(i\xi x)\,d\xi \quad ,$$

which describes the spatial structure of the beam, the incident field reduces to

$$F^i(x,y) \simeq q\left[x + \frac{\alpha_0}{\beta^{(1)}(\alpha_0)}y\right]\exp[i(\alpha_0 x - \beta^{(1)}(\alpha_0)y)] \quad . \tag{5.33}$$

5.5.2 Response of the Structure to a Plane Wave

Referring to Chap.1, when an incident plane wave $F^i(x,y,\alpha) = \exp[i\alpha x]\exp[-i\beta^{(1)}(\alpha)y]$ strikes the structure, it generates a diffracted field $F^d(x,y,\alpha)$ given in a Floquet form

$$F^d(x,y,\alpha) = u(x,y,\alpha)\exp(i\alpha x) \quad ,$$

where the function $u(x,y,\alpha)$ is periodic with period d with respect to x. Thus it can be represented by its Fourier series

$$u(x,y,\alpha) = \sum_n u_n(y,\alpha)\exp(inKx) \quad , \quad K = \frac{2\pi}{d} \quad .$$

Above the modulated region where the Rayleigh expansion (5.10) is valid, $u_n(y,\alpha) = B_n \exp[i\beta_n^{(1)}y]$. The result is that the previously studied poles $\alpha^p = k_0\delta^p$ of the S matrix appear to be poles of the functions $u_n(y,\alpha)$. Each of them is the propagation constant of a leaky wave supported by the modulated waveguide. When the incidence θ is chosen in such a way that there exists an integer n_0 so that

$$\alpha + n_0K = \mathrm{Re}\{\alpha^p\} \quad , \tag{5.34}$$

the leaky wave can be excited by the incident field , and the coupling phenomenon is found. We speak of coupling by means of the n_0^{th} order, and with the wavelengths and groove spacings generally used, $n_0 = +1$. Let us introduce $\tilde{\alpha}^p = \alpha^p - n_0K$. In the vicinity of $\tilde{\alpha}^p$, $u_n(y,\alpha)$ can be written in the form

$$u_n(y,\alpha) = \frac{v_n(y,\alpha)}{\alpha - \tilde{\underline{\alpha}}^p} \quad .$$

Near α_0, $v_n(y,\alpha)$ are slowly varying functions of α. Thus

$$u_n(y,\alpha) \simeq \frac{v_n(y,\alpha_0)}{\alpha - \tilde{\underline{\alpha}}^p} \quad , \tag{5.35}$$

and

$$F^d(x,y,\alpha) = \sum_n \frac{v_n(y,\alpha_0)}{\alpha - \tilde{\underline{\alpha}}^p}\exp[i(\alpha + nK)x] \quad . \tag{5.36}$$

An important remark needs to be made. Let us suppose that all the various media are lossless. Then the energy balance criterion is satisfied. The unitarity of the diffraction operator implies that, for all the terms corresponding to propagating waves, a zero α^z exists and is the conjugate complex of the pole α^p (otherwise the corresponding efficiency should become greater than unity if α is close to $\underline{\alpha}^p$). As a result, nothing special is observed on the efficiency curves. On the other hand, the evanescent terms are not concerned by the energy balance and have no reason to present zeros associated to α^p. Thus they may present strong resonance peaks when α is varied on the real axis and verifies (5.34). In order to confirm

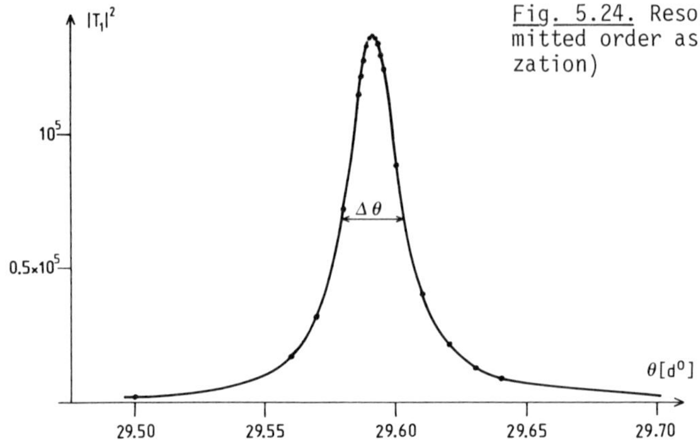

Fig. 5.24. Resonance curve of the +1 transmitted order as a function of θ(TE polarization)

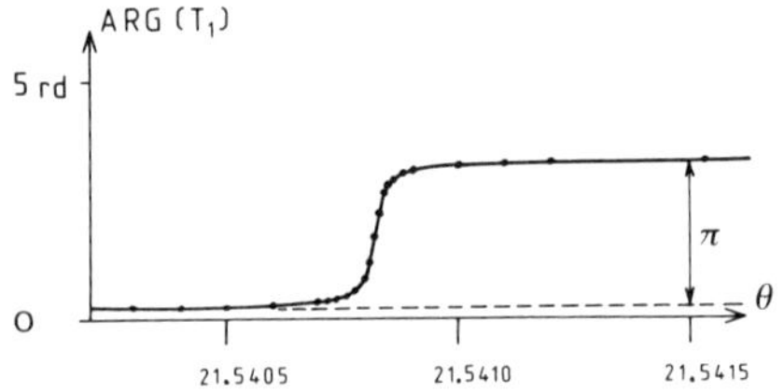

Fig. 5.25. Variations of the argument of T_1 in the vicinity of a resonance which occurs when $\theta = 21.54082^\circ$(TE polarization)

this prediction, Fig.5.24 shows the plot of the square of the modulus of the Rayleigh coefficient T_1 of the +1 transmitted order in medium 4 as a function of θ. The parameters of the structure are: $d = 0.606$ µm, $h = 0.1$ µm, $b = 0.43$ µm, $c = 1.23$ µm, $\lambda = 0.6328$ µm, $\nu_1 = 1$, $\nu_2 = 1.62$, $\nu_3 = 1.57$, $\nu_4 = 1.515$. The groove profile is a sinusoid. The computation of the poles of the S matrix gives $\alpha^p = 15.2714 + i\ 0.001883$. Then (5.34,35) predict a resonance peak around $\theta_R = 29.5915^\circ$, which perfectly coincides with the results of a direct calculation shown in Fig.5.24. From (5.35), it follows that the width at half height is classically equal to $2\ \mathrm{Im}\{\alpha^p\}$. Thus, (5.35) shows a width $\Delta\theta = 2\ \mathrm{Im}\{\alpha^p\}/k_0 \cos\theta_R$. Applied to the parameters of the curve in Fig.5.24, it gives $\Delta\theta = 0.025^\circ$, which again is in very good agreement with the direct calculation.

Other computations on different structures always give the same order of magnitude of the width $\Delta\theta$, which shows how tight the control of the incidence must be in order to get the coupling phenomenon. The result is that the resonances can be missed very easily when one varies the angle of incidence and computes the Rayleigh coefficients. But on the span of the resonance, the argument of the Rayleigh coefficient jumps by π, as seen in Fig.5.25, and implied by (5.35). This jump can be used to detect the resonance angles θ_R of the structure.

5.5.3 Response of the Structure to a Limited Beam

The response to a plane wave being known, the linearity of the problem allows us to determine the response to a superposition of plane waves in a simple manner

$$F^d(x,y) = \int_{-\infty}^{+\infty} p(\alpha-\alpha_0) \sum_n \frac{v_n(y,\alpha_0)}{\alpha - \tilde{\alpha}^p} \exp[i(\alpha + nK)x] \, d\alpha \quad . \tag{5.37}$$

Of course, in order to get a good coupling, the beam must be centered on the resonance angle θ_R, which implies that $\alpha_0 = \mathrm{Re}\{\tilde{\alpha}^p\}$. Thus, if we introduce $a'' = \mathrm{Im}\{\alpha^p\} = \mathrm{Im}\{\tilde{\alpha}^p\}$, (5.37) becomes

$$F^d(x,y) = \int_{-\infty}^{+\infty} p(\alpha-\alpha_0) \sum_n \frac{v_n(y,\alpha_0)}{\alpha - \alpha_0 - i\alpha''} \exp[i(\alpha+nK)x]d\alpha \quad . \tag{5.38}$$

As we did for the incident beam, the expression of the diffracted field can be simplified by the introduction of a convenient function. Let us define

$$r(x) = -i \int_{-\infty}^{+\infty} p(\xi) \frac{\exp(i\xi x)}{\xi - i\alpha''} \, d\xi \quad , \tag{5.39}$$

the diffracted field reduces to

$$F^d(x,y) = r(x) \; i \sum_n v_n(y,\alpha_0) \exp[i(\alpha_0 + nK)] \, x \quad . \tag{5.40}$$

The determination of the function r(x) is straightforward if we note that

$$\frac{dr(x)}{dx} = \int_{-\infty}^{+\infty} \frac{\xi p(\xi) \exp(i\xi x)}{\xi - i\alpha''} \, d\xi$$

$$= \int_{-\infty}^{+\infty} \frac{(\xi - i\alpha'')p(\xi) \exp(i\xi x)}{\xi - i\alpha''} \, d\xi + \int_{-\infty}^{+\infty} \frac{i\alpha'' p(\xi) \exp(i\xi x)}{\xi - i\alpha''} \, d\xi = q(x) - \alpha'' r(x) \; .$$

Thus r(x) is solution of the differential equation

$$\frac{dr(x)}{dx} + \alpha'' r(x) = q(x) \quad , \tag{5.41}$$

and from the classical theorem that the Fourier transform of an integrable function vanishes at infinity, r(x) verifies

$$r(-\infty) = 0 \quad . \tag{5.42}$$

Equations (5.41,42) show that r(x) is real if q(x), which describes the incident beam, is a real function.

5.5.4 Determination of the Coupling Coefficient

We call the coupling coefficient the ratio of the guided power to the power of the incident beam. The guided power is the one carried by the wave whose subscript n is equal to n_0 defined by (5.34). All the other terms in the Fourier series carry a negligible amount of energy, compared to this order. This is due to the fact that, except under very particular circumstances, only one space harmonic is excited by the incident beam.

For TE polarization, the complex Poynting vector $\underline{\Pi}^i$ of the incident wave is given by

$$\underline{\Pi}^i = \frac{\mathcal{E}^i \wedge \mathcal{H}^{i*}}{2} = \frac{\mathcal{E}^i \cdot \mathit{grad}\ \mathcal{E}^{i*}}{-2i\omega\mu} ,$$

where the star designates the complex conjugate.

If $\hat{\underline{u}}$ is the unit vector of the axis of the beam

$$\underline{\Pi}^i \cdot \hat{\underline{u}} = iCF^i\ \mathit{grad}\ F^{i*} \cdot \hat{\underline{u}}, \text{ where C is a constant.}$$

From (5.33), the incident beam can be expressed as

$$F^i(x,y) = q\left(\frac{x_1}{\cos\theta_R}\right)\ \exp(-ik_0 y_1) \quad , \text{ where}$$

$$x_1 = x\cos\theta_R + y\sin\theta_R \quad ,$$

$$y_1 = -x\sin\theta_R + y\cos\theta_R \quad .$$

In the new coordinate system

$$\underline{\Pi}^i \cdot \hat{\underline{u}} = C\ i\ F^i \frac{\partial F^{i*}}{\partial y_1} = C\ k_0 \left| q\left(\frac{x_1}{\cos\theta_R}\right)\right|^2 \quad .$$

Thus the incident power p^i is given by

$$p^i = \int_{-\infty}^{+\infty} \underline{\Pi}^i \cdot \hat{\underline{u}}\ dx_1 = C\ k_0\ \cos\theta_R \int_{-\infty}^{+\infty} |q(x)|^2\ dx \quad . \tag{5.43}$$

The calculation of the guided power p^g implies the calculation of

$$\underline{\Pi}^g \cdot \hat{\underline{x}} = iCF^g\ \mathit{grad}\ F^{g*} \cdot \hat{\underline{x}} \quad , \tag{5.44}$$

where $\hat{\underline{x}}$ is the unit vector of the x axis and

$$F^g = r(x)i\ v_{n0}(y,\alpha_0)\ \exp[i(\alpha_0 + n_0 K)x] \quad .$$

Thus

$$\frac{\partial F^{g*}}{\partial x} = -i\, v^*_{n0}(y,\alpha_0) \left[\frac{dr}{dx} - i(\alpha_0 + n_0 K) r(x)\right] \exp[-i(\alpha_0 + n_0 K)x] \quad .$$

Recalling (5.41), we get

$$\underline{\Pi}^g \cdot \hat{\underline{x}} = |v_{n0}(y,\alpha_0)|^2 \{i[r(x)q(x) - \alpha'' r^2(x)] + (\alpha_0 + n_0 K) r^2(x)\} \quad .$$

The calculation of the real part of the flux of $\underline{\Pi}^g$ through a plane perpendicular to the Ox axis gives

$$p^g = (\alpha_0 + n_0 K) r^2(x) \int_{-\infty}^{+\infty} |v_{n0}(y,\alpha_0)|^2\, dy \quad ,$$

which leads to the expression of the coupling coefficient

$$C(x) = \frac{p^g}{p^i} = \frac{(\alpha_0 + n_0 K) r^2(x) \int_{-\infty}^{+\infty} |v_{n0}(y,\alpha_0)|^2\, dy}{k_0 \cos\theta_R \int_{-\infty}^{+\infty} q(x)^2\, dx} \quad . \tag{5.45}$$

5.5.5 Application to a Limited Incident Beam

In order to illustrate the preceding considerations, and for the sake of simplicity, let us consider a limited beam with spatial structure given by

$$q(x) = 1 \text{ if } |x| \leqq \frac{\ell}{2}\ , \quad q(x) = 0 \text{ otherwise.}$$

Then (5.41) can be solved analytically and gives

$$r(x) = 0 \quad \text{if} \quad x < -\frac{\ell}{2} \quad ,$$

$$r(x) = \frac{1}{\alpha''} \{1 - \exp[-\alpha''(x + \tfrac{\ell}{2})]\} \quad \text{if} \quad |x| < \frac{\ell}{2} \quad ,$$

$$r(x) = \frac{1}{\alpha''} [1 - \exp(-\alpha''\ell)] \exp(\frac{\alpha''\ell}{2}) \exp(-\alpha'' x) \quad \text{if} \quad x > \frac{\ell}{2} \quad .$$

It is easy to see that this function presents a maximum for $x = \ell/2$. Thus the maximum coupling coefficient C_M is found for $x = \ell/2$

$$C_M = \frac{(\alpha_0 + n_0 K)[1 - \exp(-\alpha''\ell)]^2 \int_{-\infty}^{+\infty} |v_{n0}(y,\alpha_0)|^2\, dy}{\alpha''^2 k_0 \cos\theta_R\, \ell} \quad .$$

It is useful to write this expression in terms of a product of two factors

$$C_M = \frac{2[1 - \exp(-\alpha''\ell)]^2}{\alpha''\ell} \; \frac{(\alpha_0 + n_0 K) \int_{-\infty}^{-\infty} |v_{n0}(y,\alpha_0)|^2\, dy}{2\alpha'' k_1 \cos\theta_R} \quad .$$

The first factor depends on the grating coupler and on the beam. It is the function described by ULRICH [5.47] which has its maximum value 0.81 when $\alpha''\ell = 1.25$. The second factor, which is always inferior unity [5.34], only depends on the grating coupler and must be determined by computer programs. Computations show that it should not be omitted and that the maximum coupling coefficient is often far inferior to the value given by the first term. Numerical results also show that C_M is greatly decreased when one introduces small losses in the photoresist and that this decrease is dependent on the modulation (the higher the modulation, the smaller the decrease). Since in practice, the photoresist always presents some losses, it is important to choose groove depths which are not too low.

On the other hand, the optimization of the beam width can be solely conducted with the first term. The optimal width L_0 is simply given by

$$L_0 = \ell \cos\theta_R = \frac{1.25 \cos\theta_R}{\alpha''} \quad .$$

With the usual values of α'', L_0 is found to be about a millimeter, and can be less if the losses are important.

For TM polarization, the calculation of the coupling coefficient can be conducted along the same lines, except that the refractive index of the media appears in the calculation of the guided power.

A curious and interesting situation is found when the incident wave comes from the substrate in such a way that no transmitted order exists, and only one reflected order is supported by the structure. An extension of the above formalism [5.48] shows that the coupling coefficient is not dependent upon groove shape, and generally presents higher values than for the classical case. With a convenient incident beam, the upper limit 0.81 can be reached. Such a structure is designated by the name of reverse coupler.

Because of space limitations, we have not presented a detailed review of the great amount of work done on grating coupler theories and their applications to integrated optics. The reader interested in this field can refer to some recent books [5.49-51].

References

5.1 R.W. Wood: Philos. Mag. 4, 396 (1902)
5.2 J.W.S. Rayleigh: Philos. Mag. 14, 60 (1907)
5.3 J.W.S. Rayleigh: Proc. Roy. Soc. A 79, 399 (1907)
5.4 R.W. Wood: Phys. Rev. 48, 928 (1935)
5.5 L.R. Ingersoll: Phys. Rev. 17, 928 (1921)
5.6 J. Strong: Phys. Rev. 49, 291 (1936)
5.7 C.H. Palmer Jr.: J. Opt. Soc. Am. 42, 269 (1952); 46, 50 (1956)
5.8 J.E. Stewart, W.S. Gallaway: Appl. Opt. 1, 421 (1962)
5.9 J. Hagglund, F. Sellberg: J. Opt. Soc. Am. 56, 1031 (1966)
5.10 J.J. Cowan, E.T. Arakawa: Z. Phys. 97, 235 (1970)
5.11 M.C. Hutley: Opt. Acta 20, 607 (1973)
5.12 M.C. Hutley, V.M. Bird: Opt. Acta 20, 771 (1973)

5.13 M.C. Hutley, J.F. Verrill, R.C. McPhedran: Opt. Commun. 11, 207 (1974)
5.14 U. Fano: J. Opt. Soc. Am. 31, 213 (1941)
5.15 A. Hessel, A.A. Oliner: Appl. Opt. 4, 1275 (1965)
5.16 A. Wirgin, R. Deleuil: J. Opt. Soc. Am. 59, 1348 (1969)
5.17 M. Nevière, M. Cadilhac, R. Petit: Opt. Commun. 6, 34 (1972)
5.18 J.J. Cowan, E.T. Arakawa, L.R. Painter: Appl. Opt. 8, 1734 (1969)
5.19 R.H. Ritchie, E.T. Arakawa, J.J. Cowan, R.N. Hamm: Phys. Rev. Lett. 22, 1530 (1968)
5.20 R.C. McPhedran, D. Maystre: Nouv. Rev. Opt. 5, 241 (1974)
5.21 R.C. McPhedran, D. Maystre: J. Spectrosc. Soc. Jpn. 23, suppl. 1, 13 (1974)
5.22 E.G. Loewen, D. Maystre, R.C. McPhedran, I. Wilson: Jpn. J. Appl. Phys. 14, suppl. 14-1, 143 (1975)
5.23 M. Nevière, P. Vincent, R. Petit: Nouv. Rev. Opt. 5, 65 (1974)
5.24 M.C. Hutley, J.P. Verrill, R.C. McPhedran, M. Mevière, P. Vincent: Nouv. Rev. Opt. 6, 87 (1975)
5.25 D. Maystre, R. Petit: Opt. Commun. 17, 196, (1976)
5.26 D. Maystre, M. Nevière: J. Opt. (Paris) 8, 165 (1977)
5.27 M.C. Hutley, D. Maystre: Opt. Commun. 19, 431 (1976)
5.28 M. Nevière, D. Maystre, P. Vincent: J. Opt. (Paris) 8, 231 (1977)
5.29 E.G. Loewen, M. Nevière: Appl. Opt. 16, 3009 (1977)
5.30 D. Maystre, M. Nevière, P. Vincent: Opt. Acta 25, 905 (1978)
5.31 H.L. Bertoni, T. Tamir: Appl. Phys. 2, 157 (1973)
5.32 M. Nevière, R. Petit, M. Cadilhac: Opt. Commun. 8, 113 (1973)
5.33 M. Nevière, P. Vincent, R. Petit, M. Cadilhac: Opt. Commun. 9, 48 (1973)
5.34 M. Nevière, P. Vincent, R. Petit, M. Cadilhac: Opt. Commun. 9, 240 (1973)
5.35 A. Saad, H.L. Bertoni, T. Tamir: Proc. IEEE 62, 1552 (1974)
5.36 T. Tamir: Nouv. Rev. Opt. 6, 273 (1975)
5.37 T. Tamir, H.L. Bertoni: J. Opt. Soc. Am. 61, 1397 (1971)
5.38 T. Tamir, S.T. Peng: Appl. Phys. 14, 235 (1977)
5.39a V. Shad, T. Tamir: Opt. Commun. 23, 113 (1977)
5.39b H. Raether: *Excitation of Plasmons and Interband Transitions by Electrons*, Springer Tracts in Modern Physics, Vol. 88 (Springer, Berlin, Heidelberg, New York 1980)
5.40 J.M. Elson: Appl. Opt. 16, 2873 (1977)
5.41 J.P. Laude, D. Maystre: To be published
5.42 J.J. Cowan, E.T. Arakawa: Z. Phys. 237, 97 (1970)
5.43 A. Otto, W. Sohler: Opt. Commun. 3, 254 (1971)
5.44 J.A. Arnaud: "Piecewise Homogeneous Media", in *Beam and Fiber Optics* (Academic Press, New York 1976)
5.45 G. Goubau: J. Appl. Phys. 21, 1119 (1950)
5.46 D. Maystre, R.C. McPhedran: Opt. Commun. 12, 164 (1974)
5.47 R. Ulrich: J. Opt. Soc. Am. 60, 1337 (1970)
5.48 P. Vincent: Thesis, Université d'Aix-Marseille III, France (1978)
5.49 D. Marcuse: *Integrated Optics* (IEEE Press, New York 1973)
5.50 M.K. Barnowski: *Introduction to Integrated Optics* (Plenum Press, New York 1974)
5.51 T. Tamir (ed.): *Integrated Optics*, 2nd ed., Topics in Applied Physics, Vol. 7 (Springer, Berlin, Heidelberg, New York 1979)

6. Experimental Verifications and Applications of the Theory

D. Maystre, M. Nevière, and R. Petit

With 105 Figures

The reader interested in a detailed study of grating properties and grating manufacturing problems can refer to several now well-known papers [6.1-3]. We must lay stress on the fact that the properties described below are mainly those which cannot be found in classical optics text books because they are in fact consequences of the electromagnetic theory. After a lot of valuable experimental verifications, this theory is used to obtain numerous efficiency curves corresponding to the usual profiles of commercial gratings. Some hints are then given on how to use them in classical and sometimes rather unusual mountings.

6.1 Experimental Checking of Theoretical Results

6.1.1 Generalities

Taking into account the numerous tests that have been systematically used throughout the theoretical and numerical studies (convergence, energy balance and reciprocity criteria, for example), comparison between numerical data and experimental results may seem unnecessary: we agree with Maxwell's equations and the computer never gets its sums wrong! Nevertheless, as physicists, it is our duty to ascertain the relevance of the mathematical model we have used: Are infinite gratings a good representation of gratings we use in practice? Are plane waves well adapted to represent limited beams we encounter in spectrometers? Is it reasonable to trust the recently published values of refractive indices we need in the visible, UV and X-ray regions? For one's peace of mind, it is desirable that a comparison be made between experimental and theoretical efficiency curves for some gratings whose profiles and indices have been carefully checked. Indeed, such a comparison is not easy; the difficulties depend on the spectral domain. One is greatly tempted to use microwaves: in this domain, metals can be considered as perfectly conducting and, of course, the profile is accurately known, but the number of grooves is low and the representation of the incident beam by a plane wave is questionable. This objection no longer holds in the visible and a fortiori in the UV, but it is then more and more difficult to determine the profile and the indices of the materials.

The ideal compromise seems to be the far infrared. Let us recall that using wavelengths between 10 and 25 microns MADDEN and STRONG [6.4] clearly showed how efficiency curves are very dependent on the polarization. Their measurements have been used by PETIT [6.5] to test the validity of his first theoretical results. Since then, it seems that the infrared has not been used very much; this is probably because it is rather difficult to perform accurate photometric measurements in this range.

6.1.2 Microwave Region

Experimental data were already available in the fifties [6.6-8] before the birth of the electromagnetic theory of gratings. New measurements were made during the pioneering work period [6.9-13]. Here, we will speak only of a study conducted by DELEUIL, the main purpose of which was a comparison with the results of the newborn electromagnetic theory as developed in the early sixties by PETIT and WIRGIN. In DELEUIL's experimental arrangement, the basic instrument was a microwave spectrometer operating in the K_a band (7.5 mm < λ < 11.3 mm). Since it was working as an interferometer, it permitted him to eliminate the influence of power variations of the source and to suppress errors due to the nonlinearity of the detector. In other words, DELEUIL used a null method as already done by STROKE some years before in the experiment described in the last section of his thesis [6.10]. He estimated that the measurements were made with less than three percent relative error [6.13]. He noticed that, as soon as the number of grooves is greater than about twelve, gratings behave as if they were infinitely wide. Nevertheless he generally used a number of grooves greater than twenty. We reproduce here some results obtained for ruled gratings (Fig.6.1) and lamellar gratings (Fig.6.2). The profile corresponding to Fig.6.1a,b is the one for which the efficiency curve has been determined using the scalar theory by MADDEN and STRONG [6.4]. It was used in 1965 to verify the theoretical predictions of PETIT [6.5] but the theoretical data reported here have been recently computed again in our laboratory using more reliable programs (especially in the TM case which had not yet been rigorously solved in 1966). The curves in Fig.6.2 are taken from [6.13] which one can refer to for more information. Taking into account the difficulties of the measurements, the agreement between theory and experiments is pretty good; it can be said that, thanks to the microwaves, opticists clearly understood that grating efficiencies cannot be predicted without the use of Maxwell equations.

6.1.3 On the Determination of Groove Geometry and of the Refractive Index

Always associated with measurements in the optical domain are the difficulties of determining the exact form of the grooves. They can be overcome using electron microscope or mechanical techniques whose detailed descriptions does not fall within the scope of this book.

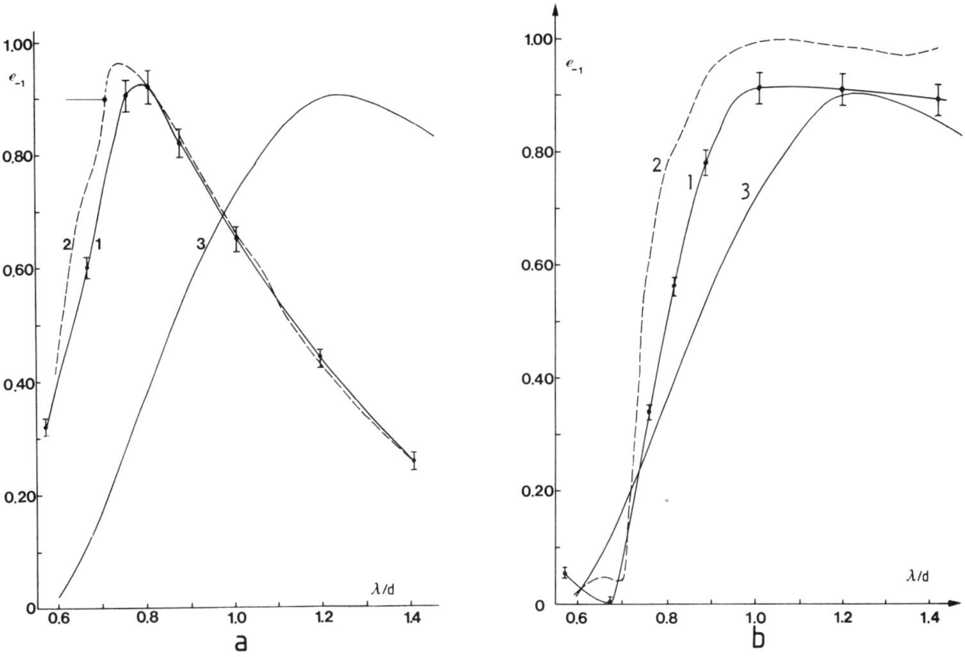

Fig. 6.1. Efficiency curves in microwaves for a metallic ruled grating in TE (a) and TM (b) polarizations. (1) experimental curves ; (2) electromagnetic theory; (3) scalar theory. The points represent the measurements with the corresponding estimated errors

In electron microscopy many observation techniques have been described, which allow the direct measurement of all the relevant profile parameters. We can quote for example "cast shadow methods" in which the grooves are parallel to the electron beam [6.14,15] and the "sandwich method" whose strange name is quickly understood when one reads its description [6.16]. The scanning electron microscope has also been used [6.17]. One of the difficulties often encountered here is to avoid the deformation of the carbon replica which resembles paper folded like an accordeon. Also worth noting is the possible use of the stereo electron microscope which, if it does not avoid possible replica deformations, gives beautiful reconstitutions [6.5,18] especially useful for qualitative examination (Fig.6.3). It must be emphasized that quantitative studies are also possible but a collaboration with specialists is then required (Fig.6.4).

Concerning mechanical techniques, the Talystep seems to be the most commonly used instrument. Its principle is extremely simple: a diamond stylus is drawn across the grating surface and its vertical movement is converted to an electrical signal by the use of an electro-mechanical transducer. After amplification, the signal is displayed as a pen recorder trace. The instrument has a vertical resolution of about 1 nm which is more than adequate in the visible region. Its use may be limited by the finite size and included angle of the stylus which prevents it

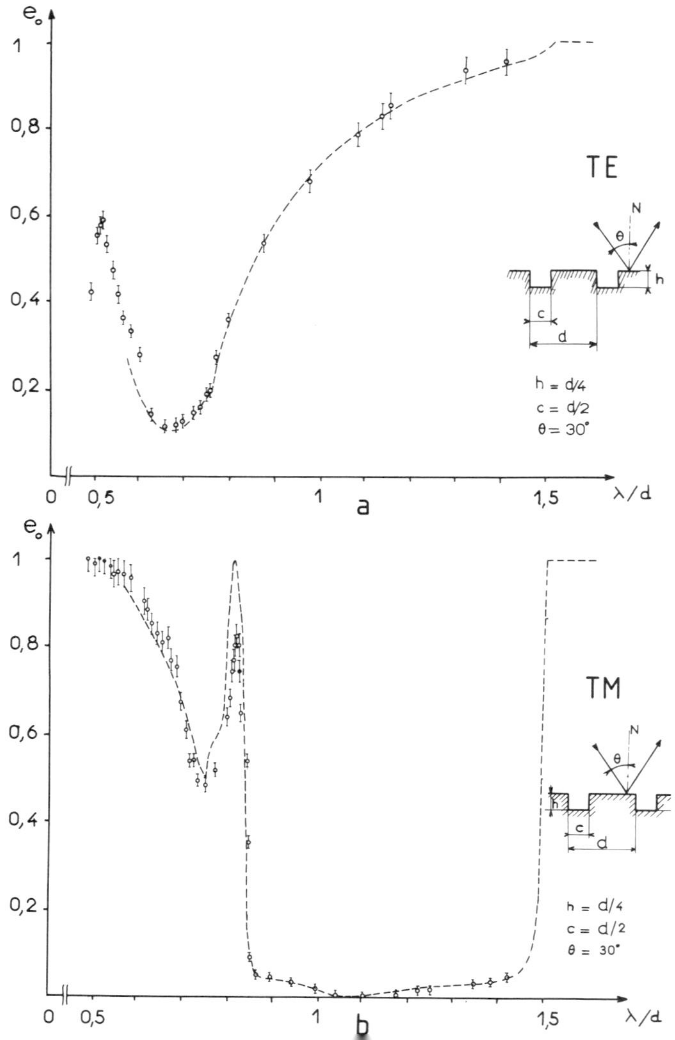

Fig.6.2a,b. Zero-order efficiency vs λ/d for a lamellar perfectly conducting grating. (---) Theoretical results; (Ι̊)Experimental measurement with its estimated error; (a) TE polarization; (b) TM polarization

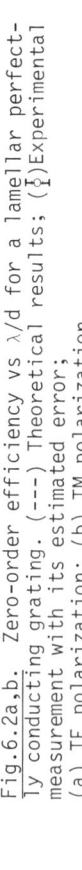

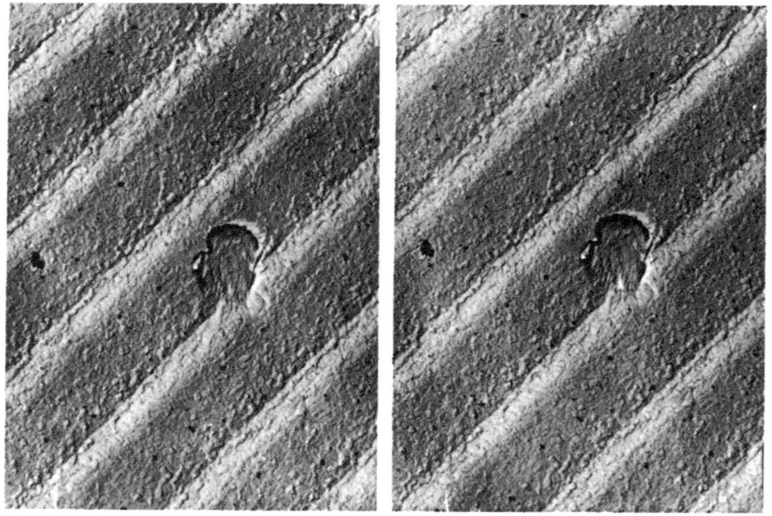

Fig.6.3. Stereoscopic photographs of a 600 groove/mm grating. When placed in the common focal plane of two identical lenses, they give an effect of relief. The big defect can be used to make the two images coincide

from reaching the bottom of deep grooves. After some convenient improvements like those suggested by VERRILL [6.20], the Talystep however turns out to be able to record grating profiles with a slope as high as 45° and to give useful measurements on gratings with up to 1200 l/mm. Referring to [6.3], comparison of scanning electron microscope and Talystep measurements on the same grating sometimes agrees to within about 1° concerning the blaze angle. Of course, the Talystep is the right device to record shallow sinusoidal profiles [6.21].

In this section, we are concerned with profile determination with a view to verify theoretical predictions. Nowadays, since the electromagnetic theory of gratings is well established, one may hope to obtain the groove profile from the efficiency curves. This is one of the most important purposes of theoretical work now in progress on inverse diffraction [6.22].

The determination of complex refractive indices of grating materials will not be developed here. The values used in our calculations are often given by HUNTER [6.23] or taken from the American Institute of Physics Handbook [6.24]. Additional data have been provided by ROBIN [6.25], LUKIRSKII et al. [6.26], IRANI et al. [6.27]. Whereas in the optical domain, the data from several authors are generally in good agreement, strong discrepancies occur in the XUV and X-ray domains where even the concept of refractive index is questionable. Since the behavior of grat-

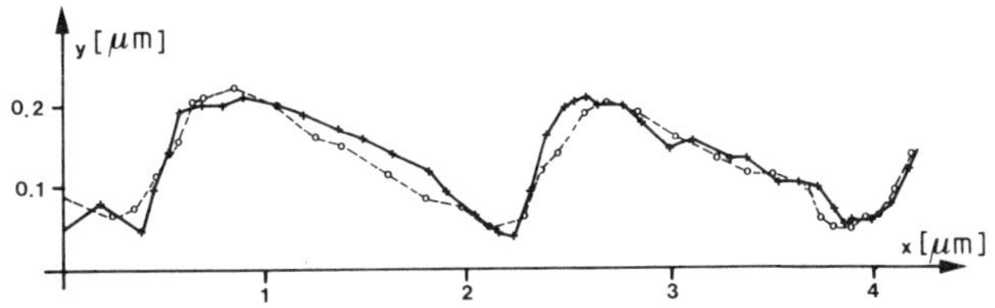

Fig. 6.4. Profile restitution obtained by VOLKMANN [6.19] (C. Zeiss, Oberkochen) from two photographs similar to those of Fig.6.3. The two curves were drawn by two different observers

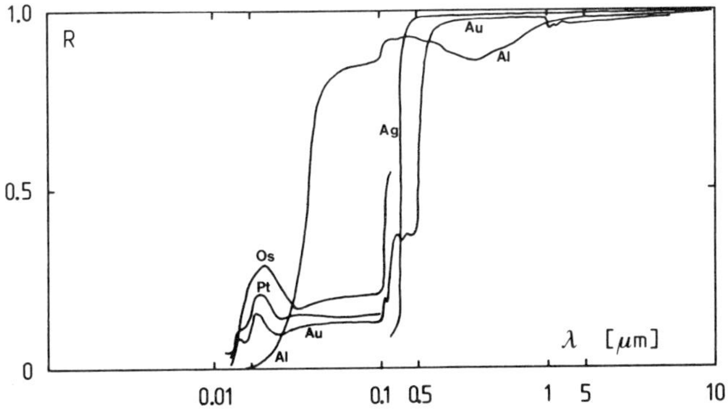

Fig. 6.5. Normal incidence reflectivity of different metals from X-rays to infrared region

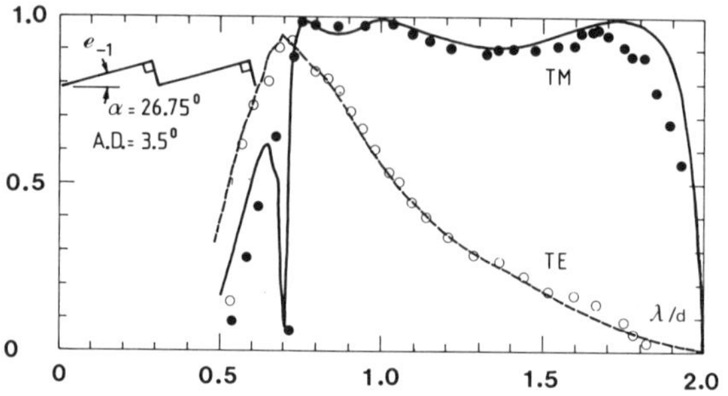

Fig. 6.6. Efficiency curves of a 26.75° blaze angle echelette grating in the infrared, for an angular deviation (AD) equal to 3.5° between incident and -1 order diffracted waves

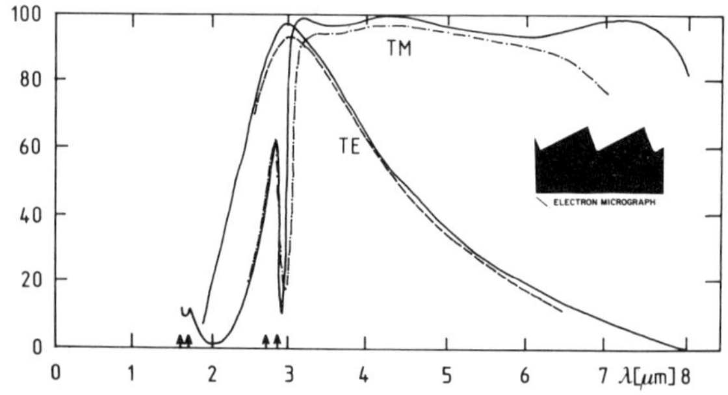

Fig. 6.7. Efficiency curve of a 240 l/mm, 26.75° blaze angle, aluminum, echelette grating used in a near Littrow mounting and in the infrared region

ing is strongly dependent on these refractive indices, the quality of these data plays a fundamental role in the success of grating theories.

The choice of the grating material is, first of all, conditioned by its reflectivity. In near UV, visible and infrared regions, metals are good reflectors and several possibilities are offered to experimenters. On the other hand, in vacuum UV, only one good reflector, namely aluminum, can be found. At shorter wavelengths (XUV), the reflectivities of all materials fall below 30%, and culminate at a few percent in the X-ray region. Figure 6.5, drawn from many data provided by the preceding authors, shows in outline the reflectivity behavior of the principal materials used in grating technology. If, at first glance, aluminum has the best reflectivity under 0.3 μm, one must keep in mind the problems linked with the aluminum oxide which forms on the unprotected material. Above 0.2 μm, the oxide layer is transparent and does not significantly affect the reflectivity of the surface. This does not hold any longer under 0.2 μm, where gratings are generally protected from oxidizing by a thin layer of magnesium fluoride. Unfortunately, under 0.11 μm, the loss problem arises in MgF_2 and aluminum must be abandoned below 0.09 μm. There, gold generally takes preference, although platinum and sometimes osmium can be used with success.

6.1.4 Infrared

Except for the very near infrared domain, the reflectivity of all materials is very close to unity and all metals may be assumed to be perfectly conducting. Thus the first formalisms developed are sufficient to predict the grating behavior in this range. Figure 6.6 shows, for example, calculations made at wavelengths between 2.7 and 13.3 μm and the corresponding experimental data for a 150 groove/mm echelette grating[1]. The correspondence is almost perfect provided that groove geometry is close to the one assumed and that the small reflectance losses are taken into account by taking measurements relative to a mirror made of the grating material. The strong TM anomaly characteristic of 26.75° blaze angle is well predicted by the theory. Small discrepancies only appear at grazing incidence ($\lambda/d > 1.7$). Other examples, not only concerning the first but the second and third orders, can be found in [6.28]. The same agreement is found. Comparisons on a 240 groove/mm echelette grating (Fig.6.7) taken from [6.29] lead to the same conclusions. The groove profile shown on the electron micrograph attests to the quality of the grating, which, in the infrared region, guarantees a good comparison between theory and experiments.

6.1.5 Visible Region

The conclusions are quite different as it appeared as soon as the first comparisons were achieved in 1973. Figure 6.8 taken from [6.18] relates the first observed discrepancies. Curve 1 gives the measured efficiencies [6.30] of a 830 groove/mm holographic silver grating as a function of the angle of incidence θ and for a fixed wavelength. From a Talystep determination, the groove shape showed itself to be sinusoidal with groove depth equal to 0.20 μm. Assuming such a shape, calculations conducted with infinite conductivity theories (Curve 2) completely failed to predict the observed efficiency for TM polarization. This failure seemed quite astounding keeping in mind that the reflectivity of silver is 0.98. Any attempt to explain the mismatch by a distorsion of the profile failed, too. On the other hand, the first calculations based on the new-born finite conductivity theory had already predicted the strong influence of finite conductivity on the TM efficiency curves [6.31]. Indeed, the use of the finite conductivity theory closed the debate (see Curve 3 in Fig.6.8 b). Not only the TM polarization anomalies were well predicted, but calculations also predicted strong absorption peaks on the total diffracted energy curve (Curve 4), which sometimes falls to about 0.5. These phenomena looked very strange at the time since the silver plane reflectance is close to unity and the grating we are dealing with has a modulation less than 1/6. They have recently been shown to be linked with the excitation of surface waves (see Chap.5). Another merit of these curves was to exhibit a relative symmetry which has been interpreted by the extension of the reciprocity theorem to finite conductivity gratings (see Chap.2 and [6.32]).

After this first comparison, a systematic study [6.33-35] was undertaken which established the inadequacy of perfect reflectivity models for holographic gratings

---

1 Ninety degrees apex angle ruled grating

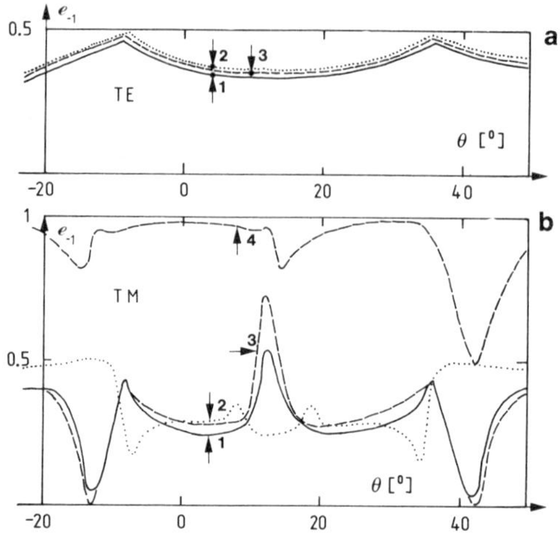

Fig. 6.8. Efficiency of a 830 l/mm sinusoidal holographic silver grating in the visible (λ = 0.521 μm), as a function of the incidence for TE (a) and TM (b) polarization

Fig. 6.9. -1 order efficiency and total energy curves (DE) for a 810 l/mm, 190 nm groove depth sinusoidal grating for different surface materials (TM polarization), (a) experimental measurements; (b) finite conductivity theory results

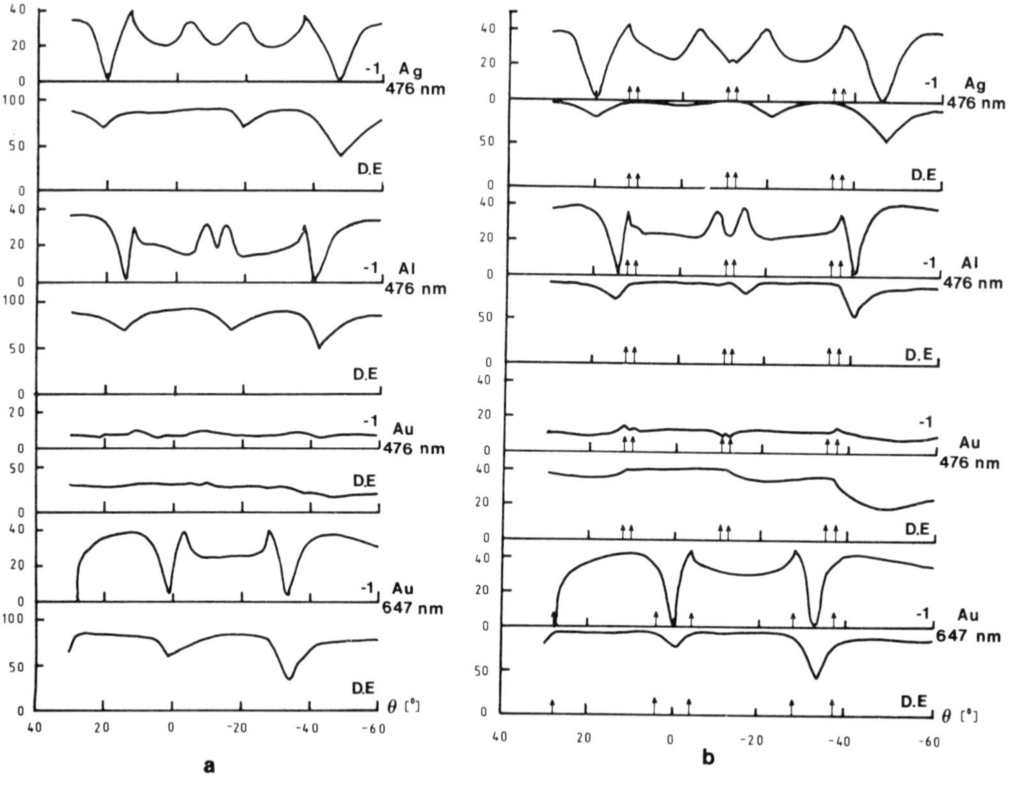

in the visible region. Finite conductivity models gave quantitative agreement as well for efficiency curves (even near anomalies) as for total diffracted energy (DE) curves including near absorption peaks (Fig.6.9). Similar investigations were conducted for blazed gratings [6.28,29]. Examples of comparison are given for a gold grating (Fig.6.10), and an aluminum grating (Fig.6.11). There is no need for

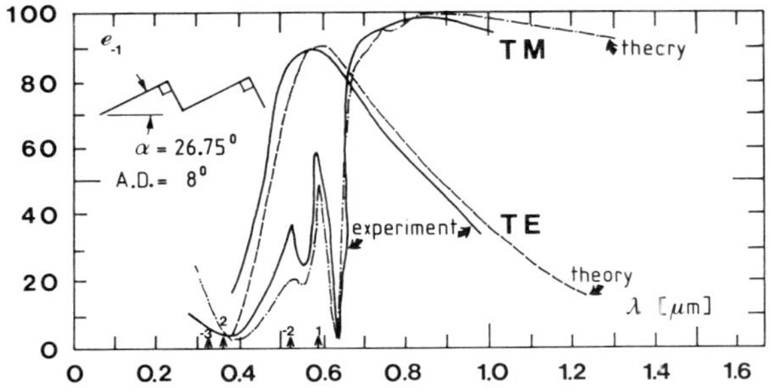

6.10. Efficiency of a 1200 l/mm, 26.75° blaze angle gold echelette grating in the visible, in a near-Littrow mount. Theoretical curves and measurements

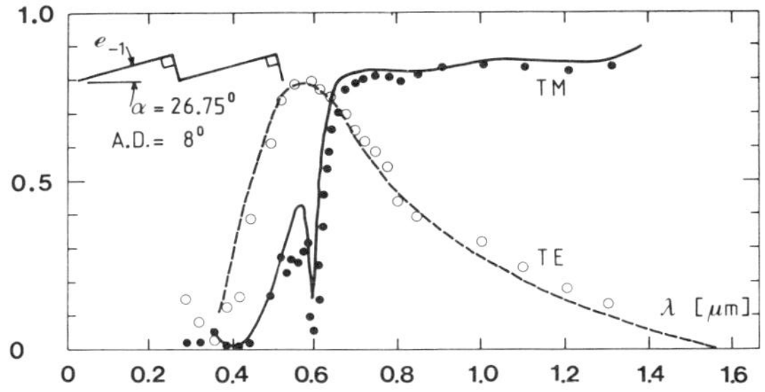

Fig.6.11. Efficiency of a 1200 l/mm, 26.75° blaze angle aluminum echelette grating in the visible and in a near-Littrow mount. Theoretical curves and experimental points

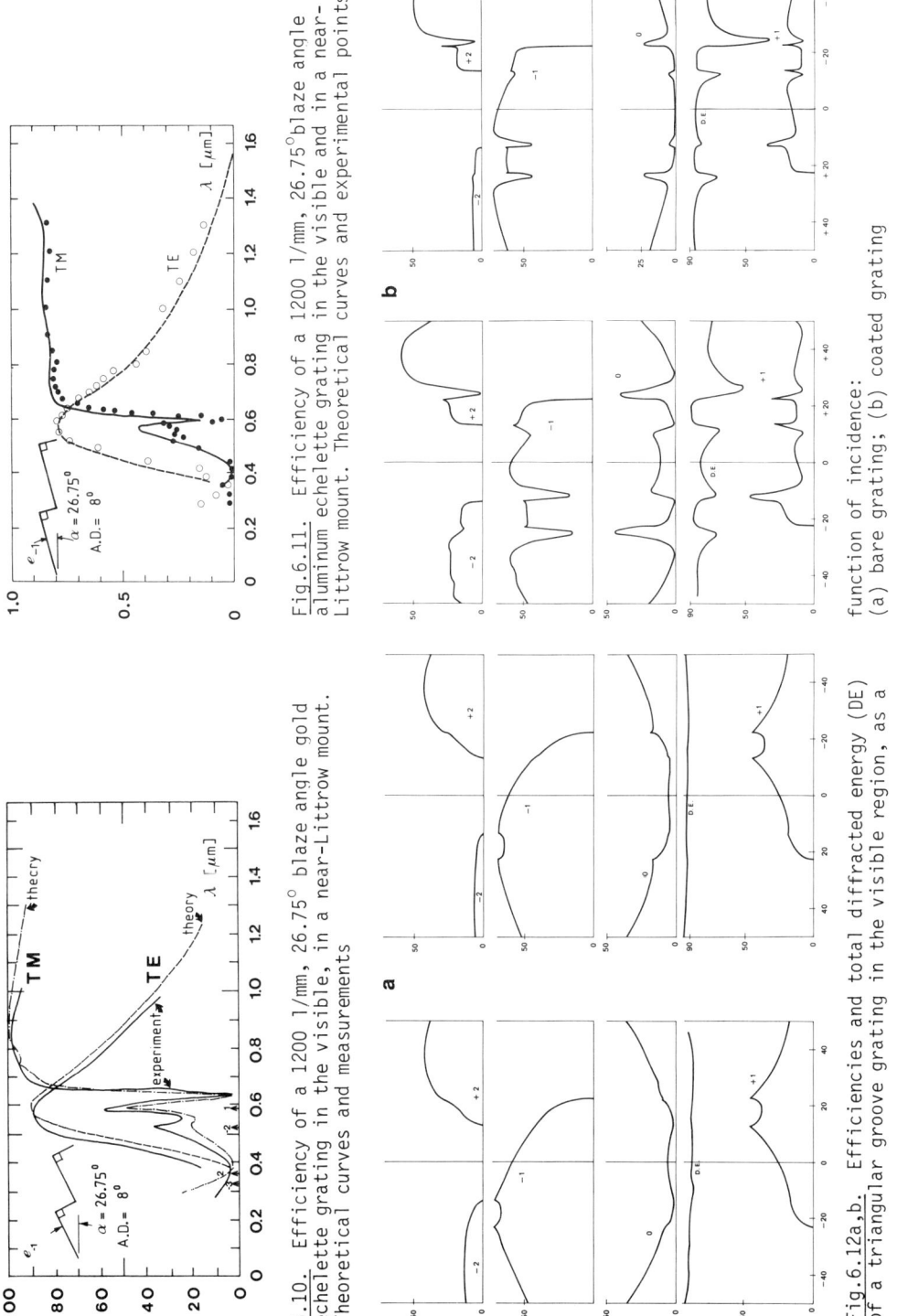

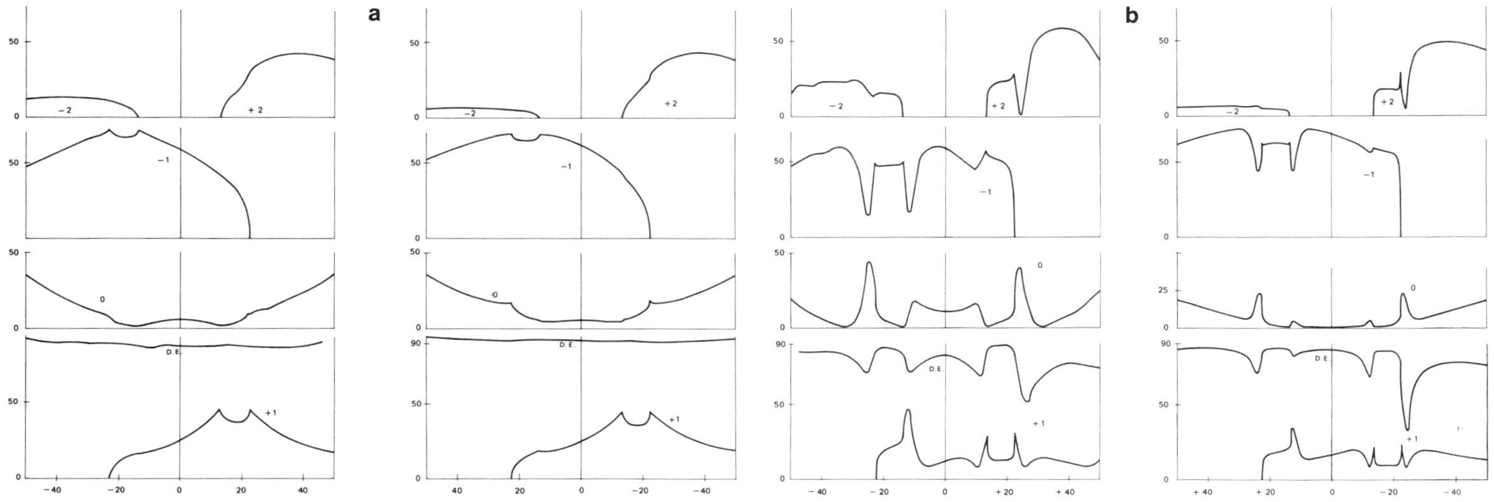

Fig.6.12a,b. Efficiencies and total diffracted energy (DE) of a triangular groove grating in the visible region, as a function of incidence: (a) bare grating; (b) coated grating

comment. Many other examples can be found in [6.28,29], which all confirm the accuracy of finite conductivity theories for predicting grating behavior in the visible region.

In addition to the study of bare gratings, comparisons between theory and experiments were made to know the influence of a dielectric coating. Big changes in efficiency curves, already predicted by calculations assuming perfect reflectivity [6.36] were in fact observed in experimental results [6.37]. Figure 6.12a,b, taken from [6.38], shows TE efficiencies of a 1264 grooves/mm, 23° blaze angle, 110° apex angle ruled aluminium grating illuminated at 0.488 μm wavelength, as a function of the angle of incidence. Figure 6.12a is related to the bare grating, while Figure 6.12b is related to the same grating coated with a 0.1 μm layer of silicon monoxide. In both figures, experimental data are on the left, and theoretical results on the right. All the observed anomalies introduced by the SiO layer are predicted, even the slightest ones, as well as the sharp absorption peaks in the total energy curve. It may seem curious that, through the coating, the smooth TE efficiency curves are turned into distorted ones which recall TM efficiency behavior. This fact has been explained in Chap.5 by the introduction of surface (leaky) waves in the film waveguide formed by the dielectric overcoating, which have similar properties to surface plasmon waves found for TM polarization.

The strong energy absorption peaks found in theoretical and experimental results give rise to the following question: is it possible to find a grating, with a convenient geometry, able to absorb in totality a given wavelength under a convenient incidence? The reader who read Chap.5 knows that the answer is affirmative. In TM polarization, a bare grating answers the problem, while a similar result is obtained with a dielectric coated grating used under TE polarization. Figure 6.13, taken from [6.39] shows that this theoretical prediction has been thoroughly confirmed by experiments, the residual reflected energy being no more than 0.5%. Moreover, the predicted change of π on the phase of the reflected wave near the absorption peak has been observed with an interferometer [6.39]. A similar investigation [6.40] was conducted for TE polarization on a dielectric coated aluminum grating (Fig.6.14). This time the residual reflected energy was found to be about 5%, and closely met the predictions.

In order to complete the series of comparisons, let us give some details concerning the invariance theorem established in Sect.3.7 for perfectly conducting gratings. Measurements were performed in natural light and conical diffraction mounting for -1 order efficiencies and for a 610 l/mm, 34° blaze angle aluminum grating, under incidence $\theta = 13^{\circ}$. They show that the invariance property of efficiency is, of course, no longer verified, since the reflectivity of the surface varies with the incidence and wavelength. Nevertheless, the efficiency variations are of the same order of magnitude as the changes in reflectivity, as shown in Table 6.1 in which ϕ designates the angle between the incident beam and the cross-section plane of the grating. Thus the invariance theorem can still be applied

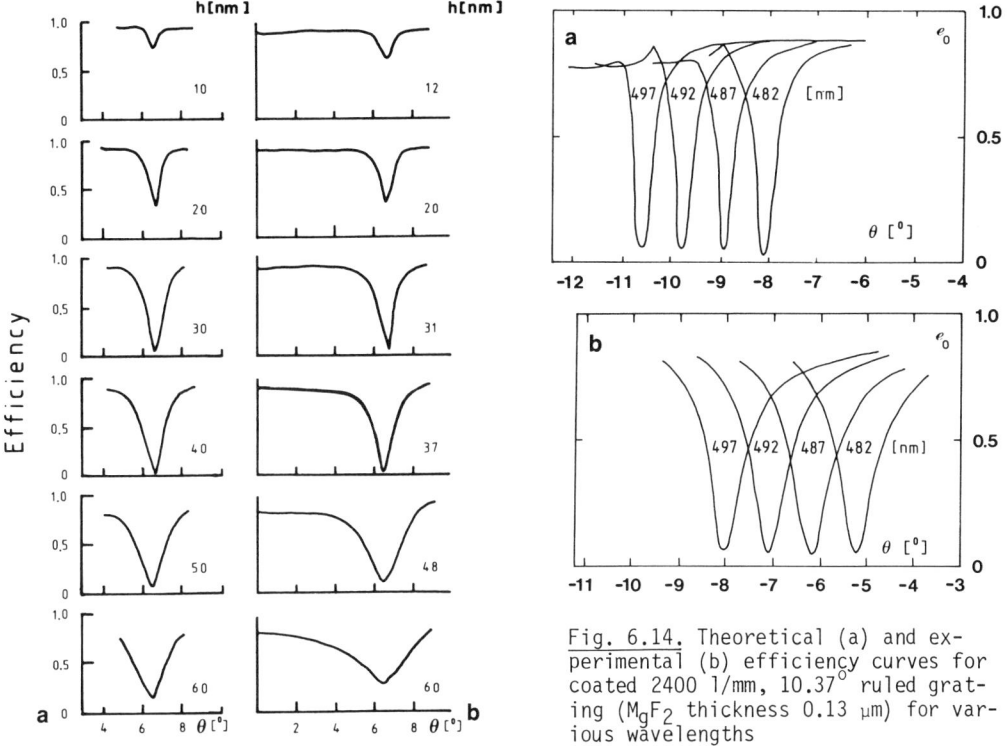

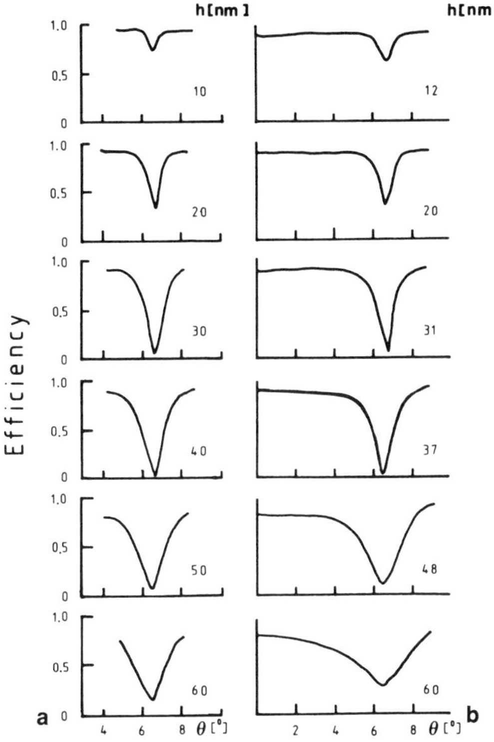

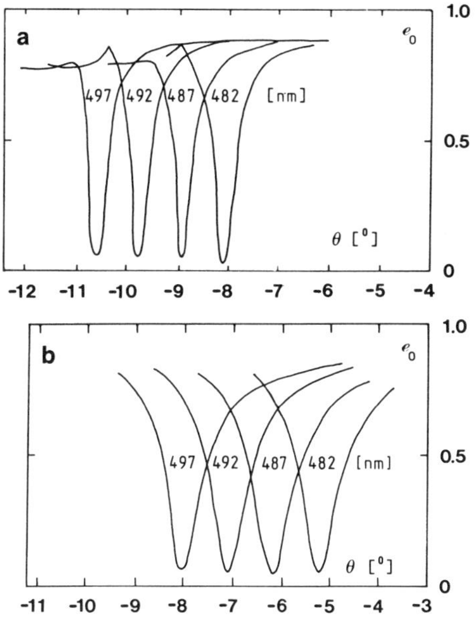

Fig. 6.14. Theoretical (a) and experimental (b) efficiency curves for coated 2400 l/mm, 10.37° ruled grating (M_gF_2 thickness 0.13 μm) for various wavelengths

Fig. 6.13. Theoretical (a) and experimental (b) zero-order efficiency curves for bare holographic gratings (1800 l/mm) of various groove depths h, used with TM polarized light (λ = 0.647 μm). The values of h are given in nm on each figure

with satisfactory accuracy provided that one takes into account the correction due to the reflectivity of the surface. To conclude the last two sections, let us try to specify the wavelength under which finite conductivity theories must be used. Of course, one can think that as soon as the reflectivity of a grating material exceeds about 0.9, the perfectly conducting model applies. Following Fig.6.5, this would occur above 0.3 μm for silver and aluminum, and above 0.5 μm for gold. The preceding study (Fig.6.8) clearly shows that this is not so. Extensive calculations have shown that it is only above 4 μm that the infinite conductivity model can be safely used, as can be observed in [Ref.6.18, Fig.24]. Below 4 μm, it is safer to use finite conductivity theories although infinite conductivity models are still able to provide the necessary insight for most practical problems. The difference between the two predictions depends on polarization. For TE polarization and for most commercial gratings, finite conductivity generally results in multiplying the efficiencies by the reflectivity. But for highly modulated gratings, the efficiency reduction often reaches twice as much. For TM polarization however, the effect of finite conductivity is to introduce a completely new kind of grating anomalies, linked with plasmon resonances, and to shift and change the shape of existing anom-

Table 6.1. Comments on the invariance theorem (from [6.18])

λ_0 [μm]	ϕ [deg]	e_{-1}
0.7	21	0.62
0.65	30	0.60
0.6	37	0.62
0.55	43	0.64
0.5	48	0.66
0.45	53	0.68
0.4	58	0.64
0.35	62	0.61
0.3	66.5	0.60

alies. Outside the anomaly region, no simple rule can be found, except that finite conductivity always reduces grating efficiencies.

6.1.6 Near and Vacuum UV

These regions, which extend from visible to 0.1 μm, are characterized by the fact that there exists one known material, and only one, namely aluminum, which presents a high reflectivity. This implies the use of thin overcoatings of magnesium fluoride. But provided that these overcoatings are thin enough (25 nm) their optical interaction is negligible. Thus, efficiencies almost as high as in the visible domain can be obtained, and comparison with experiments gives about the same degree of concordance, generally a little worse. As an example, Fig.6.15 shows only 15%

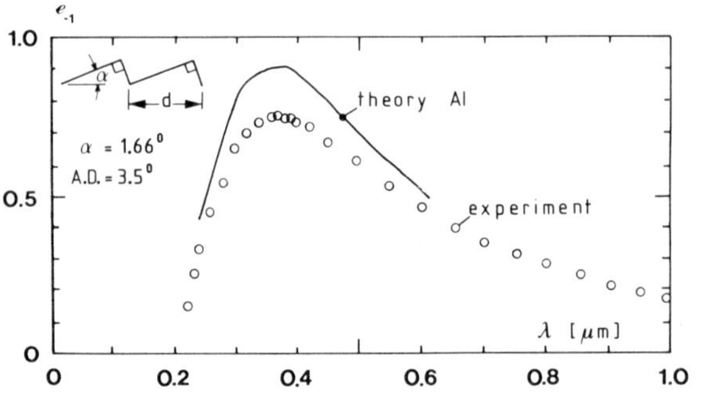

Fig. 6.15. Theoretical and experimental efficiency curves for a 158 l/mm aluminum ruled grating in near UV

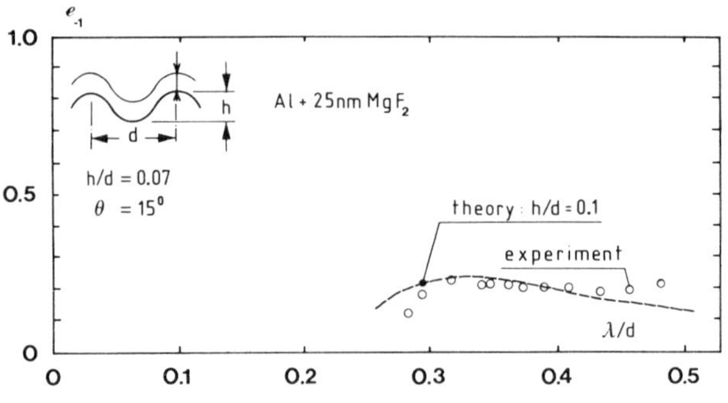

Fig. 6.16. Theoretical and experimental efficiency curves for a 2400 l/mm aluminum dielectric coated sinusoidal grating in near UV (h/d = 0.07, θ = 15°)

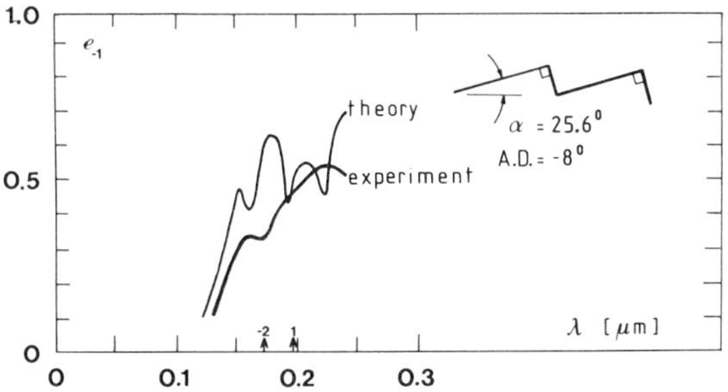

Fig. 6.17. Theoretical and experimental efficiency curves, illuminated with natural light, for a 3600 1/mm ruled grating in UV

discrepancy in the blaze peak region of a shallow 158 groove/mm echelette aluminum grating. It is important to notice, however, that such a grating used in this spectral region operates at λ/d ratios much lower than in the visible region. If one tries to operate at the same λ/d ratio as in the visible range, which implies the use of 2400 and even 3600 mm^{-1} groove frequencies, one is confronted with problems concerning the quality of the grating ruling. Figure 6.16 shows a comparison between theory and experiments for a 2400 groove/mm aluminum sinusoidal grating, coated with the standard 25 nm MgF_2 overcoating which still gives a good agreement. In Fig.6.17, relative to a 25.6° blaze angle, 3600 groove/mm aluminum grating used with 8° deviation between incident beam and -1 order, the concordance becomes qualitative only.

6.1.7 XUV Domain

The main feature of this region, which extends from 30 to 100 nm, is that the reflectivities of all materials fall to very small values as the wavelength decreases. Thus one is often forced to deal with high angles of incidence in order to get utilizable efficiencies. Figure 6.18 shows experimental and theoretical efficiencies for a 1200 groove/mm, 52 nm groove depth, sinusoidal gold grating used under 80° incidence. The agreement is pretty good taking into account the increasing influence of micro-irregularities in this domain, which scatter light outside the spectral orders. The same comparison (Fig.6.19) for an echelette grating gives a good concordance for the zero order, but only a qualitative one for the first order. This is not surprising if one keeps in mind the difficulties of correctly ruling a 2.6° blaze angle grating. Despite the groove quality problem, the ruled grating presents higher maximum efficiencies than the sinusoidal one, a fact which can be understood if one considers that, for the blaze wavelength, the -1 order is specularly reflected on the large facet. Even for the echelette grating used at the optimum, the -1 order efficiency remains small, due to the fact that an important part of the energy is carried out by the zero order, which is of no interest for spectroscopy. A possible means to avoid that, is to use an incident beam propagating out of the cross-section plane (off-plane mounting). Among the numerous possi-

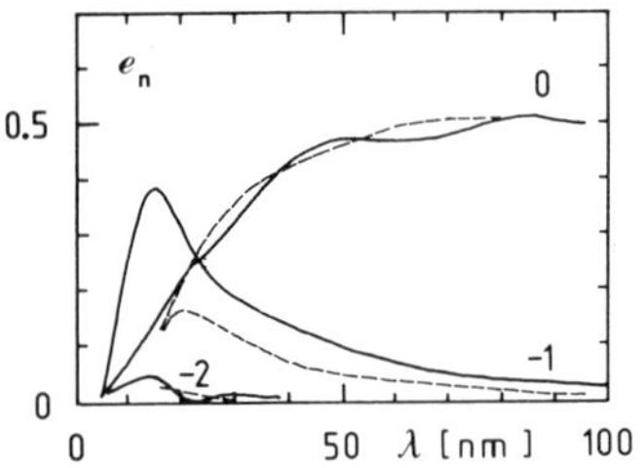

Fig. 6.18

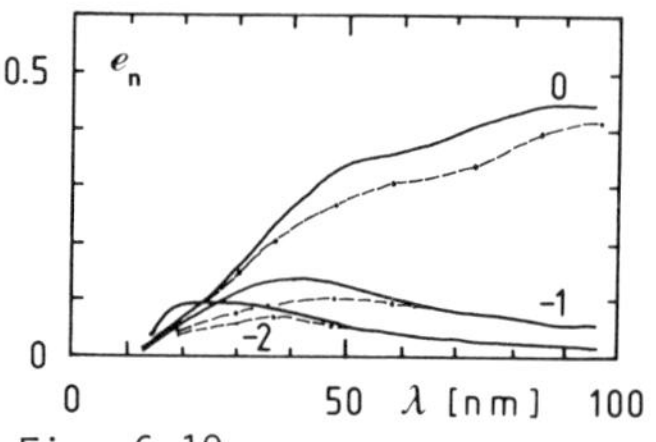

Fig. 6.19

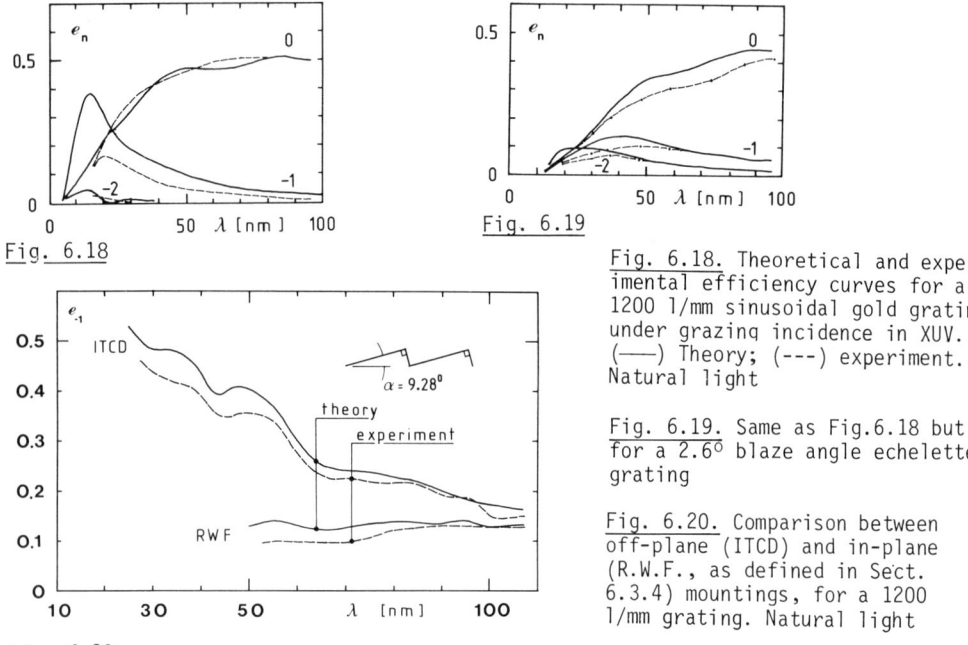

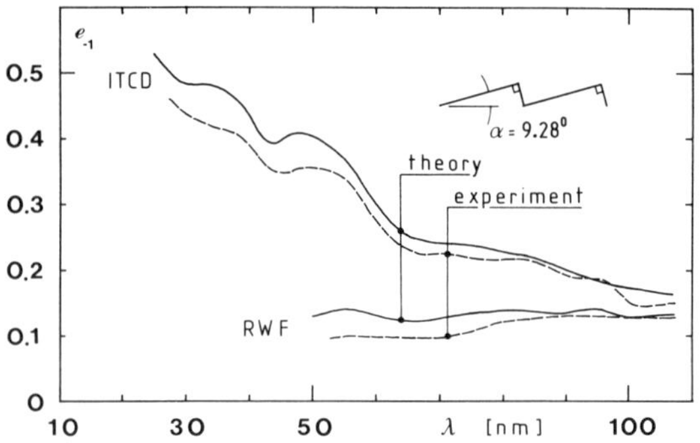

Fig. 6.20

Fig. 6.18. Theoretical and experimental efficiency curves for a 1200 l/mm sinusoidal gold grating under grazing incidence in XUV. (——) Theory; (---) experiment. Natural light

Fig. 6.19. Same as Fig.6.18 but for a 2.6° blaze angle echelette grating

Fig. 6.20. Comparison between off-plane (ITCD) and in-plane (R.W.F., as defined in Sect. 6.3.4) mountings, for a 1200 l/mm grating. Natural light

bilities, a special mounting, devised from theoretical studies, turns out to be particularly efficient. This is the so-called ITCD (Invariance Theorem Conical Diffraction) or GMS (Generalized Maréchal and Stroke) configuration, in which the incident and -1 order beams propagate in the plane of the small facet of an echelette grating. More details about it will be given hereafter (see Sect.6.4.2). Figure 6.20 shows the increase in efficiency both seen on theoretical and measured data, when one goes from a classical in-plane mounting to the ITCD mount. In order to appreciate the quality of the agreement, it is worth noting that for the smallest wavelengths, the grating supports about 200 diffracted orders. However, the experimental results here confirm that the electromagnetic theory still gives reliable results. Due to the low λ/d ratio, one may think that scalar theories could provide the same predictions. As will be explained below, this point of view is completely wrong, because of the high angles of incidence which are used.

6.1.8 X-Ray Domain

The preceding remarks concerning in-plane mountings apply again, with more strength. Here the ITCD mount is the only one which can give more than 10% efficiencies, inplane values being about 1% even at very grazing incidences. In this domain, the data for the refractive indices of gold depend on the experimental conditions. Thus it is impossible to expect a quantitative concordance. However, Fig.6.21 compares a set of theoretical results [6.41] with the measurements of WERNER [6.42] for an extremely off-plane mounting. A 5° blaze angle, 3600 groove/mm gold grating is illu-

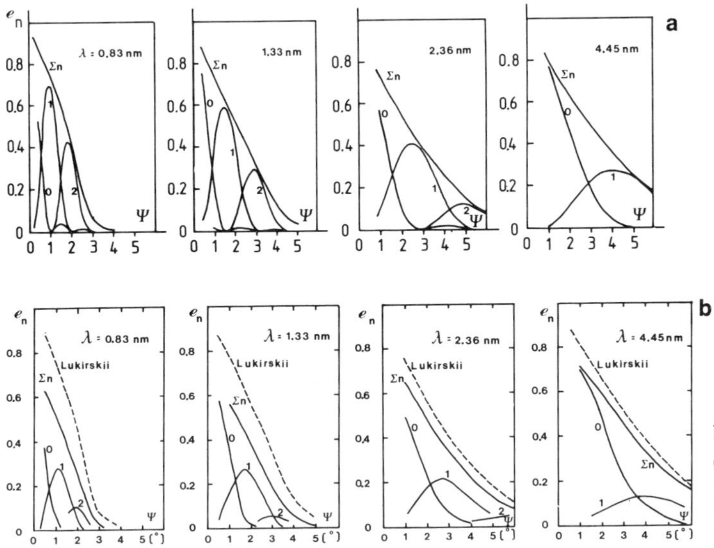

Fig. 6.21. Comparison of theoretical (a) and experimental (b) results in the X-ray domain. Σ_n is the total diffracted energy

minated by a beam located in a plane parallel to the grooves and perpendicular to the grating surface. The curves are plotted as functions of the grazing angle $\psi = \pi/2 - \phi$, for four different wavelengths. The total energy reflected by the grating shows a perfect agreement with the reflectance measured by LUKIRSKII et al. Concerning the zero, first and second order efficiencies, both theory and experiments exhibit the same features, the maxima being located exactly at the same grazing angles. The experimental values are about half the predicted ones. But with the reserves previously pointed out concerning the refractive indices used in the calculations, on top of which come surface irregularities, the comparison may be considered as satisfactory. Recent comparisons not yet published on echelette and lamellar gratings have shown discrepancies not superior to 10% at wavelength 44 Å, which was completely unexpected in this spectral domain.

6.2 Systematic Study of the Efficiency of Perfectly Conducting Gratings

The properties of perfectly conducting gratings have been more widely investigated than those of the metallic ones. This can be explained by two reasons. First, numerical programs dealing with metallic gratings in the visible region are very few in number and have been recently elaborated. Second, the model of the perfectly conducting grating is especially interesting because the efficiency of such a grating only depends on θ, λ/d and on the shape of the cross section. In other words, for a fixed incidence θ, the grating whose profile is given by $y/a = f(x/a)$, illuminated with wavelength λ/a generates the same efficiencies on all the spectral orders as the grating $y = f(x)$ illuminated with wavelength λ. Thus, for a given shape of profile, only two parameters, namely λ/d and θ, actually have an influence on the efficiencies. This will enable us to perform systematic studies of the most commonly used commercial gratings in Littrow or near-Littrow mounts. Knowing λ/d, λ will be supplied by Table 6.2, at least for usual spatial frequencies. Unfortunately, this property does not hold any longer for metallic gratings since the optical index ν varies with the wavelength. Thus, for metallic gratings, λ and d must be separately considered as parameters.

Furthermore, the numerical studies on perfectly conducting gratings are valuable. Indeed, we know that this model is very well adapted when the wavelength exceeds 4 μm and that, between 0.1 μm and 4 μm, it gives good indications, at least for TE polarization.

Of course, a thorough investigation of the properties of any type of perfectly conducting grating cannot be envisaged. As in the studies already published, we will consider mainly currently used commercial grating, viz., ruled gratings, sinusoidal holographic gratings and symmetric lamellar gratings. However, an empirical equivalent rule will enable us to compare the properties of these three types of gratings, and to extend our conclusions to a wide class of other gratings. A second limitation will concern the type of mounting. In practice, we deal with the nth order Littrow mount[2] in which

$$2 \sin \theta = -n \frac{\lambda}{d} \quad . \tag{6.1}$$

Generally, the mountings used by opticists are close to Littrow. The deviation $D = \theta_{-m} + \theta$ (with the notations defined in Chap.1) is not null but small. Then, (6.1) becomes

$$\sin\left(\theta - \frac{D}{2}\right) \cos \frac{D}{2} = -n \frac{\lambda}{d} \quad . \tag{6.2}$$

Some general ideas concerning the influence of a small or large deviation on the efficiency obtained in a Littrow mount will be given. Some curves given in the following come from studies already published [6.28,29,43] but most have been drawn especially for this book using the computer programs described in [6.44] (ruled and sinusoidal gratings) and [6.45] (lamellar gratings).

6.2.1 Systematic Study of Echelette Gratings in -1 Order Littrow Mount

The properties of such gratings are certainly the best known, at least for two reasons. First, because some decades ago, the ruled grating was practically the only one to be manufactured and even today, it remains extensively used. Secondly,

2 If θ is positive, and using our notations, n is necessarily negative. Nevertheless the expression "first-order Littrow mount" is very often used instead of "-1 order Littrow mount".

Table 6.2. Wavelength, in μm, as a function of λ/d, and the number N (lines/mm)

N \ λ/d	0.1	0.2	0.3	0.4	0.5	0.6	0.7	0.8	0.9	1.0	1.1	1.2	1.3	1.4	1.5	1.6	1.7	1.8	1.9	2.0
3600	0.03	0.06	0.08	0.11	0.14	0.17	0.19	0.22	0.25	0.28	0.31	0.33	0.36	0.39	0.42	0.44	0.47	0.50	0.53	0.56
2400	0.04	0.08	0.13	0.17	0.21	0.25	0.29	0.33	0.38	0.42	0.46	0.50	0.54	0.58	0.63	0.67	0.71	0.75	0.79	0.83
1800	0.06	0.11	0.17	0.22	0.28	0.33	0.39	0.44	0.50	0.56	0.61	0.67	0.72	0.78	0.83	0.89	0.94	1.00	1.06	1.11
1200	0.08	0.17	0.25	0.33	0.42	0.50	0.58	0.67	0.75	0.83	0.92	1.00	1.08	1.16	1.25	1.33	1.42	1.50	1.58	1.67
900	0.11	0.22	0.33	0.44	0.56	0.67	0.78	0.89	1.00	1.11	1.22	1.33	1.44	1.56	1.67	1.78	1.89	2.00	2.11	2.22
600	0.17	0.33	0.50	0.67	0.83	1.00	1.17	1.33	1.50	1.67	1.83	2.00	2.17	2.33	2.50	2.67	2.83	3.00	3.17	3.33
300	0.33	0.67	1.00	1.33	1.67	2.00	2.33	2.67	3.00	3.33	3.67	4.00	4.33	4.67	5.00	5.33	5.67	6.00	6.33	6.67

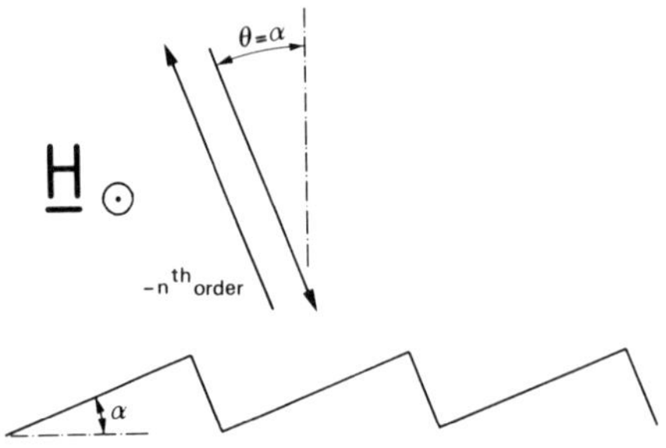

Fig. 6.22. Theorem 1

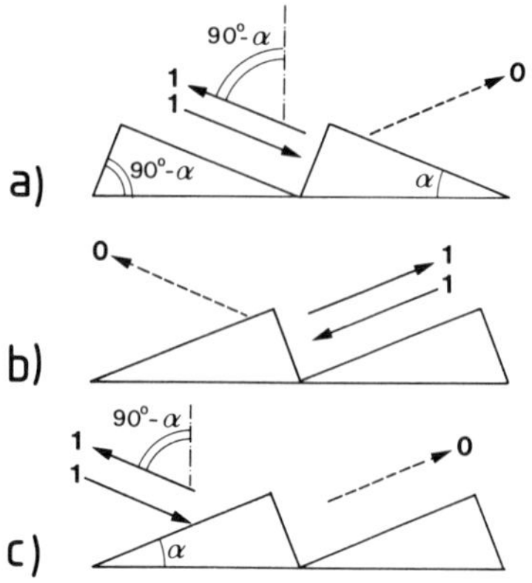

Fig. 6.23. Theorem 2

some very important analytical properties (see below) allow one to predict, in the TM case of polarization, the so-called blaze wavelength, for which the efficiency reaches unity. Probably due to this latter property, the name of "blazed gratings" is frequently employed for ruled gratings. However, we will see later that holographic and lamellar gratings possess comparable blaze properties, at least in the resonance region where few orders are diffracted. Thus, in our opinion, the term "blazed grating" is now inappropriate.

Let us begin to state two theorems which analytically predict the efficiency of a grating in particular circumstances.

Theorem 1: In TM polarization, when an echelette grating, with blaze angle α, is illuminated under the incidence $\theta = \alpha$ in -n order Littrow mount (Fig.6.22), the diffracted energy is concentrated in the -n order [6.46]: the efficiency e_{-n} is equal to unity, all the others vanish.

Theorem 2: In the -1 order Littrow mount and for TM polarization, when an echelette grating is illuminated under the incidence $\theta = 90^\circ - \alpha$ (Fig.6.23c), the diffracted energy is concentrated in the -1 order, provided that $\alpha < 70.53^\circ$. The efficiency e_{-1} is equal to unity, all the others vanish [6.47].

Theorem 1 is a consequence of the uniqueness theorem. Clearly, when (6.1) holds, the superposition of the incident wave and a diffracted wave propagating in the opposite direction satisfies both Maxwell equations and boundary conditions. The demonstration of theorem 2 can be outlined with the help of Fig.6.23. Starting from Theorem 1 (Fig.6.23a), we get Fig.6.23b through a symmetry. The reciprocity theorem (Sect.1.2.5) ensures the vanishing of e_0 in Fig.6.23c. As far as only two diffracted orders are propagating, the energy balance imposes $e_{-1} = 1$. This happens when $2\lambda/d > 1 + \sin\theta$ or, taking into account (6.1), when $\theta > \sin^{-1}(1/3)$. This last value will be denoted hereafter by $\theta_c = 19.47^\circ$. Obviously $\theta > \theta_c$ implies $\alpha < \alpha_c = 90 - \theta_c = 70.53^\circ$.

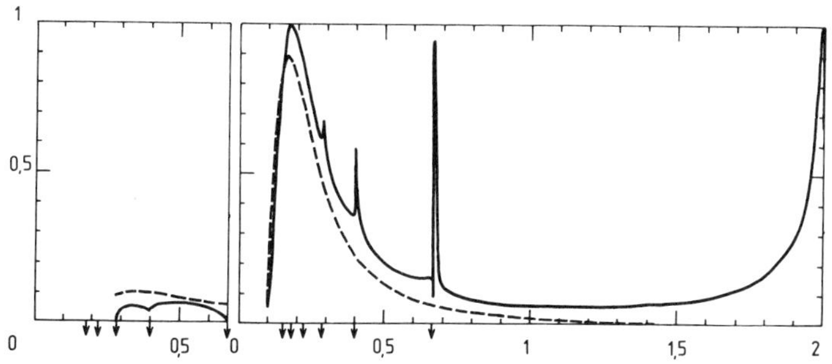

Fig.6.24.
Echelette grating.
$\alpha = 5^{\circ}$. (-) TM polarization
(---) TE polarization

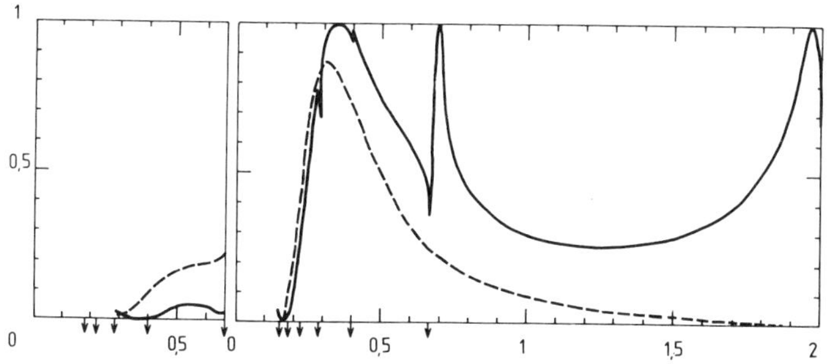

Fig.6.25.
Echelette grating. $\alpha = 10^{\circ}$

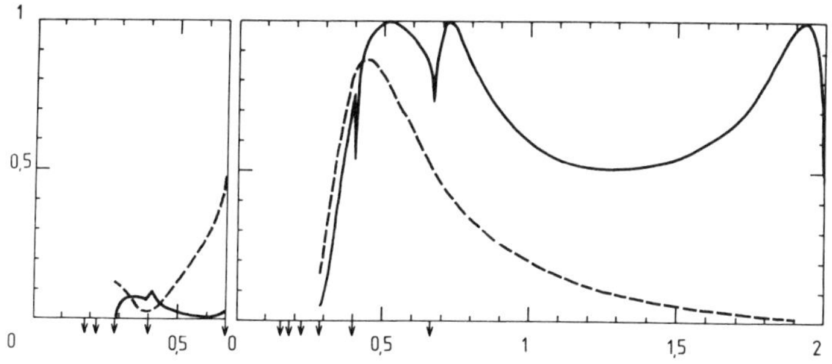

Fig. 6.26.
Echelette grating. $\alpha = 15^{\circ}$

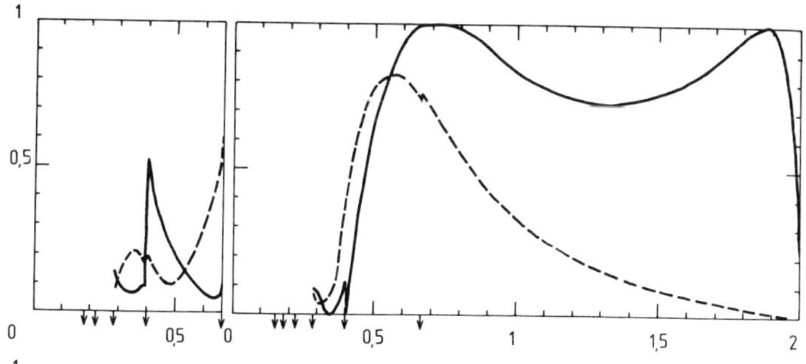

6.27.
Echelette grating. $\alpha = 20^{\circ}$

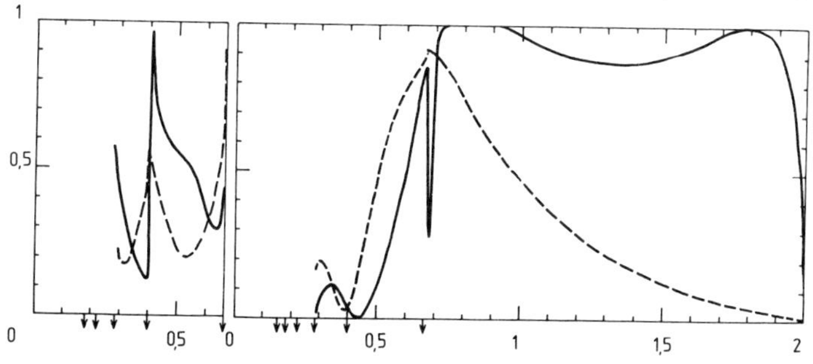

Fig.6.28.
Echelette grating. $\alpha = 25^{\circ}$

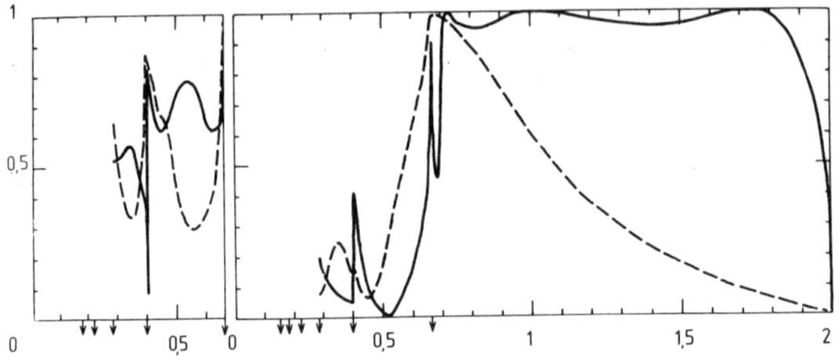

Fig. 6.29. Echelette grating. $\alpha = 30^\circ$

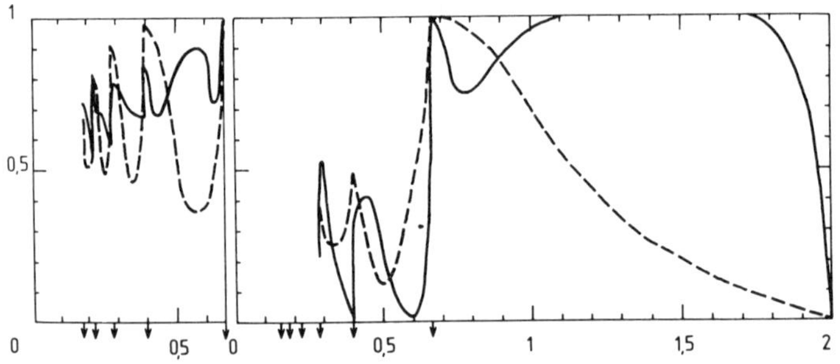

Fig. 6.30. Echelette grating. $\alpha = 35^\circ$

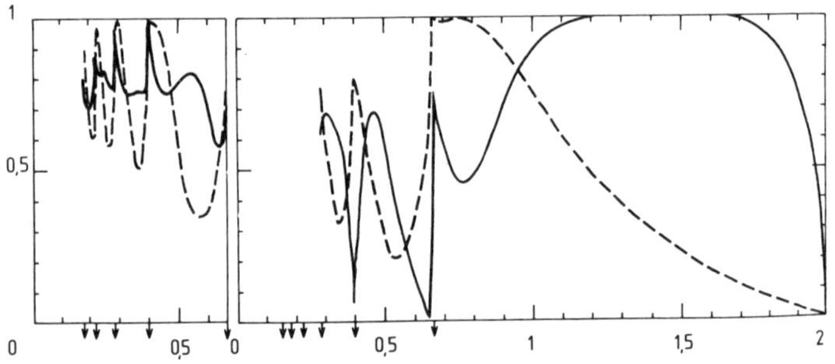

Fig. 6.31. Echelette grating. $\alpha = 40^\circ$

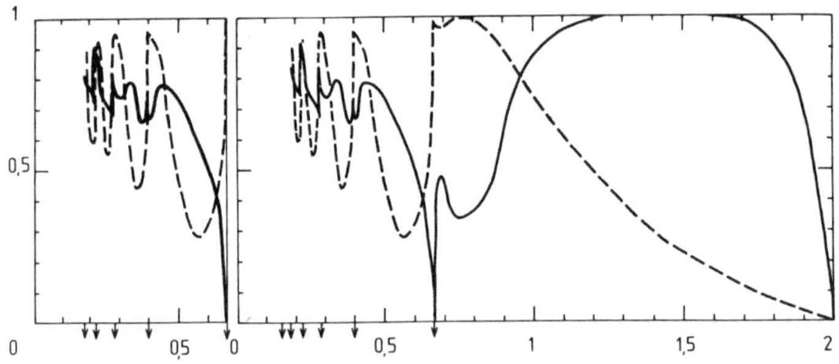

Fig. 6.32. Echelette grating. $\alpha = 45$

Let us turn now to the curves (Figs.6.24-32) which give the efficiency e_{-1} of gratings as a function of λ/d when the blaze angle α, in degrees, is known. To use the curves we have to bear in mind that two gratings having complementary blaze angles generate the same efficiency provided that $\theta > \theta_c$, i.e., $\lambda/d > 2/3$. For a fixed value α_0 of α, the curve drawn in the right-hand rectangle gives, not only the efficiency for the blaze angle α_0, but also for the blaze angle 90 - α_0 as long as λ/d is greater than 2/3. For the latter (90 - α_0), the efficiency curve in the complementary range of wavelengths ($\lambda/d < 2/3$) is plotted in the left-hand rectangle.

The first remark to be made is that the two above theorems are well-verified in the TM case: the two perfect blazings are always observed for the predicted wavelength λ_b and λ_b' such that

$$\frac{\lambda_b}{d} = 2 \sin\alpha \quad , \quad \frac{\lambda_b'}{d} = 2 \cos\alpha \quad .$$

Generally, a third perfect blazing is observed in the vicinity of the first Rayleigh wavelength $\lambda_R^1 = 2d/3$ which unfortunately, is also characterized by the strongest anomaly[3] . Nevertheless, it must be noticed that this anomaly disappears for $\alpha = 20^\circ$, when the blaze wavelength λ_b predicted by theorem 1 is close to λ_R^1. The other anomalies are located at the other Rayleigh wavelengths λ_R^n. Their importance decreases when n increases and, as observed above for λ_R^1, the anomaly at wavelength λ_R^n disappears when $\lambda_R^n \simeq \lambda_b$.

In TE polarization, a maximum value exceeding 83% is always observed for a wavelength λ_c which, for the lowest blaze angles, practically coincides with λ_b. This is not surprising since $\lambda_b/d = 2 \sin\alpha$ being small, we are practically in the scalar region and the efficiency in TE polarization is close to unity, i.e., the efficiency reached in TM polarization. The value of λ_c first increases with the blaze angle. When it reaches λ_R^1, it remains stationary whereas the associated efficiency progressively increases and reaches unity for a blaze angle $\alpha \simeq 34^\circ$. If the blaze angle goes on increasing, the perfect blazing still holds, but λ_c is shifted towards the right. It is worth noting that efficiency curves in TE polarization are much more regular than in TM polarization, in particular in the vicinity of the Rayleigh wavelengths.

From these curves and also from previous calculations, it is interesting [6.28] to classify the echelette gratings into six groups depending on the blaze angle or, in other words, on the modulation depth (see Table 6.3). For the gratings of the first group, the behavior is unusually simple: polarization effects are negligible, at least for incident angles close to α. Hence, the peak in TM, TE and natural light, is always found where $\lambda = \lambda_b = 2d \sin\alpha$, and approaches 100%. Anomalies can be neglected, especially around λ_b. Fifty percent efficiency and above is obtained over a wavelength region from 0.67 to 1.8 times the blaze wavelength, i.e., more than an octave.

In the group II ($5^\circ < \alpha < 10^\circ$), polarization effects begin to arise especially at Rayleigh wavelengths. The TE peak, having an efficiency in the order of 90%,

---

3 The Rayleigh wavelengths are associated with the appearance of new propagating orders when λ decreases. In Littrow mount, they are defined by $\lambda_R^n = 2d/(2n+1)$. On efficiency curves they are indicated by small arrows.

Table 6.3. Classification of usual gratings

	I very low modulation	II low modulation	III medium modulation		V high modulation	VI very high modulation
				IV		
Echelette gratings (α)	0°-5°	5°-10°	10°-18°	special low anomalies 18°-22°	22°-38°	38°-45°
Sinusoidal gratings (h/d)	0.-0.05	0.05-0.15	0.15-0.25		0.25-0.4	> 0.4
Symmetrical lamellar gratings (h/d)	0.-0.05	0.05-0.10	0.10-0.20		0.20-0.30	> 0.3

is shifted towards wavelengths shorter than λ_b. The previously noted rule of thumb for the range of 50% minimum efficiency still holds here.

Gratings of group III exhibit stronger polarization effects. The 50% efficiency range shifts slightly upwards, going from 0.72 to twice the blaze wavelength λ_b. Here the reduction and the shift of the TE efficiency peak from λ_b are more pronounced.

Group IV occupies a unique position. As noticed above, the anomaly at the Rayleigh wavelength λ_R^1 is completely suppressed. One observes that this minimal anomaly TM behavior is accompanied by the lowest first order TE peak, about 82%. The range for above 50% unpolarized efficiency again moves upwards slightly, from 0.75 to 1.9 times the blaze wavelength.

In the former groups, the efficiencies obtained in the domain $\lambda < \lambda_R^1$ with $\alpha < 45^\circ$ were better than those observed for the complementary blaze angle (left side of Fig.6.24-32).

The outstanding characteristic of group V gratings is a very strong anomaly at λ_R^1, which practically coincides with the maximum TE efficiency. The perfect blazing for TE case is obtained for blaze angles superior to about 35°. A second, and very useful attribute, is the existence of a wide angular band in TM polarization over which very high efficiencies are observed. The effect of high TE efficiency is to reduce in practice the effect of the TM anomaly, so that the 50% efficiency band extends from 0.6 to nearly twice the nominal blaze wavelength. However, it must be remembered that the upper limit is not always attainable in spectrometers. An interesting and surprising remark is that for the highest blaze angles of this region (see Fig.6.30), it becomes more efficient to use the grating by illuminating the small facet. In this case, the natural light efficiency remains superior to 50% even if $\lambda < \lambda_R^1$.

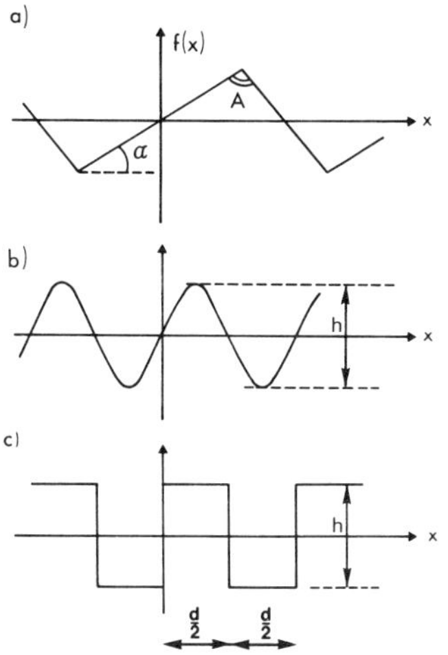

Fig. 6.33a-c. Three types of commercial gratings. (a) ruled, (b) sinusoidal, (c) symmetrical lamellar profile

This remark holds for the group VI as well. It seems that the better efficiency curve for $\lambda < \lambda_R^1$ is found for $\alpha \simeq 50^o$. In the complementary wavelength region, the shift between TE and TM perfect blazings becomes very important. Hence, the peak of unpolarized light efficiency does not exceed 85%.

6.2.2 An Equivalence Rule Between Ruled, Holographic and Lamellar Gratings

The aim of this section is to show that an approximate but very simple equivalence rule allows us to simplify considerably the understanding of some grating properties and to deduce fundamental properties of sinusoidal, lamellar and ruled gratings (Fig.6.33) from those of echelette gratings previously described.

The equivalence rule only concerns the gratings whose profiles have a center of symmetry, i.e., for which the function f(x) can be expanded in a sine series

$$f(x) = \sum_{n=1}^{\infty} b_n \sin(nKx) \quad .$$

It can be stated as follows: As long as there are only two propagating diffracted orders, the three profiles in Fig.6.33 lead to the same efficiency provided that their fundamentals have the same amplitude b_1. This rule which has been suggested by a numerical study [6.49] is perfect in the limit case $h/d \to 0$ [6.49] and gives good predictions for all commercial gratings. It can be visualized in Fig.6.34. which is simply a representation of b_1 for a ruled grating as a function of the blaze angle α, for various apex angles.

In the grating nomogram (Fig.6.34), two abscissa axes are drawn both graded in the ratio h/d. The upper one concerns sinusoidal gratings and the lower one symmetrical lamellar gratings: the depth of a lamellar gratings is $\pi/4$ times that of the associated sinusoidal grating. The ordinate axis is graded in terms of

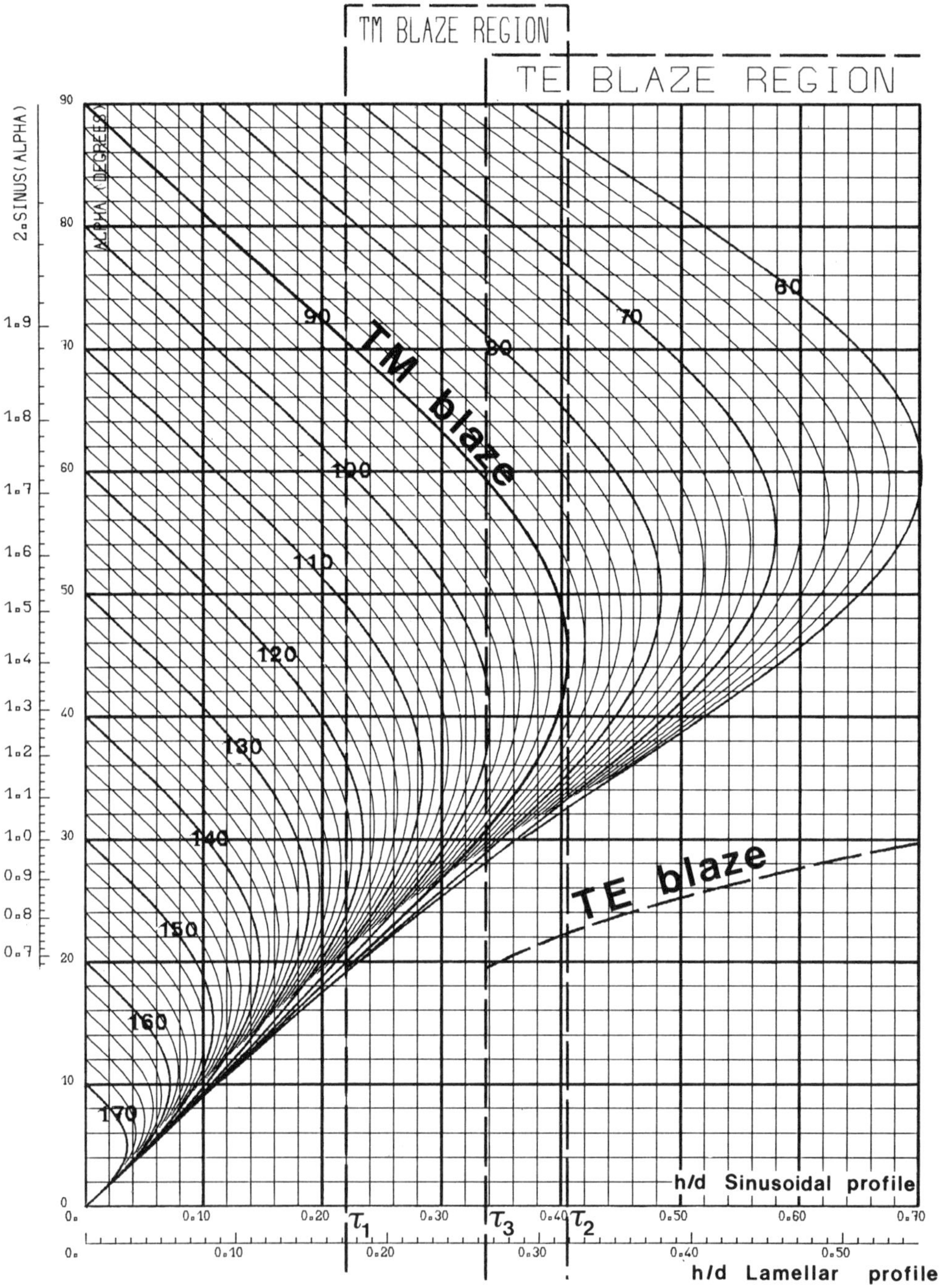

Fig. 6.34. The grating nomogram

blaze angle α. Each curve corresponds to a given apex angle A which is varied with an increment of 2°. A ruled grating (A,α) is represented by a point in the plane whose abscissa gives the normalized depth (h/d) of the associated sinusoidal and lamellar gratings. All the curves have an axis of symmetry perpendicular to the α axis. This means that two ruled gratings having apex angle A and, respectively blaze angle α and π-A-α have the same fundamental. Thus, from the equivalence rule, such gratings should have the same efficiencies. Indeed they do, as can be rigorously established using the reciprocity theorem and the conservation of energy. Let us notice that ruled gratings with an apex angle less than 90° should have a blaze angle between 90° -A and 90°. If this rule is not observed, the function f(x) is not defined in a unique manner and the profile cannot be represented by a Fourier series. A thorough numerical verification of the validity of the equivalence rule is described in [6.48]. For example, Fig.6.35 shows the efficiency curve in Littrow mount of some gratings "equivalent" to a sinusoidal grating having a groove depth to period ratio equal to 0.23. From many similar computations, the following "rule of thumb" can be drawn: for gratings having a depth-to-period ratio less than 1/3, (that is for practically all commercial gratings), the efficiencies of " equivalent" gratings seldom differ more than 10% both for infinitely conducting gratings and metallic gratings and whatever the polarizations. The discrepancies are always observed near the Rayleigh wavelength λ_R^1 and indisputably, it is the efficiency of the lamellar grating which deviates most from the rule, especially in TE polarization. However, even in this extreme case, the shape and position of the efficiency peak are practically unaltered.

Let us notice finally that the nomogram enables us to know without calculations the blaze wavelength of an echelette grating for TM polarization and in Littrow mounting. Such a grating is represented by a point lying on the curve A = 90° which runs from the origin to the upper left corner. From Theorem 1 stated in Sect.6.2.1, it follows that a first perfect blazing in TM polarization is obtained for $\theta = \alpha$. Thus, this angle is equal to the ordinate of the point in the nomogram. Theorem 2, in the same section, predicts a second perfect blazing, when $\alpha < 70.53^\circ$, at the angle of incidence $\theta = 90-\alpha$. These angles can be found at the ordinates of

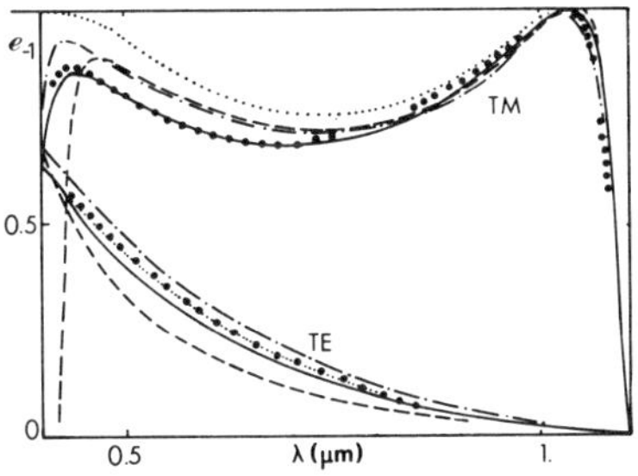

Fig. 6.35. Efficiency of 1800 l/mm perfectly conducting gratings having the same fundamental. (——) Sinusoidal grating, h/d = 0.23; (---) lamellar grating, h/d = 0.1806; (....) echelette grating, A = 90°, α = 69.5°; (-·-·-) ruled grating, A = 74°, α = 86°; (····) ruled grating, A = 120°, α = 28°

the second intersection of the curve $A = 90^\circ$ with the vertical passing through the representative point of the grating. Once more, this property applies to gratings such as $\alpha < 70.53^\circ$, and so situated below the upper intersection of the curve $A = 90^\circ$ with a vertical dashed line (corresponding practically to $h/d = 0.22$ for a sinusoidal profile). Knowing the angles of incidence θ for which the efficiency equals 100%, one often prefers to know the corresponding wavelengths $\lambda = 2d \sin\theta$. This is the reason for which we have drawn a second ordinate axis graded in 2 $\sin\alpha$. This axis makes it possible, having located the blaze angle θ on the α axis, immediately to obtain the blaze wavelength to period ratio λ/d, and so λ by using Table 6.2. We will see in the next section that the blaze wavelengths for TE polarization can also be deduced from the nomogram.

6.2.3 Systematic Study of the Efficiency of Holographic Gratings in -1 Order Littrow Mount

It is well known that many holographic gratings, generated by exposing photoresist in interference fringe fields, have profiles very close to sinusoidal ones. Furthermore, numerical calculations have shown that sinusoidal gratings have interesting efficiency curves and that the adoption of closely related profiles, such as symmetrical trapezoidal ones, does not significantly modify the efficiencies [6.50]. Although during the last years, techniques have been elaborated to diversify the profiles of holographic gratings [6.51,52], our study will be restricted to the sinusoidal model, in which the efficiency in Littrow mount, as a function of λ/d, only depends on the groove depth over groove spacing ratio h/d. The efficiency curves are given in Figs.6.36-43. Again it seems useful to distinguish five regions of groove depth to groove spacing ratios, which roughly correspond through the grating nomogram to those defined for ruled gratings (see Table 6.3).

Before giving detailed comments on these regions, let us show that the use of only the grating nomogram allows us to make interesting predictions about the blazing properties of sinusoidal gratings. If $h/d < \tau_2 = 0.403$, a sinusoidal grating can be associated to two echelette gratings with complementary blaze angles α and $90-\alpha$, to which correspond, for TM polarization, two normalized blaze wavelengths, namely 2 $\sin\alpha$ and 2 $\cos\alpha$. As long as the equivalence rule holds (i.e., $\lambda/d > 2/3$, which, through the nomogram, corresponds to $h/d > \tau_1 = 0.22$), one can expect for these two wavelengths an efficiency close to 100%. In fact only one peak, the one which corresponds to an incidence less than 45°, generally holds the attention of spectroscopists. It must be pointed out that it is precisely this peak which is affected when h/d falls under τ_1. This is why the strip $\tau_1 < h/d < \tau_2$ is called TM blaze region on the nomogram.

If $h/d > \tau_2$ it is no longer possible to associate an echelette grating with a given sinusoidal one. Consequently no perfect blazing is expected, a fact which is

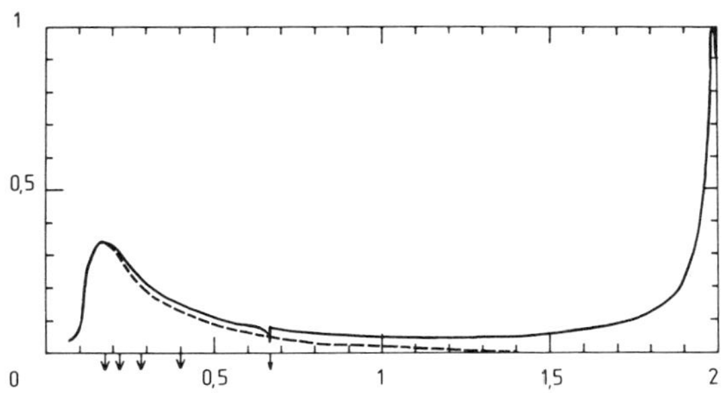

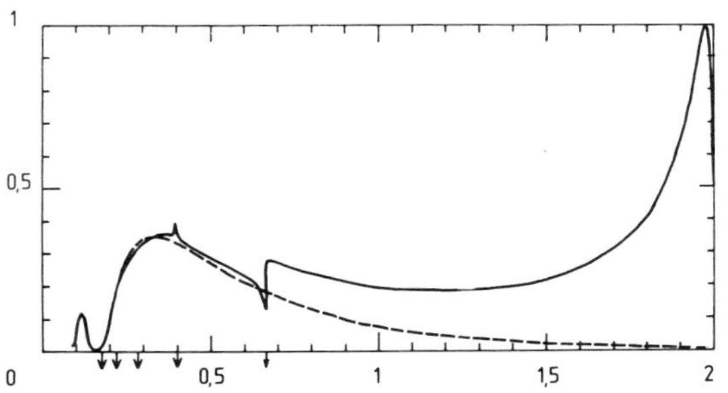

Fig.6.36. Sinusoidal grating. h/d = 0.05. (—) TM polarization (---) TE polarization

6.37. Sinusoidal grating. h/d = 0.10

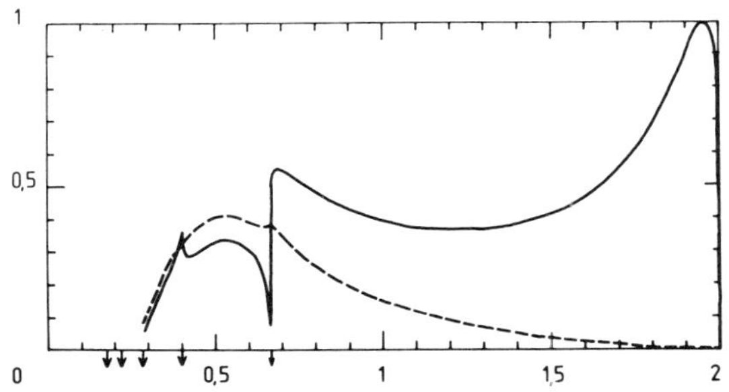

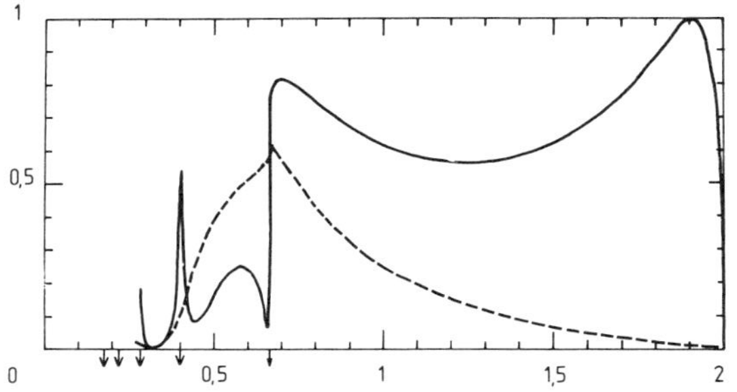

6.38. Sinusoidal grating. h/d = 0.15

Fig.6.39 Sinusoidal grating. h/d = 0.20

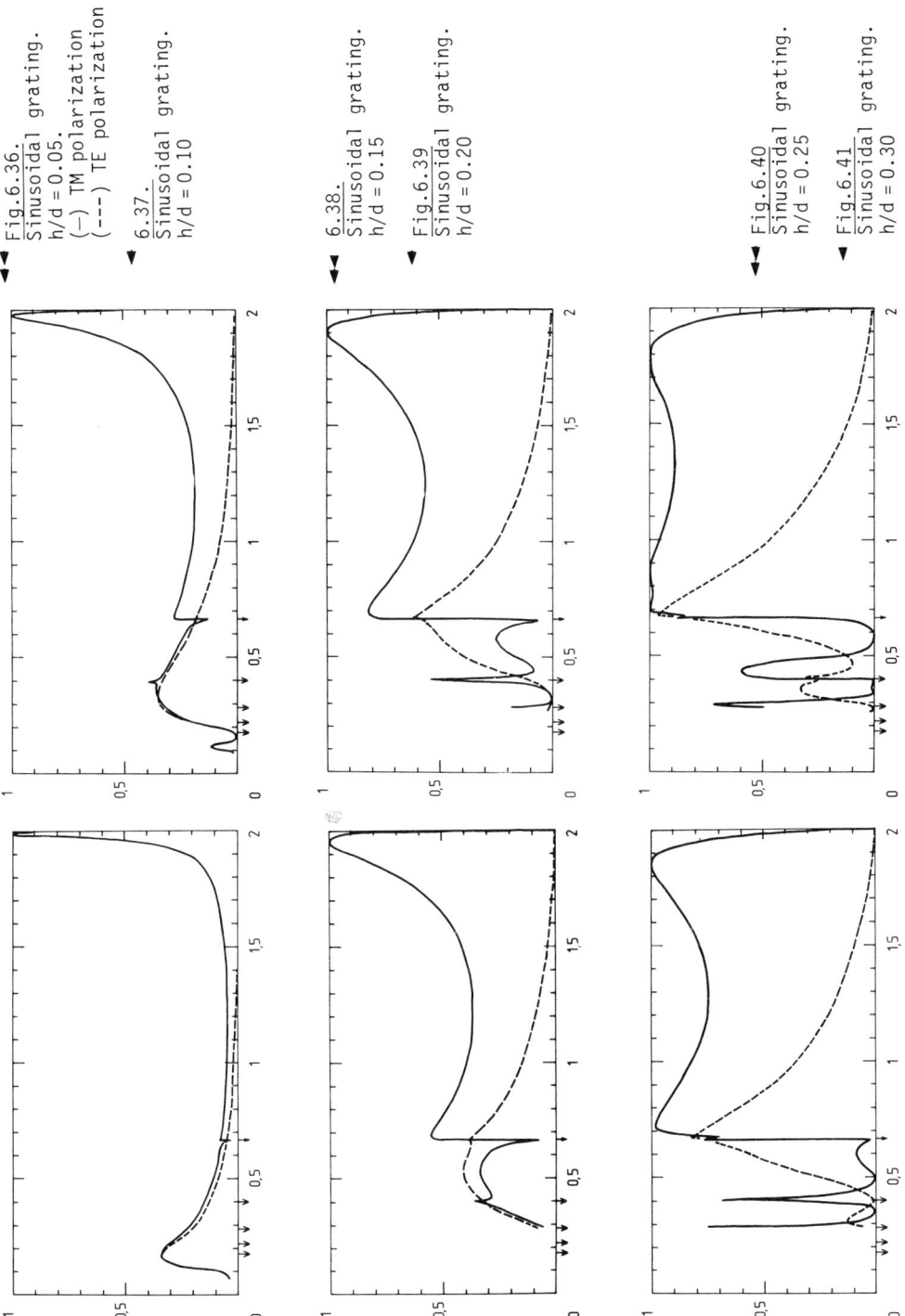

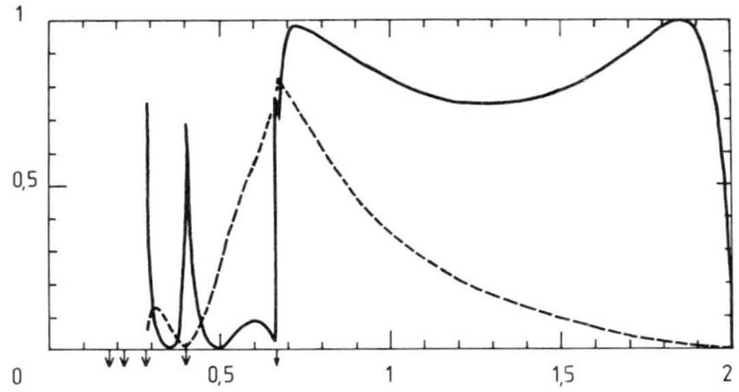

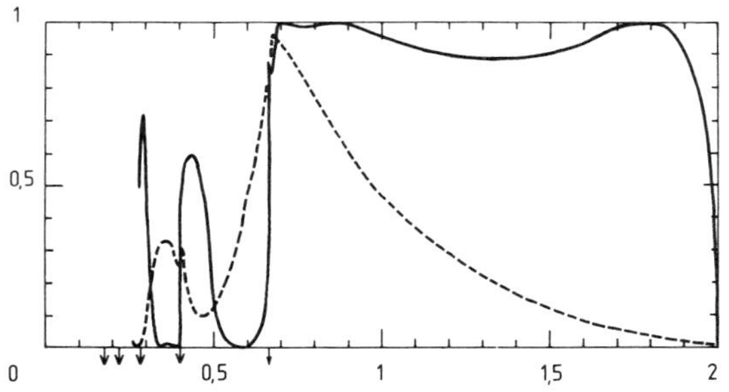

Fig.6.40 Sinusoidal grating. h/d = 0.25

Fig.6.41 Sinusoidal grating. h/d = 0.30

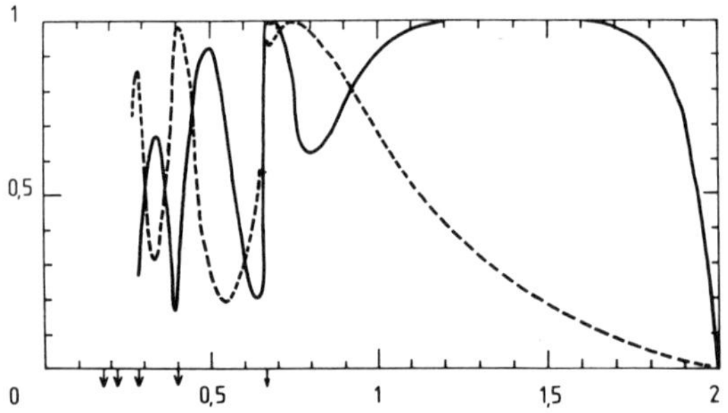

Fig. 6.42
Sinusoidal grating.
h/d = 0.40

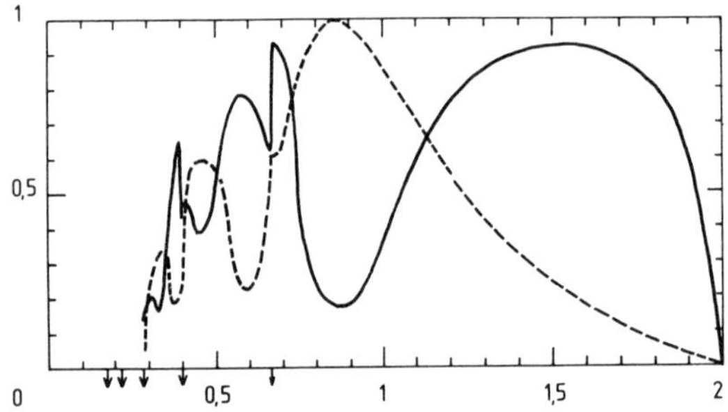

Fig. 6.43
Sinusoidal grating.
h/d = 0.50

indeed observed in the rigorous results reported in Fig.6.43. However, the perfect blazing reappears [6.53] for highly modulated gratings ($h/d \simeq 1$), which seems of little interest from a practical point of view.

Before leaving the TM polarization, a curious phenomenon must be emphasized. A very shallow grating ($h/d \ll \tau_1$) is equivalent to an echelette grating with a blaze angle close to 90°. Consequently, its efficiency curve must exhibit a perfect blazing under grazing incidence as predicted by Theorem 1 of Sect.6.2.1. All these predictions are confirmed by the computed efficiencies plotted in Figs.6.36-43. For TE polarization, we can expect a perfect blazing for $\tau_3 < h/d < \tau_2$ where τ_3 roughly corresponds to the critical blaze angle of 35° above which a perfect blazing is obtained for echelette gratings. This prediction is confirmed by the efficiency curves, but the exact value 0.34 of τ_3 given by a numerical study is slightly smaller than that predicted above. It is worth noting that, contrary to what happens for TM polarization, the perfect blazing still holds for $h/d > \tau_2$, the associated incidence angle being shifted towards the right. For this reason, the region $h/d > \tau_3$ is called "TE blaze region". In this region, a dashed line, drawn from rigorous computations, gives for sinusoidal gratings the variation of the incidence associated with the TE blaze peak as a function of h/d. Of course, this line can be used to find the location of the perfect blazing of ruled and lamellar gratings. For example, a grating with 40° blaze angle and 70° apex angle, is associated with a sinusoidal one with $h/d \simeq 0.49$. From the dashed line a perfect blazing is expected for an incidence close to 25°.

Let us now give the main features of the grating behavior in each of the five regions of groove depth.

Since with the present state of the art it is not practical to make ultrafine pitch gratings (i.e., > 6000 l/mm frequency), it is often necessary in the UV region to operate at low λ/d ratios. The set of curves consequently show that getting usable efficiency calls for operation with very low modulation depths (group I). For this type of grating (Fig.6.36), the perfect blazing obtained in grazing incidence for TM polarization does not seem to be usable in spectroscopy. Below λ_R^1, such gratings can be considered to lie completely in the scalar domain [6.54]. Scalar theory leads to efficiency curves whose shape is independent of h and always have a peak value of 34% for $\lambda = 3.4$ h [6.55]. Electromagnetic theory, for all practical purposes, leads to identical results. Since the peak efficiency of ruled gratings in the same scalar domain is 100%, holographic gratings appear to operate at a disadvantage of approximately 3 to 1.

As with echelette gratings, important polarization effects (Fig.6.37,38) begin to arise in the low modulation region (group II). The peak predicted by the scalar theory still holds, but the efficiency slowly increases for TE polarization. The Rayleigh anomaly for $\lambda = \lambda_R^1$ begins to develop. A second peak for TM polarization appears immediately above λ_R^1, whose value reaches about 50% for $h/d = 0.15$ (Fig.6.38). Since the perfect blazing only appears for $\lambda/d \simeq 2$, the efficiency for unpolarized light remains low (less than 25% for $\lambda/d \simeq 1.2$).

In the medium modulation region (group III), all the predictions of scalar theory fail (Figs.6.38-40). Stronger polarization effects can be seen and the efficiency peak near $\lambda = \lambda_R^1$ increases for both TM and TE polarization. The perfect blazing is shifted towards shorter wavelengths. The efficiency for TM polarization generally exceeds 50% for $\lambda > \lambda_R^1$ but remains very low for shorter wavelengths.

It is worth noting that the Rayleigh anomaly never disappears when the modulation increases. Consequently, a "special low anomaly" region (group IV) does not exist for sinusoidal gratings.

It is certainly in the high modulation region (group V) that the behavior of sinusoidal gratings and that of echelette gratings look the most alike. Indeed, in this region, the equivalence rule applies for the two perfect blazings of echelette gratings in TM polarization. Practically all the remarks already stated for echelette gratings apply, at least for wavelengths superior to the Rayleigh wavelength λ_R^1 under which the efficiency remains very weak. Without doubt, this region can be considered as the best from the point of view of efficiency.

It is very difficult to give some common features for the gratings of group VI. We will merely point out that, for the lowest groove depths ($h/d \simeq 0.4$), the efficiency curves are very similar to those obtained for corresponding echelette gratings, except again for $\lambda < \lambda_R^1$. For this last domain of wavelength, it must be recalled that the efficiency of echelette gratings was very high, especially when illuminated on the small facet. This feature does not hold for sinusoidal gratings with either TE or TM polarization.

In conclusion, a comparison of the properties of echelette and sinusoidal gratings requires us to distinguish the domain $\lambda < \lambda_R^1$ from its complementary. In the former, the echelette grating always gives a better efficiency curve. In the latter, the properties are very similar, except perhaps in the vicinity of λ_R^1.

6.2.4 Systematic Study of the Efficiency of Symmetrical Lamellar Gratings in -1 Order Littrow Mount

Until recently, rectangular groove gratings occupied a rather minor role in the application of diffraction gratings. They are now in regular use, especially at the two extremes of the optical spectrum. One is for X-rays [6.56], the other in the infrared for beam sampling mirrors [6.57,58]. The development of ion etching procedures in recent years [6.59] together with new methods for generating fine pitch patterns in photoresist [6.60-61] have made it worthwhile to investigate a possible spread into other spectroscopic applications, when typical wavelength to groove spacing ratios are 0.25 to 2. The efficiency curves in Littrow mount are given in Figs.6.44-51 for usual groove depths.

If $\lambda > \lambda_R^1$, one can roughly say that the properties of symmetrical lamellar gratings are very close to those of the corresponding sinusoidal and echelette ones. Almost all the considerations about the behavior of sinusoidal gratings continue

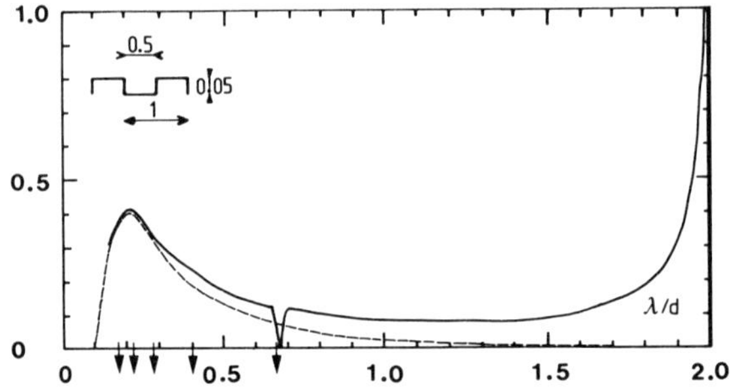

Fig. 6.44. Lamellar grating. h/d = 0.05. (——) TM polarization; (----) TE polarization

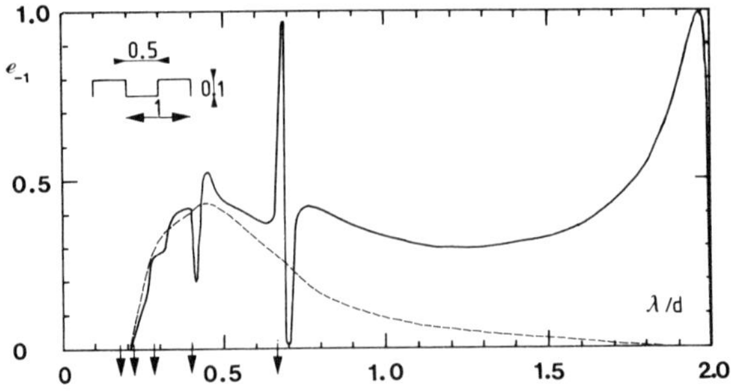

Fig. 6.45. Lamellar grating. h/d = 0.10

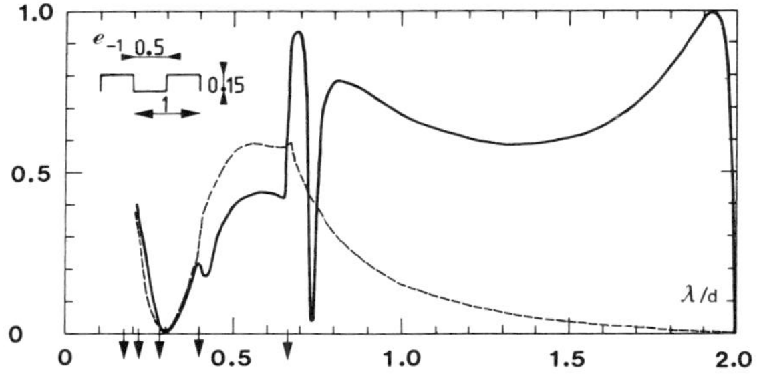

Fig. 6.46. Lamellar grating. h/d = 0.15

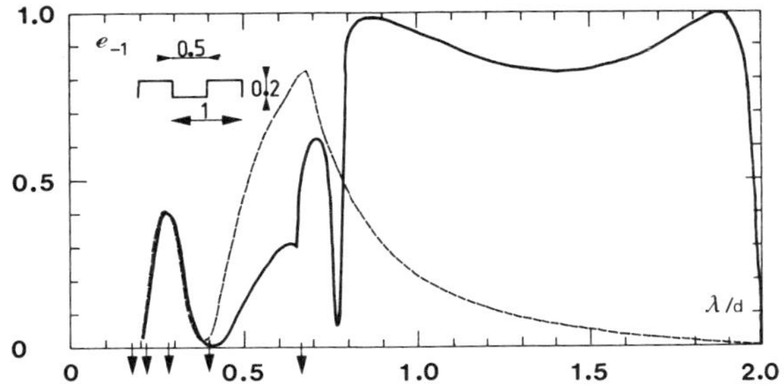

Fig. 6.47. Lamellar grating. h/d = 0.20

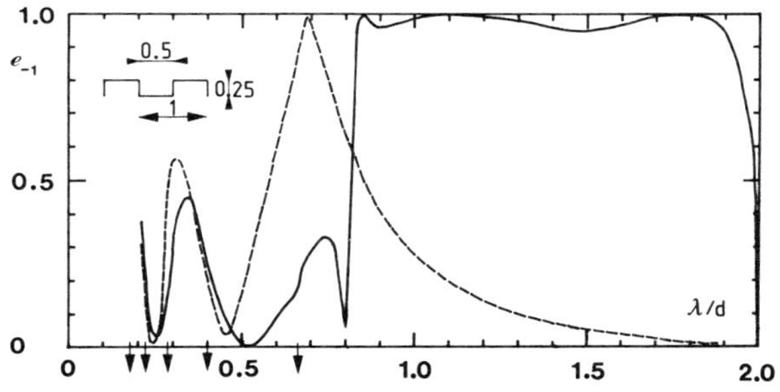

Fig. 6.48. Lamellar grating. h/d = 0.25

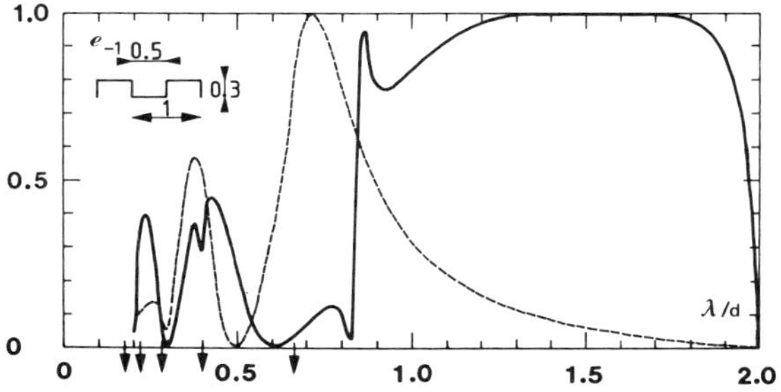

Fig. 6.49. Lamellar grating. h/d = 0.30

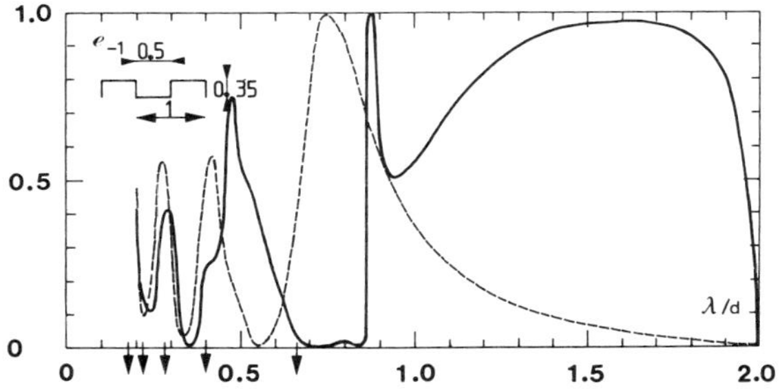

Fig. 6.50. Lamellar grating. h/d = 0.35

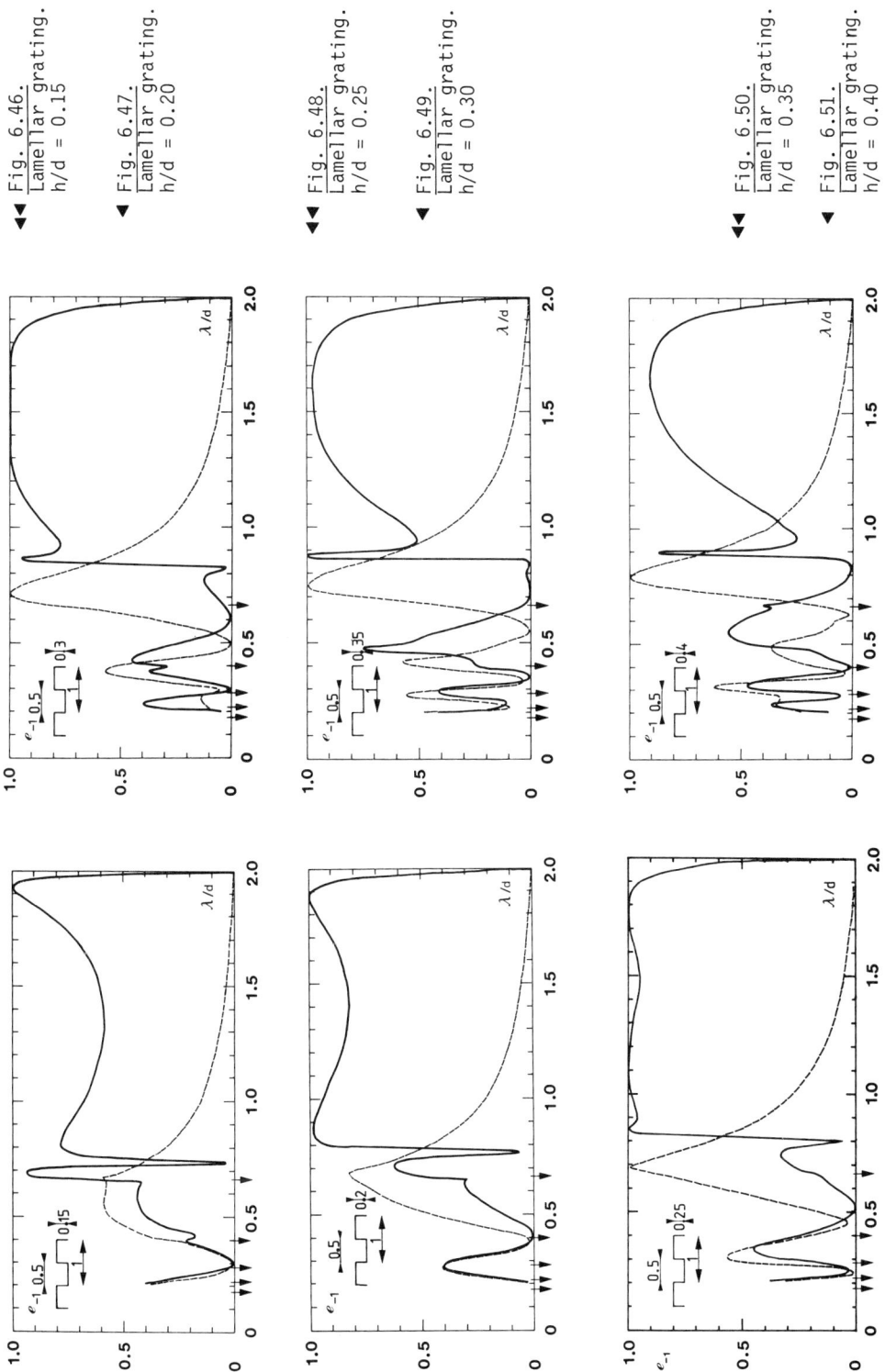

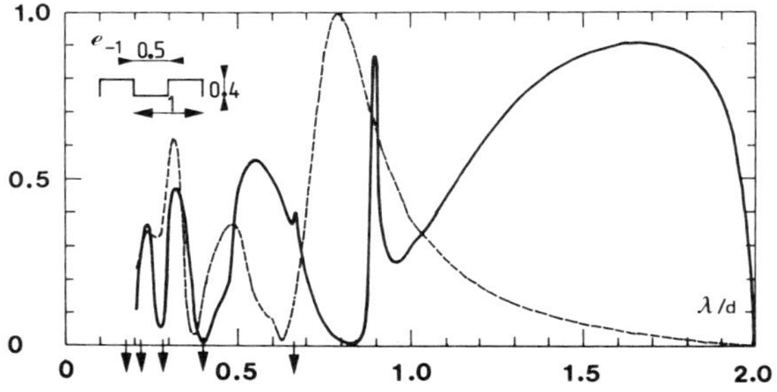

Fig. 6.51. Lamellar grating. h/d = 0.40

to hold, the different groups being well defined in Table 6.3. However, two important points must be stated. First, the TM polarization curves of lamellar gratings are less interesting than those of sinusoidal ones because of a spectacular anomaly, which arises for a wavelength very close to λ_R^1 for low groove depths and is shifted towards the right for higher modulation. Secondly, the half-width of the efficiency peak in TE polarization is shorter for lamellar gratings than for echelette and sinusoidal ones. These two facts involve a significantly marked separation between the TE and TM efficiency peaks, even in group V. For these reasons, symmetrical lamellar gratings appear to be less interesting than holographic and echelette gratings for spectroscopic applications[4].

On the other hand, when $\lambda < \lambda_R^1$, certain lamellar gratings (those of group I and II) have a better efficiency than the corresponding sinusoidal ones. Indeed, both scalar theories [6.55] and numerical results (see Figs.6.44,45) show that 40.5% efficiency is reached for $\lambda = 4h$. Moreover, as shown by a work not reported here [6.55] the symmetrical profile gives a better efficiency than other lamellar profiles, a conclusion which holds even in the electromagnetic domain [6.62], at least for usual groove depths.

6.2.5 Influence of the Apex Angle

Since the diamonds used by manufacturers often happen to have a facet angle different from 90°, one has to be aware of the troubles that one can encounter when the apex angle A differs significantly from 90°. Figure 6.52 shows the striking decrease of efficiency observed when the apex angle of a 20.5° blaze angle grating is increased from 90° (grating G) to 120° (grating G'). In fact the efficiency in unpolarized light drops from 93% to 65%! This can be explained using the grating nomogram. Indeed, we can see in Fig.6.34 that, when the apex angle becomes equal to 120°, the representative point of a ruled grating with $\alpha = 20.5^\circ$ leaves the "TM blaze region". Moreover, the nomogram allows us to find a grating G" having an apex angle of 120°, but a blaze angle different from α, whose efficiencies are close to those of the echelette grating G. This grating G" which is represented by the intersection of the curve $A = 120^\circ$ with the vertical line passing through the point ($A = 90^\circ$, $\alpha = 20.5^\circ$) has a blaze angle of 28°. It can be seen in Fig.6.52 that such a grating (G") gives an efficiency curve better than the first one (G'). For TE polarization, the curve is identical to that of the echelette grating G. In TM polarization, grating G" is better than grating G' for all wavelengths. Another example of application of the nomogram to answer questions which can be asked by designers of ruled gratings may be found in [6.48].

---

4 It will be shown in Sect.6.4.1 that this conclusion must be reconsidered for very deep grooves ($h/d > 0.5$).

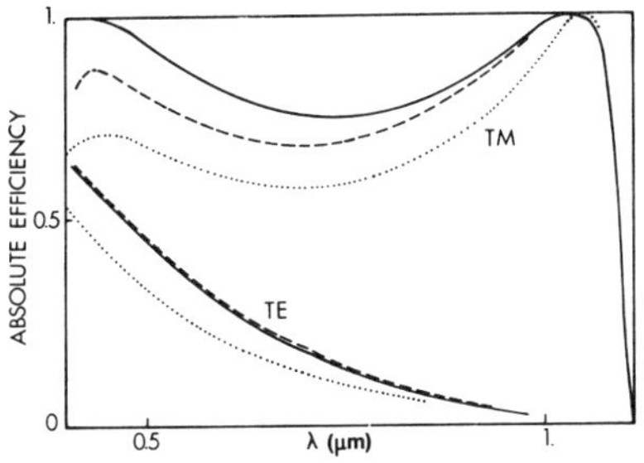

Fig. 6.52. Efficiency of 1800 lines/mm perfectly conducting ruled gratings: (——) Grating G; (....) Grating G'; (----) Grating G"

Finally, the grating nomogram makes it possible to classify the ruled gratings into several categories, depending whether or not they have about 100% efficiency peaks in TE or TM polarization. For example one can see that a ruled grating with apex angle greater than 123° cannot present a near perfect blazing whatever the polarization. Such a blazing is observed only in TM polarization if $100^{\circ} < A < 123^{\circ}$.

6.2.6 Influence of a Departure from Littrow

Except for laser tuning applications, gratings are never used in pure Littrow mounting. Monochromator mounts usually operate with constant deviation between incident and diffracted beams. Spectrographs generally work at constant angles of incidence, but in most practical systems their behavior can be fairly well diagnosed from constant deviation curves. The main studies on this topic have been described in [6.28,43]. The effects, for the same departure from Littrow, vary from negligible to significant, depending on the modulation depth. The general rule is that the higher the modulation, the more pronounced the effect. No results are shown for the very low modulated gratings (group I) for which the effect is too small to report, at least in the vicinity of the efficiency peak situated in the scalar domain.

For low modulated gratings, Figs.6.53, 54 show the effect of 45° angular deviation (AD): For echelette gratings, the only significant difference is the shift of the anomalies linked with the splitting of the pass-off wavelengths, as shown in Fig.6.53. For sinusoidal gratings (Fig.6.54), the grating behavior actually improves somewhat when AD = 45°. The TE peak increases and is shifted slightly to shorter wavelengths. In TM polarization, anomalies appear to smooth out, and efficiency increases at high λ/d values.

We have observed that the effects of a Littrow departure for medium blaze echelette gratings (group III) are similar to those described for low modulated gratings. On the other hand, for medium and high modulated sinusoidal gratings (groups III and V), we can see in Fig.6.55 the rather drastic effect due to a departure from Littrow to AD = 45°: the TE efficiency drops considerably, with no change in peak location; the span of TM high efficiency contracts at both ends because the limiting 90° diffraction angle is reached for $\lambda/d = 1.707$.

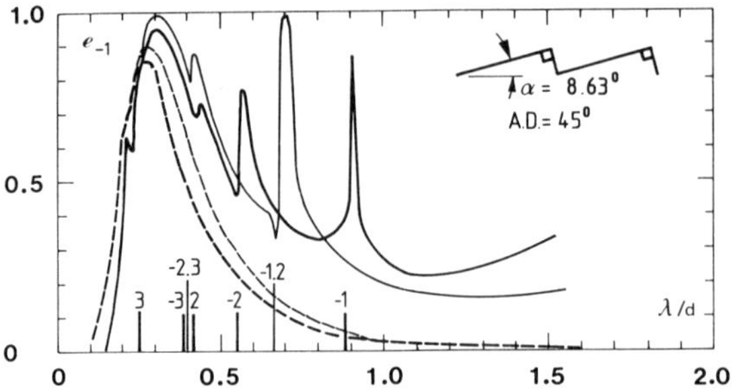

Fig. 6.53. Influence of 45° departure from Littrow for echelette gratings (Groups I and II). Efficiency curves and pass-off wavelengths for Littrow are drawn lightly

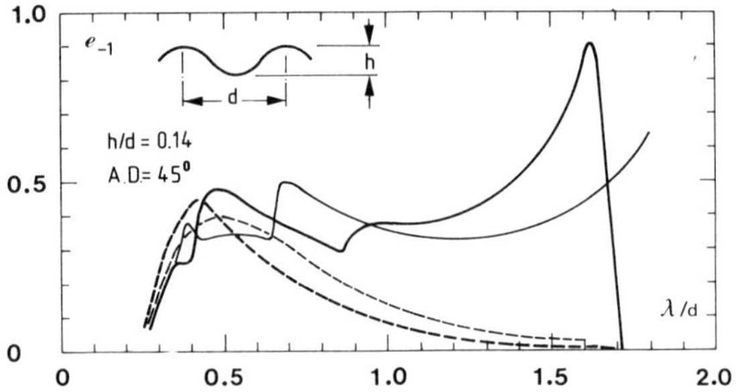

Fig. 6.54. Influence of 45° departure from Littrow for sinusoidal gratings (Groups I and II)

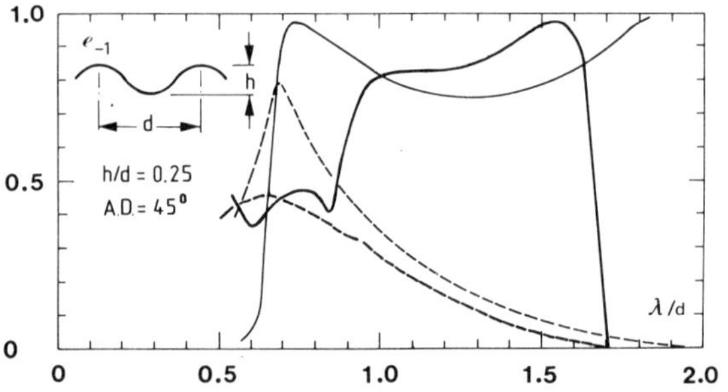

Fig. 6.55. Influence of 45° departure from Littrow for sinusoidal gratings (Group III)

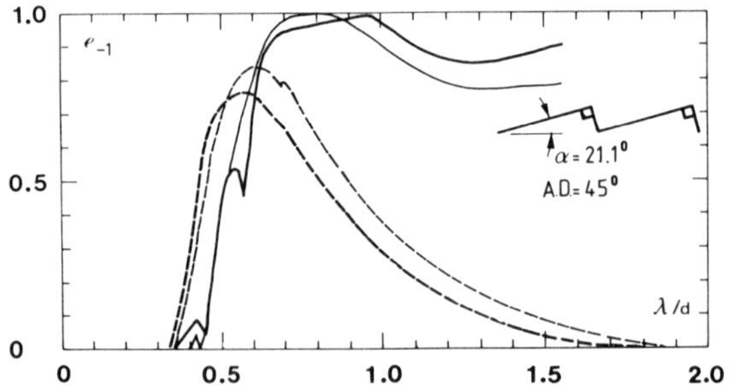

Fig. 6.56. Influence of 45° departure from Littrow for echelette gratings (Group IV)

The special low anomaly blazed gratings (group IV) maintain their properties quite well for AD values up to 15°. Beyond that, anomalies start to grow, to the extent shown in Fig.6.56 for AD = 45°. In other words, low anomaly properties are not automatically maintained. TE efficiency is also reduced slightly and shifted somewhat in the direction of shorter wavelengths.

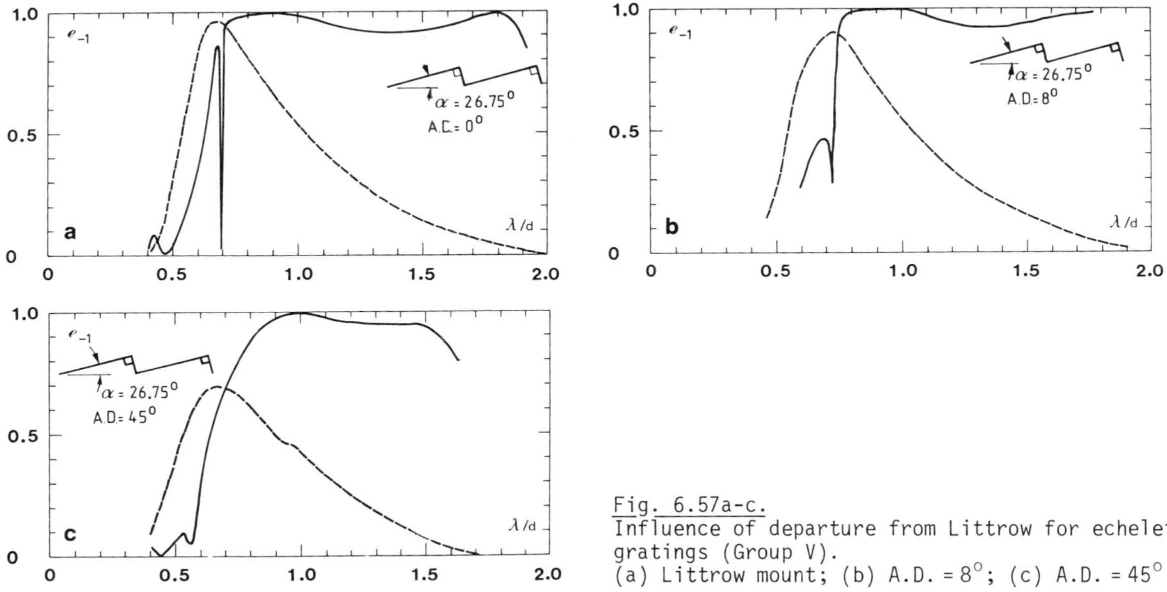

Fig. 6.57a-c.
Influence of departure from Littrow for echelette gratings (Group V).
(a) Littrow mount; (b) A.D. = 8°; (c) A.D. = 45°

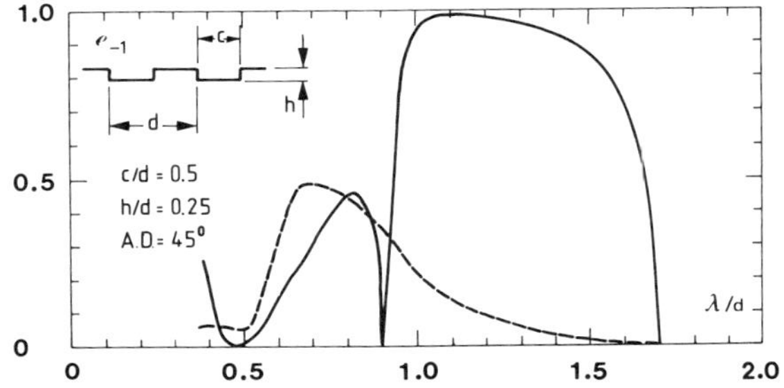

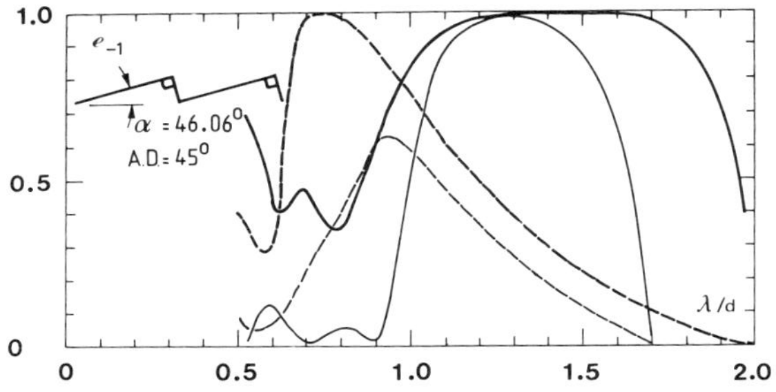

◀◀
Fig.6.58.
Influence of departure from Littrow for lamellar gratings (Group V)

◀
Fig.6.59.
Influence of departure from Littrow for blazed gratings (GroupVI)

For echelette gratings of group V, the efficiency peaks are located in a domain where the equivalence rule is valid, and the remarks already made for sinusoidal gratings of this group hold. It is evident from Fig.6.57 that the effect of small AD (8°) is to smooth out the principal TM anomaly at low expense in TE polarization. When AD becomes larger (45°), the TM anomaly is virtually suppressed but at the expense of significant shrinkage of the TM peak and of the entire TE curve. This suppression recalls the vanishing of the anomalies at $\lambda = \lambda_R^1$ observed in Littrow mounting for gratings of group IV. Figure 6.58 compared with Fig.6.48 shows that the effect of the departure from Littrow is the same for symmetrical lamellar gratings as for sinusoidal gratings.

Finally, for very high modulated gratings (group V), departure from Littrow is accompanied by similar but much stronger effects and, in addition, the TE peak is shifted towards longer wavelengths (Fig.6.59). This conclusion applies to ruled, sinusoidal or lamellar gratings.

6.2.7 Higher Order Use of Gratings

It has been noted from many numerical and experimental results that the efficiency peaks of gratings used in $-m^{th}$ order Littrow mount roughly occur at the same $m\lambda/d$ values. Therefore, the peak wavelength always tends towards the scalar region as the order m increases. Higher orders are used sometimes to increase the wavelength coverage of a single grating, or to increase the dispersion, since the latter is proportional to the tangent of the angle of diffraction. Of course, in these mountings, more than two orders are diffracted, thus the equivalence rule does not work.

Let us first consider the echelette gratings. Figures 6.60,61 [6.43] show Littrow mount efficiency curves for -2 and -3 orders and for a range of commercial blaze angles α. It must be noticed that for low and medium modulations (order -2) and even for high modulations (order -3), the curves for TE polarization are almost identical, except for high incidences. According to theorem 1 of Sect.6.2.1, the efficiency reaches 100% in TM polarization for $\theta = \alpha$. With increasing blaze angles, the location of the TE peak moves further away from the TM peak. For the -2 order Littrow mount, and for TM polarization, and contrary to what happens in -1 or -3 orders, the anomaly does not disappear when it coincides with the nominal blaze wavelength.

Figure 6.62 shows the efficiency curve in Littrow -2 of a low modulated sinusoidal grating. As expected from scalar thoeries [6.55], a 23% efficiency (related with the Bessel function J_2) is obtained when $J_2'(2\pi h/\lambda) = 0$, i.e., for $\lambda = 2.06$ h. Let us recall that in the general case of -m order Littrow mount, the scalar theory predicts that an efficiency of $|J_m(\omega)|^2$ is obtained when $\omega = 2\pi h/\lambda$ is such that $J_m'(\omega) = 0$. The result is a decrease of the peak efficiency when going to high orders.

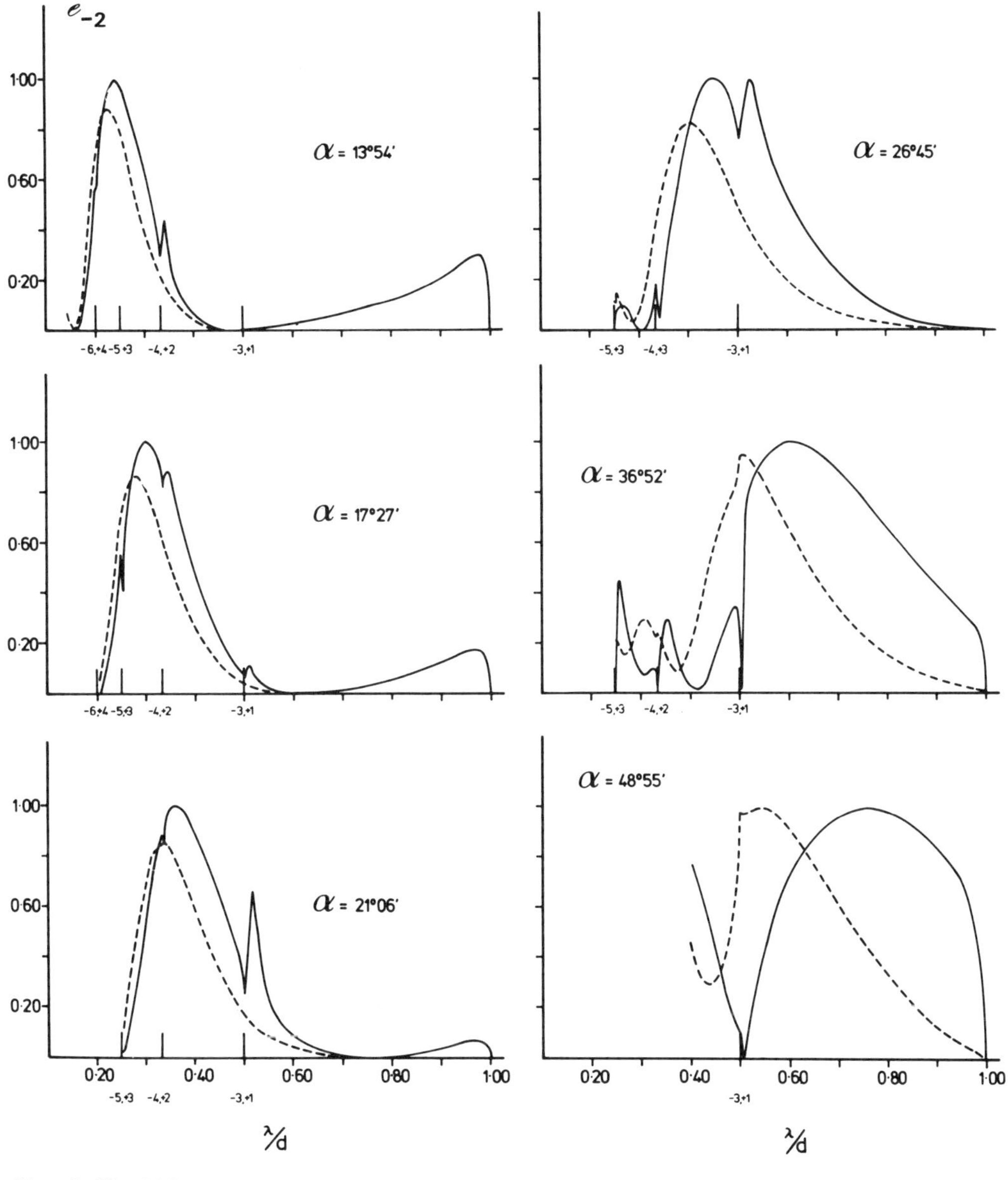

Fig. 6.60. Efficiency in -2 order Littrow mount for various blaze angles. (——) TM polarization; (----) TE polarization

We show in Fig. 6.63 the efficiency curve of a h/d = 0.22 sinusoidal grating. The low efficiency and strong anomalies make such a grating unattractive in second order. This remark also holds when using -3 order [6.43].

In contrast, highly modulated gratings do give useful efficiencies in first, second and third order Littrow mounts (Fig.6.64).

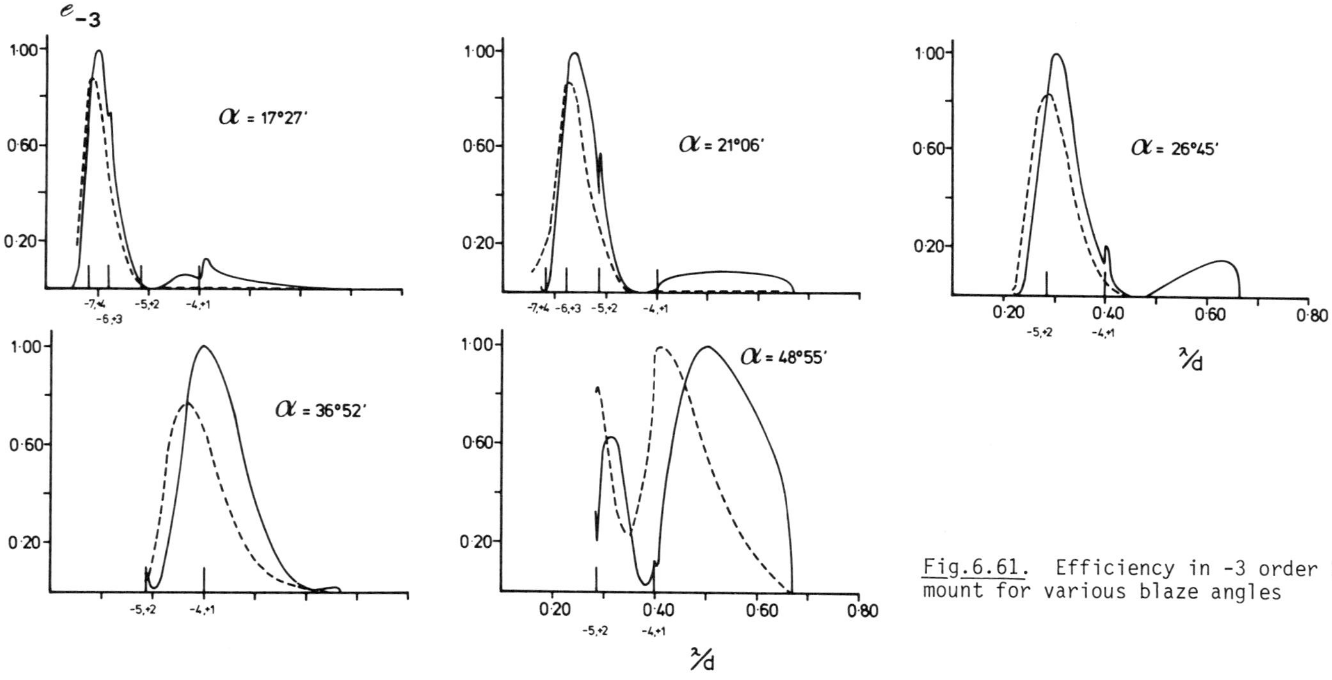

Fig.6.61. Efficiency in -3 order Littrow mount for various blaze angles

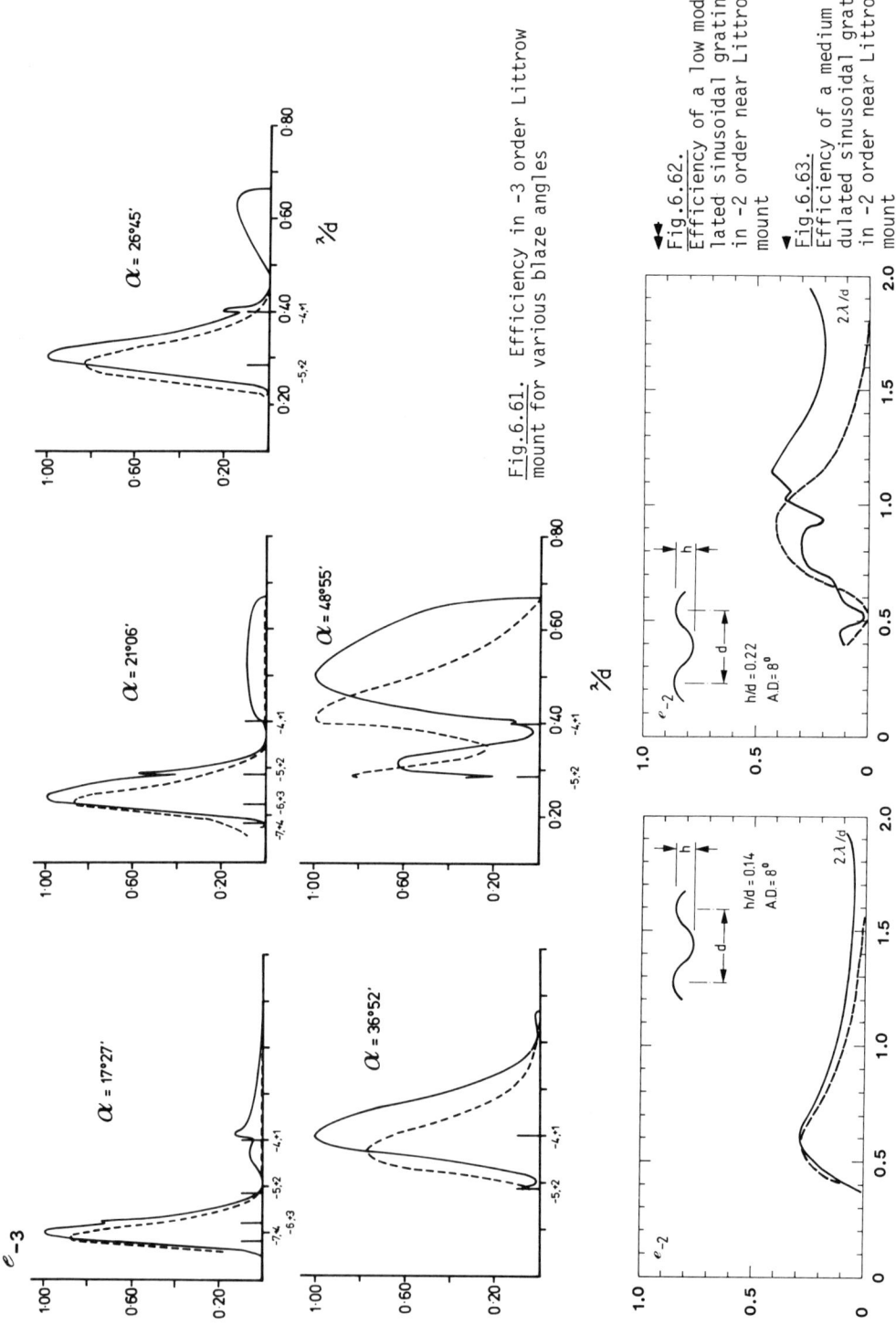

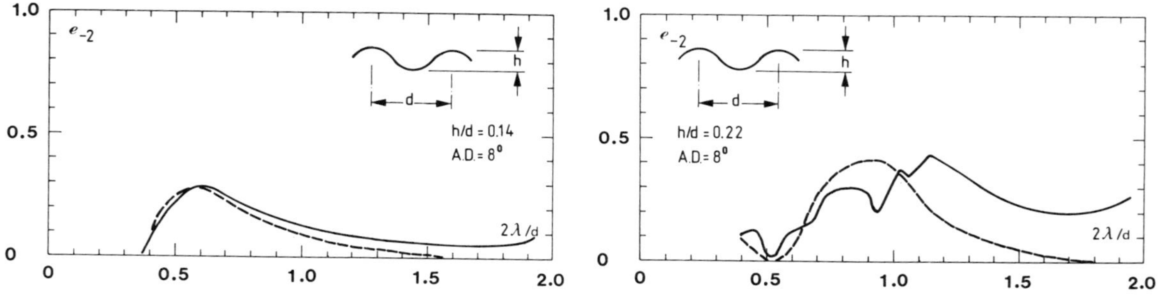

◄◄ Fig.6.62. Efficiency of a low modulated sinusoidal grating in -2 order near Littrow mount

◄ Fig.6.63. Efficiency of a medium modulated sinusoidal grating in -2 order near Littrow mount

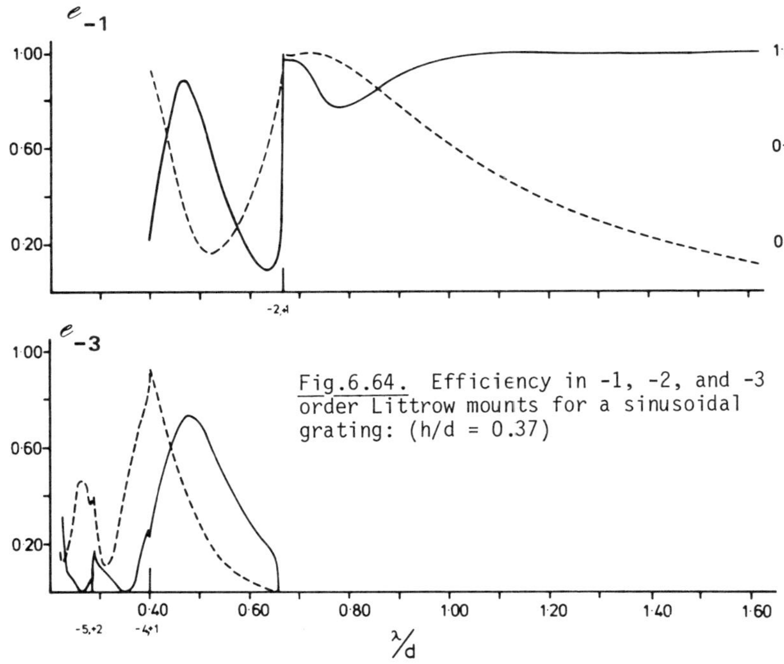

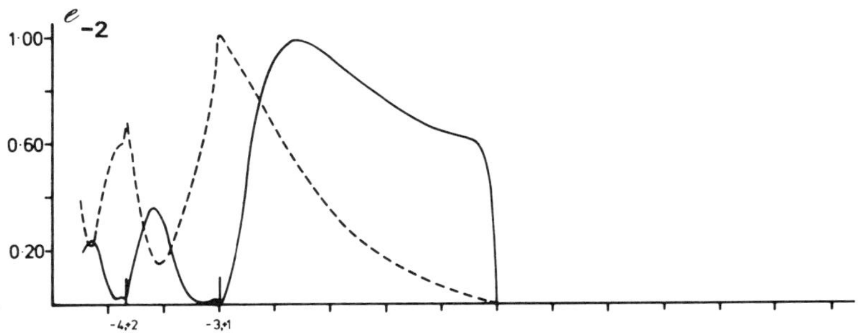

Fig.6.64. Efficiency in -1, -2, and -3 order Littrow mounts for a sinusoidal grating: (h/d = 0.37)

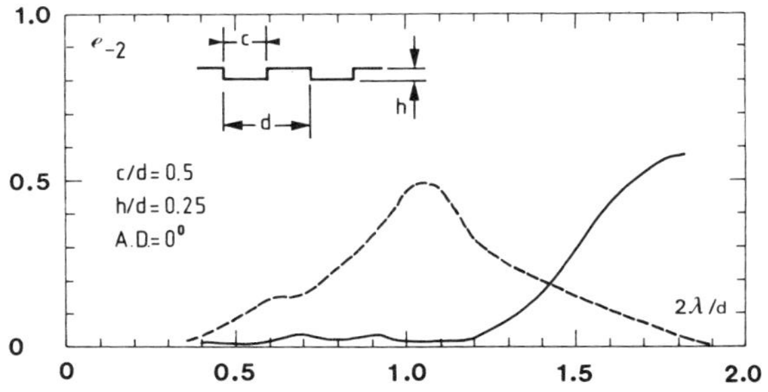

Fig. 6.65. Efficiency in -2 order Littrow mount for a symmetrical lamellar grating

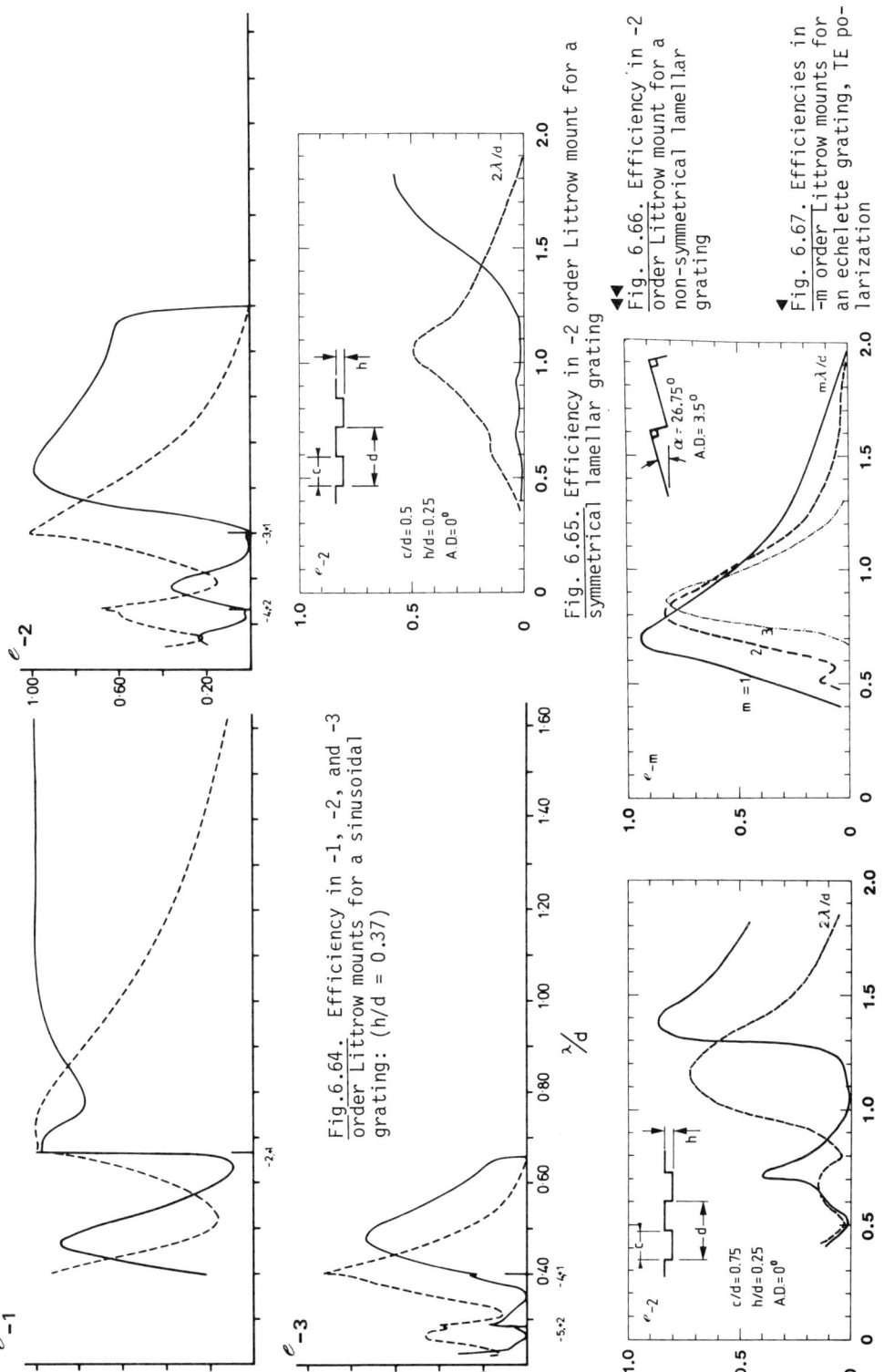

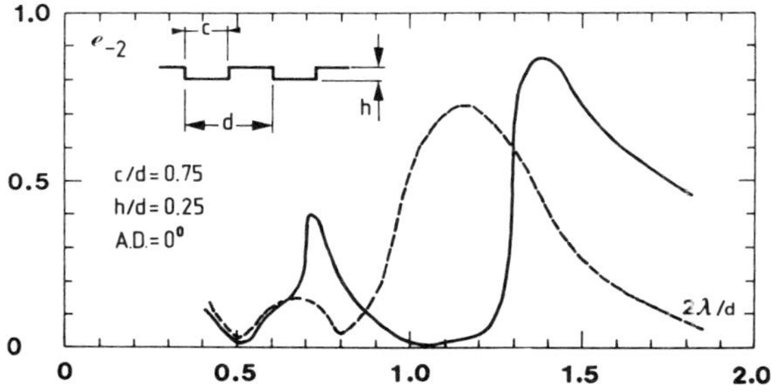

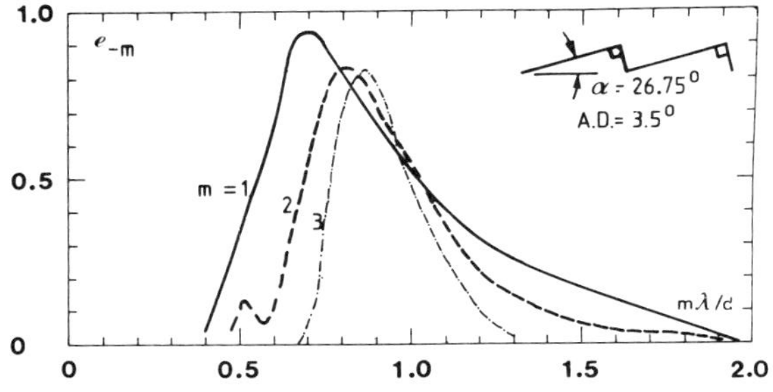

◀◀ Fig. 6.66. Efficiency in -2 order Littrow mount for a non-symmetrical lamellar grating

◀ Fig. 6.67. Efficiencies in -m order Littrow mounts for an echelette grating, TE polarization

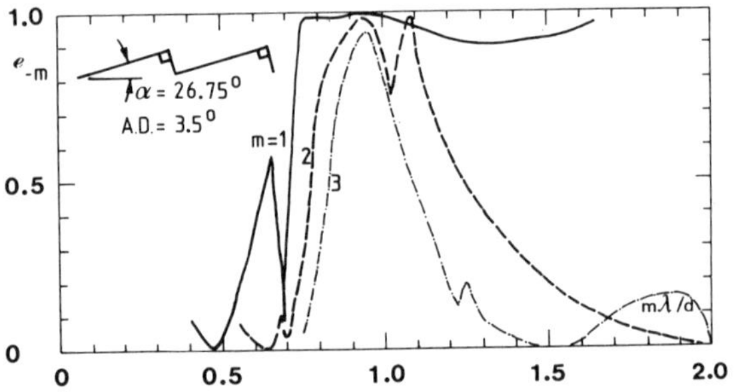

Fig. 6.68. Efficiencies in -m order Littrow mounts for an echelette grating, TM polarization

For lamellar gratings, and if c/d = 0.5, it is known that scalar theories predict a zero efficiency in the -2 order [6.63,55]. This is confirmed by the electromagnetic theory provided that h/d < 0.02. However, since the scalar approach assumes zero groove height, it is not surprising that some energy appears when h/d = 0.05. When h/d = 0.25, efficiencies up to 50% are calculated (see Fig.6.65). It is interesting to notice that the nonsymmetrical lamellar grating behaves better than the symmetrical one. Figure 6.66 shows that a rather strong increase of efficiency occurs when the width of the valley becomes three times the width of the "plateau".

In Figs.6.67,68 taken from [6.28], and experimentally confirmed, we have plotted for an echelette grating the efficiencies in -1, -2 and -3 order Littrow mounts, as functions of mλ/d ; One can see that, when m increases, the blaze width decreases and that TE and TM efficiencies approach each other.

6.3 Finite Conductivity Gratings

As already explained in Sect.6.1.5, the influence of the finite conductivity of the grating material is negligible above 4 μm. On the other hand, its influence may be fundamental under 1 μm, the 1 to 4 μm region being a transition zone where the necessity of taking into account the finite conductivity depends on the required accuracy.

6.3.1 General Rules

For TE polarization, the introduction of the finite conductivity does not affect the shape of the efficiency curve. As long as the groove depth does not exceed 0.25 d the efficiency is simply multiplied by the reflectance. For higher modulated gratings the efficiency reduction is greater than the one due to the reflectivity, reaching twice as much for h/d ≃ 0.5. In the latter case a shift of the maximum efficiency may occur, resulting in a crossover of the two curves given by infinite and finite conductivity theory [6.64].

For TM polarization the effect is very strong. The number, location and strength of anomalies are affected [6.30,31,65,66] resulting in a complete change in efficiency curves. Figures 6.69,70, taken from [6.66], clearly show this fact ; strong oscillations occur mainly at the smallest wavelengths, but beyond λ/d = 0.8, efficiency curves are smooth ; in this region, the role of finite conductivity is again in proportion to the value of the reflectivity. Due to the shift of anomalies to longer wavelengths, the width of the TM blaze plateau is somewhat reduced.

The introduction of a deviation between beams, which is common practice in most spectroscopical imstruments, leads to the same changes as for infinite conductivity theory. It results in a reduction of the blaze peak for TE polarization, and a nar-

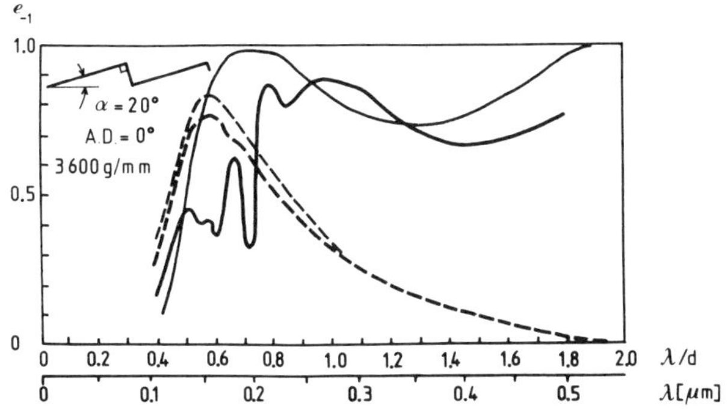

Fig. 6.69. Computed -1 order Littrow mount efficiency curves for a 20° blaze angle echelette grating. Infinite conductivity efficiencies are shown as light curves. Finite conductivity efficiencies, heavy lines, hold for aluminum

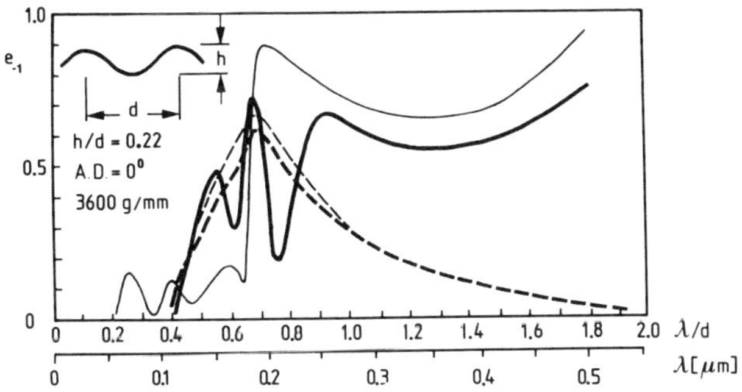

Fig. 6.70. Computed efficiency curves for a sinusoidal grating. Other conditions as in Fig.6.69

rowing of the TM blaze plateau at the two ends. When departure from Littrow and finite conductivity effects both occur, the changes in efficiency curves may be drastic.

A special attention must be devoted to the low λ/d ratio region. When λ/d decreases, one goes to the scalar domain which is characterized by low anomalies and polarization effects. Thus finite conductivity has the same influence on TE and TM curves, i.e., a simple reduction in proportion to the reflectance, even in a spectral domain where the latter is very small. For a given grating, this fact is generally found when used in higher orders.

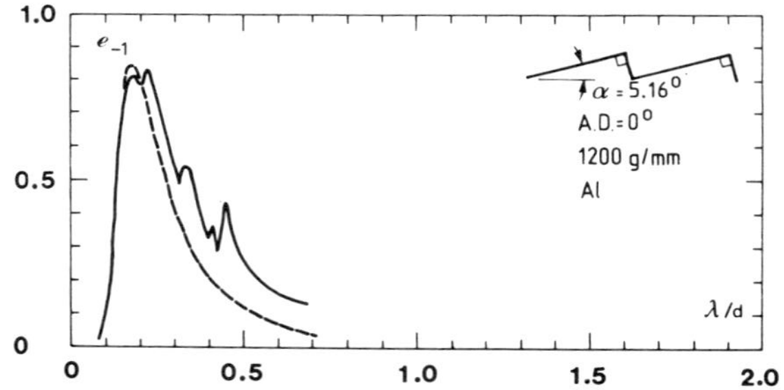

Fig.6.71

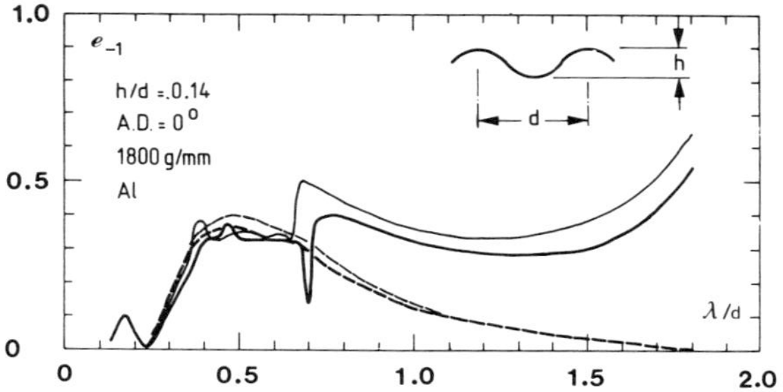

Fig.6.72

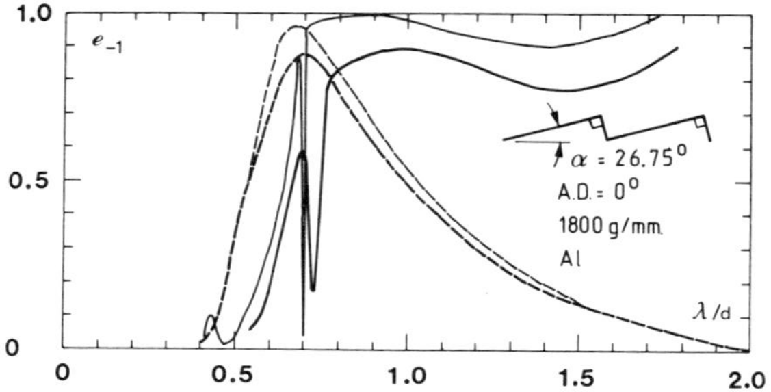

Fig.6.73

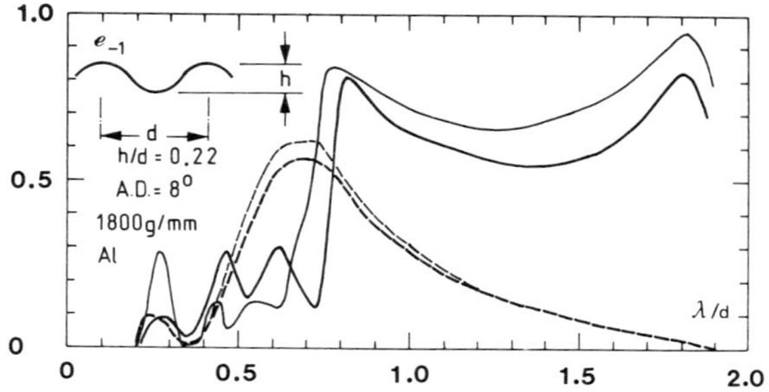

Fig.6.74

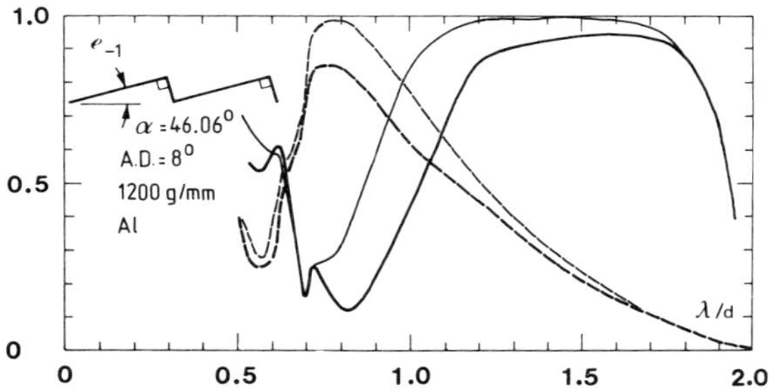

Fig.6.75

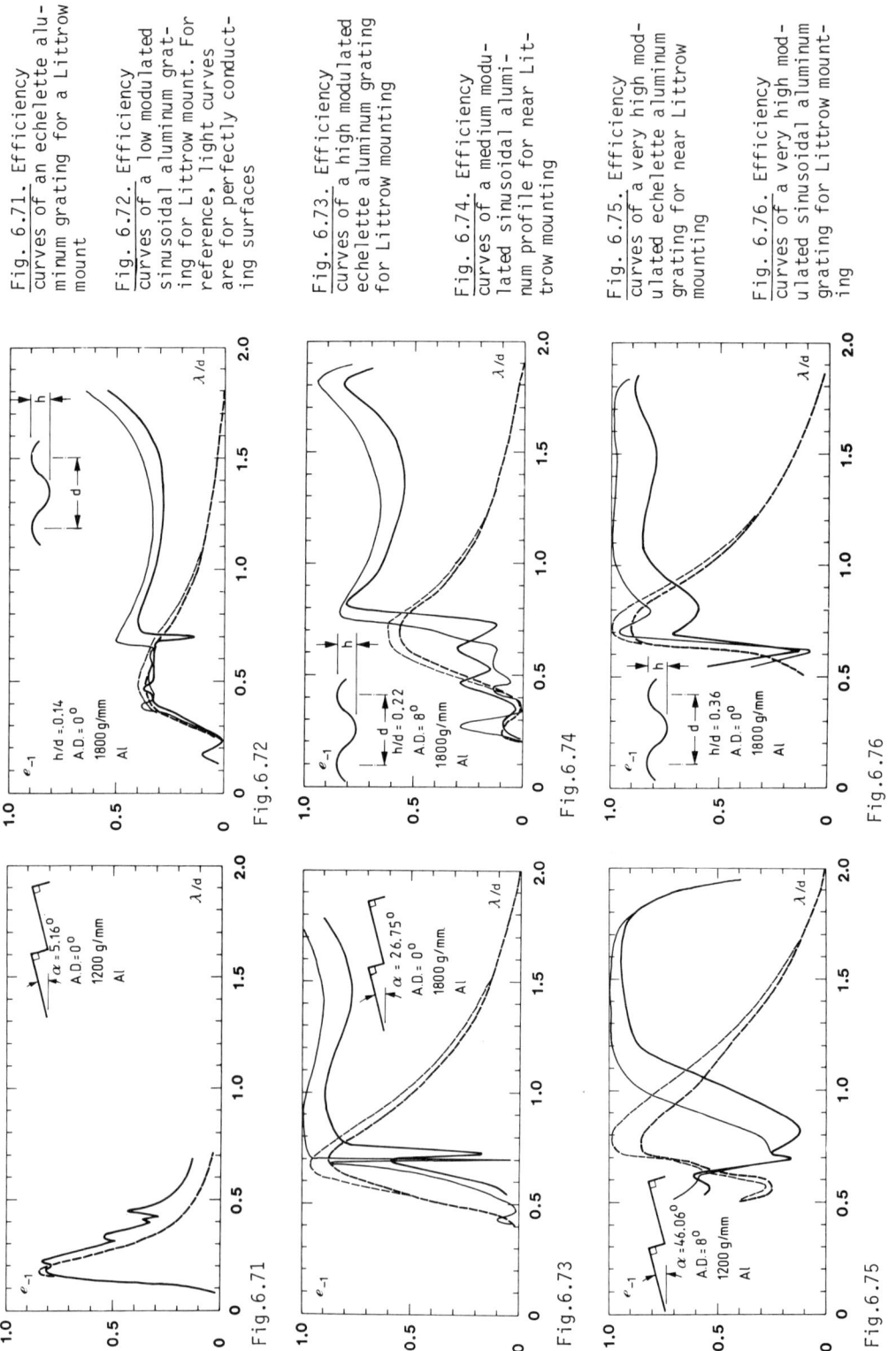

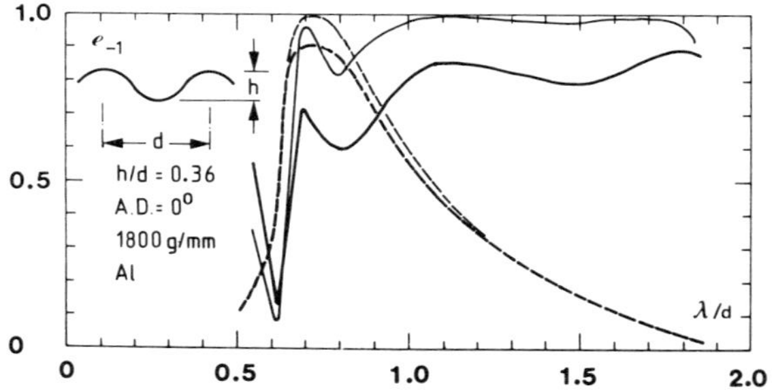

Fig.6.76

Fig. 6.71. Efficiency curves of an echelette aluminum grating for a Littrow mount

Fig. 6.72. Efficiency curves of a low modulated sinusoidal aluminum grating for Littrow mount. For reference, light curves are for perfectly conducting surfaces

Fig. 6.73. Efficiency curves of a high modulated echelette aluminum grating for Littrow mounting

Fig. 6.74. Efficiency curves of a medium modulated sinusoidal aluminum profile for near Littrow mounting

Fig. 6.75. Efficiency curves of a very high modulated echelette aluminum grating for near Littrow mounting

Fig. 6.76. Efficiency curves of a very high modulated sinusoidal aluminum grating for Littrow mounting

6.3.2 Typical Efficiency Curves in the Visible Region

In order to illustrate the preceding rules, let us give, as examples, some typical efficiency curves for blazed and holographic gratings drawn using the finite conductivity theory. Because of space limitations, only an example of low, medium, and high modulated gratings, taken from [6.28], are shown. The interested reader can refer to the original papers [6.28,66]. Figure 6.71 is related to a 5°10' blaze angle 1200 groove/mm aluminum echelette grating used in Littrow mount. Due to the low modulation, which implies a low λ/d ratio near the blaze peak, TM anomalies are weak and TE and TM curves fall close to each other. The infinite conductivity curves, not reported for the sake of clarity, would present a very similar shape. Figure 6.72 shows the same features for a low modulated holographic grating, although a pronounced anomaly appears, due to plasmon resonance effects, resulting in a light reduction of the width of the blaze plateau. Figure 6.73 shows the effect of finite conductivity on a high modulated echelette grating; the TM anomaly is shifted and weakened. Outside the anomaly region, the efficiency is decreased by a variable amount that can reach twice the reflectance loss. Similar comments apply to Fig.6.74, related to a sinusoidal profile. Figures 6.75,76, related to very high modulated echelette and sinusoidal gratings, show that the influence of finite conductivity increases with groove depth; in the case of the echelette grating, the decrease reaches about twice the reflectance drop for TE polarization

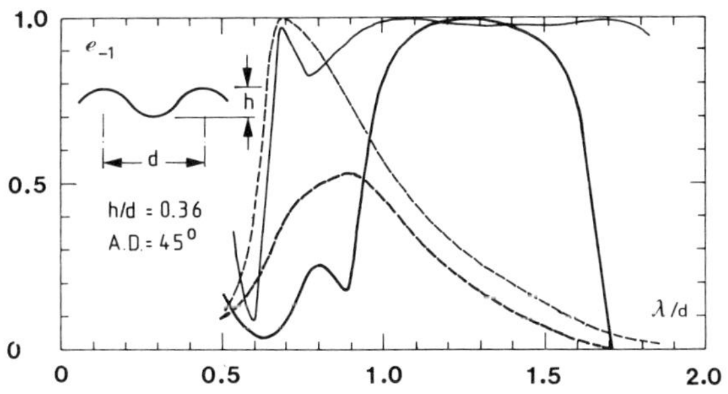

Fig. 6.77. Efficiency of a perfectly conducting high modulated sinusoidal grating with AD = 45°. For reference, light curves show Littrow efficiencies

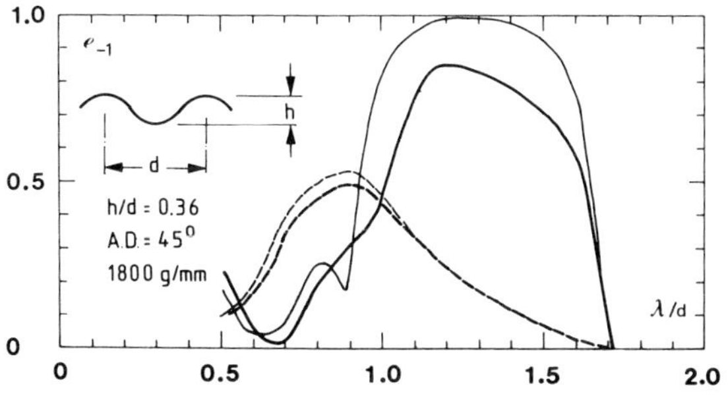

Fig. 6.78. The same as Fig.6.77 but for an aluminum grating. Light curves are the same as the heavy lines of Fig.6.77

and sometimes more for TM polarization ; a significant shift of the TE blaze peak begins to appear.

The paper by MAYSTRE et al. [6.66], which provides a set of curves for 3600 lines/mm aluminum gratings, clearly shows that the influence of finite conductivity increases with the groove frequency.

To conclude, Figs. 6.77,78 illustrate the drastic influence of both departure from Littrow and finite conductivity when they act simultaneously ; in Fig.6.77, and in TE polarization, the reduction reaches 50%, while an important shift of the blaze peak to longer wavelengths is found. For TM polarization, the span of high efficiency contracts at both ends because the limiting 90° diffraction angle is reached for $\lambda/d = 1.707$, but still reaches 100%. When the finite conductivity is included, (Fig.6.78), the TM blaze plateau loses most of its interest for that high modulated grating.

6.3.3 Influence of Dielectric Overcoatings in Vacuum UV

Dielectric overcoatings are mostly used in vacuum UV ($0.1 < \lambda < 0.2$ μm), where aluminum oxide presents important losses, to prevent aluminum gratings from oxidizing. They are also sometimes used on top of silver gratings in the very near infrared region. With the recent development of tunable lasers, stacks of layers, with alternately low and high refractive indices, are used to enhance the efficiency of gratings used as wavelength selectors. The same goal is looked for, when one uses them as beam sampling mirrors for high power lasers (see Sect.6.4.6)in the infrared region. Although it is sometimes possible to get a worthwhile increase in efficiency [6.67], the use of overcoatings must be made very carefully. As explained in Chap. 5, above a cut-off thickness, guided modes can be supported by the dielectric coated grating, producing a new kind of grating anomalies. The result is to transform smooth TE efficiency curves into irregular curves which look like those found for TM polarization, as shown experimentally by HUTLEY et al. [6.37].

As a typical example of dielectric overcoatings, Fig.6.79, taken from [6.65], shows for two thicknesses the influence of a layer of magnesium fluoride on top of an aluminum grating used in TE polarization (TM plane gives similar results). As we can see, both thicknesses improve the maximum efficiency, but for different wavelengths. The two curves cross at 1100 Å and 1410 Å. Between these values, which include the important Lyman α region, the thinner overcoating is clearly to be preferred. The advantage is 20% at 1216 Å. At wavelengths longer than 1410 Å, the thicker coating is preferred.

In the case of holographic gratings, the effect of the 250 Å MgF_2 overcoating is to give the same 10% enhancement near the peak as that seen for the echelette grating. Both gratings have their usefulness destroyed if the thickness of MgF_2 is made 1000 Å.

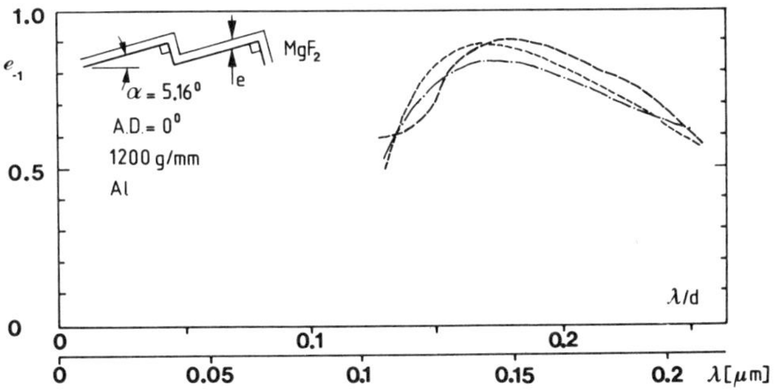

Fig.6.79

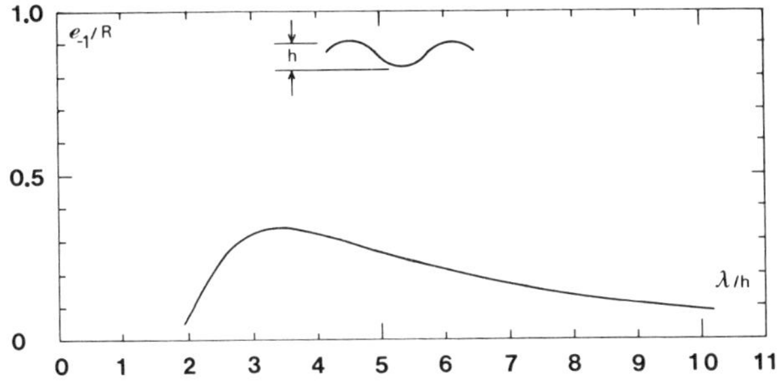

Fig.6.80

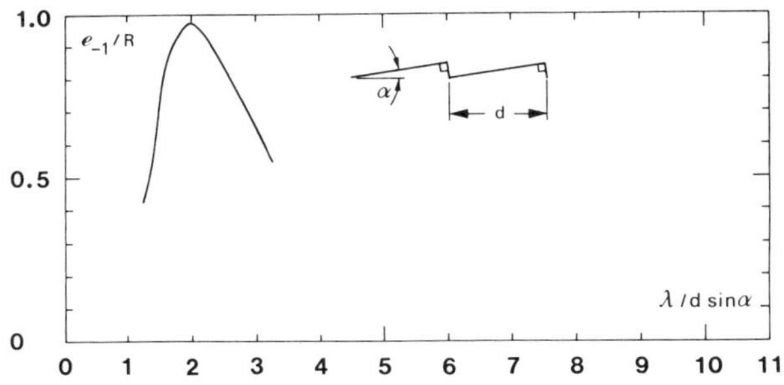

Fig.6.81

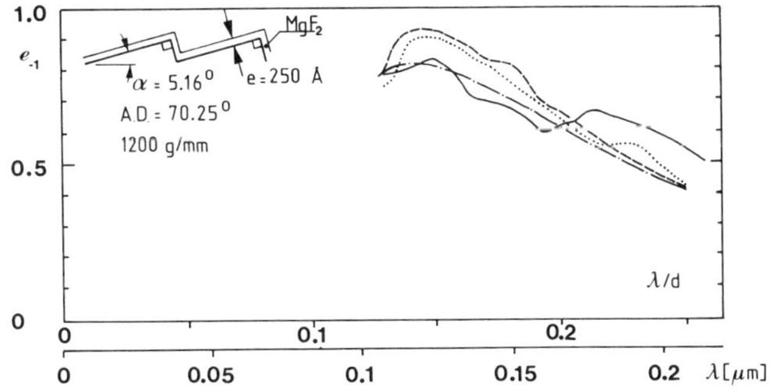

Fig.6.82

Fig. 6.79. Influence of a MgF_2 overcoating in TE polarization. (—·—·—·—) e = 0; (----) e = 250 Å; (— — —) e = 400 Å

Fig. 6.80. Universal relative efficiency curve of a sinusoidal grating used in the scalar domain. R is the reflectance of the grating material

Fig. 6.81. The same as Fig.6.80 but for an echelette grating

Fig. 6.82. Efficiency in a 70.25° deviation mounting for bare and 250 Å MgF_2 coated gratings. (——) Aluminum bare grating for TM polarization; (—·—·—) Aluminum bare grating for TE polarization; (····) Aluminum coated grating for TM polarization; (----) Aluminum coated grating for TE polarization

For these VUV gratings, it turns out that 100 Å layers of MgF_2 have no optical influence; however they provide the chemical protection required. In this spectral domain, most commercial gratings work at very low λ/d ratios. As a result, they behave in a scalar or near scalar manner. Efficiency curves are almost free of anomalies and polarization effects, and can all be deduced from universal efficiency curves, whatever their groove depth, groove spacing, and material may be. The efficiency over reflectance ratio turns out to be purely a function of the ratio of wavelength to groove depth. Figures 6.80,81 show the two universal relative efficiency curves for sinusoidal and echelette gratings. In this scalar region, sinusoidal gratings appear to operate at a disadvantage since their theoretical efficiency ceiling is 34%, compared with 100% for the echelette grating! This conclusion is moderated by the fact that in the UV domain, most grating are produced on a concave blank [6.68a].

It is important to appreciate the fact that for sinusoidal gratings, the blaze peak always occurs at a wavelength equal to 3.4 times the groove depth h, regardless of groove frequency. Thus, for a given blaze wavelength, only the groove depth is to be determined. The groove spacing can be chosen arbitrarily, provided that the grating works in the scalar region.

If the scalar behavior has the initial merit of limiting the reduction in efficiency due to finite conductivity to the amount of reflectance losses, a second useful attribute is to diminish efficiency reduction due to departure from Littrow. Figure 6.82 illustrates the effects on an echette grating used in a Seya-Namioka mount[5]. This well-known mount, which operates with angular deviation of 70.25°, is especially important because it provides the best practical possibility for monochromator-type scanning through rotation of a concave grating. Thus, it represents the largest departure from Littrow one is likely to encounter, short of going to grazing incidence. For the 5.16° blaze angle shown in Fig.6.79, one can see the 18% reduction in blaze wavelength which follows geometrically from the 70.25° departure from Littrow. Despite this large angle, we observe for a bare grating less than 5% drop in peak efficiency compared with Fig.6.79, which is explained by the low groove depth. Similar conclusions were found for holographic gratings. Note that, in Fig.6.82, the effect of the 250 Å MgF_2 coating is not only to increase efficiency 15%, but also to eliminate the small TM anomalies. The boundary of the scalar domain is not sharp. Although under near normal incidence a safe bound for λ/d can be taken as equal to 0.2 for many current gratings, it can be extended up to 0.4 in the case of low modulations ($h/d \leqq 0.1$). When the incidence increases, TE and TM curves clearly separate for $\theta = 40^{\circ}$, due to the difference in reflectivities of the grating surface. On top of that, at high angles of incidence ($\theta > 80^{\circ}$), polarization effects of even the perfectly conducting grating appear [6.55]. Then the efficiencies become unpredictible using simple rules.

---

5 Locally, concave gratings, such as those used in Seya-Namioka mounting, can be considered as plane ones.

6.3.4 The Use of Gratings in XUV and X-Ray Regions (λ < 1000 Å)

The characteristic of this region is that the reflectance of any surface material being generally small (Fig.6.5), high efficiencies cannot be expected [6.68b]. Moreover, gratings are used such that the λ/d ratio is, in general, quite small. This means that the grating generates a large number of diffracted orders, and that the calculation of efficiencies often requires a large amount of computer time. One of the essential problems is that of choosing the best mounting for a given grating. Thanks to a grant from the C.N.E.S. (Centre National d'Etudes Spatiales, France), the authors had the opportunity to make a systematic study [6.69,70] for two types of profile forms (sinusoidal and echelette). Using mainly 3600 lines/mm gold gratings, they have considered three wavelengths frequently encountered in space optics (744, 584 and 304 Å).

In the case of echelette gratings, for each wavelength, the blaze angle α and the angle of incidence θ were varied in five degree steps over all their possible ranges. For each order n, the parameters corresponding to the maximum efficiencies have been noted. This led to the conclusion that these parameters always correspond to a configuration close to that of Fig.6.83, which we call RWF (Reflected Wave on Facet) mounting. In this configuration, which satisfies the classical blazing condition deduced from scalar theories, the incident and n^{th} diffracted beam are symmetrical with respect to the normal on the facet. Important blaze effects can be observed; of course, the total diffracted energy is low (< 20%), but it can be easily concentrated in one diffracted order (relative efficiencies typically being about 90%). According to the values of the wavelength and the order, the optimal angular deviation was found to be close to zero or as large as 110°. In the latter case, i.e., for large deviation, polarization effects are substantial [6.55]. In order to understand the superiority of the RWF mount, Fig.6.84 shows that, when one leaves the corresponding blazing condition, a rapid drop in efficiency occurs for both polarizations.

Since this RWF mount holds a privileged position in XUV, it may look particularly interesting to try to derive an approximate but simple formula which allows one to predict grating efficiencies for this mount. From a reasoning based on geometrical optics and developed in [6.70], a phenomenological formula was devised which predicts echelette grating efficiency e from surface reflectivity

$$e = R(\theta'') \cdot \tau, \quad \tau = \min\left(\frac{\cos\theta'}{\cos\theta}, \frac{\cos\theta}{\cos\theta'}\right) \quad . \tag{6.3}$$

$R(\theta'')$ denotes the Fresnel reflection under incidence θ''. The significance of angles θ, θ' and θ'' is made clear in Fig.6.83. This formula, tested against many electromagnetic computations, gives a simple but accurate tool for predicting grating efficiencies. Although it was devised to be exact in the limit $\lambda/d \to 0$, numerical calculations have shown that it still gives reliable results (within 20%

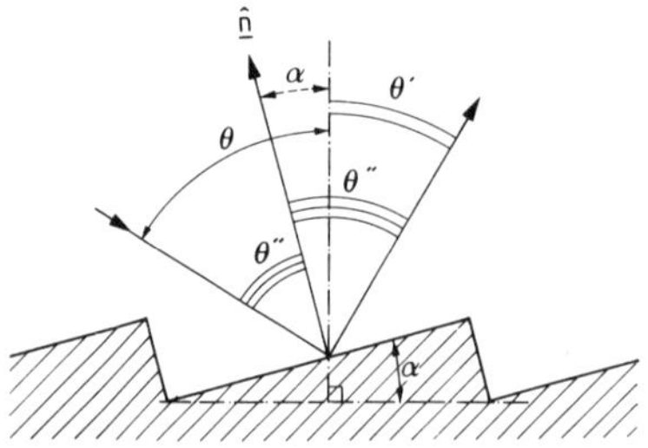

Fig. 6.83. The RWF mounting

Fig. 6.84. Variation of the efficiency of a gold grating in the +2 order at the vicinity of a RWF mount. The incidence θ is constant, as well as the wavelength λ, to avoid dispersion effects. The grating parameters are: $\alpha = 32.4^\circ$, $\theta = 12.4^\circ$, $\theta'' = 20^\circ$, $\theta' = -52.4^\circ$, $\lambda_0 = 700$ Å

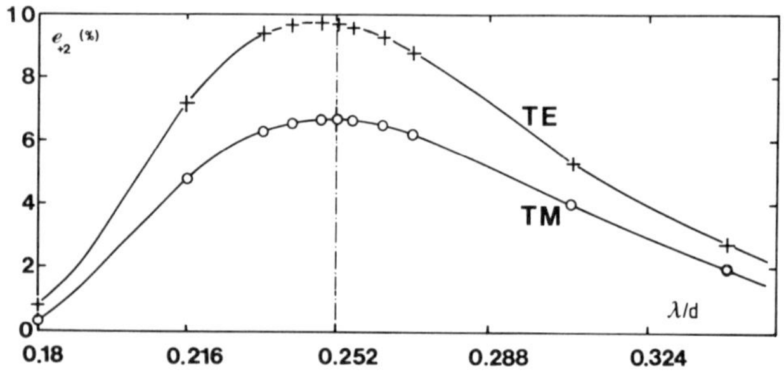

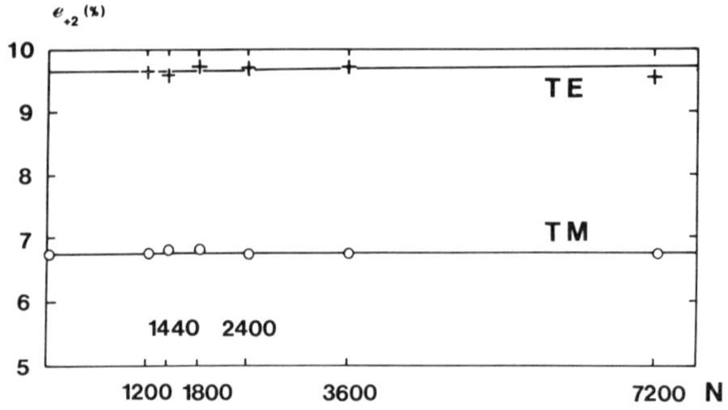

Fig. 6.85. Efficiency of a gold echelette grating in a RWF mount, as a function of the number N of lines per mm: $\alpha = 74.4^\circ$, $\theta = 4.44^\circ$, $\theta'' = 20^\circ$, $\lambda = 70$ nm. The selected order varies from 1(N=7200) to 6(N=1200)

accuracy) even when $\lambda/d = 1/2$. Figure 6.85 plots its predictions when one increases the number n of the spectral order which is selected, and simultaneously decreases the groove frequency N, in such a way to keep the nN product constant. The formula predicts unchanged efficiencies, while rigorous calculations gives only minor changes even for 7200 grooves/mm, for which, with the 700 Å wavelength used, the λ/d ratio reaches 0.504 !... .

For sinusoidal gratings, similar electromagnetic studies have been performed. The essential conclusion is that, in XUV, the sinusoidal is a notably worse performer than the echelette grating. For example, for $\lambda_0 = 744$ Å, if the groove depth lies between 0 and 1400 Å, it is not possible to obtain an efficiency in a given order superior to 10% whatever the angle of incidence may be. On the other hand, efficiencies larger than 20% can be obtained with echelette gratings.

Before ending this section, let us give some comments on polarization effects and efficiency behavior when, leaving the RWF condition, the λ/d ratio decreases and the incidence increases. When one abandons near normal incidence, differences between the TE and TM reflectances begin to appear for $\theta = 20^{\circ}$ and constitute the first cause of polarization effects, particularly important when $\theta = 40^{\circ}$ [6.55]. For near grazing incidence ($\theta > 60^{\circ}$) a specific grating phenomenon arises in that even a perfectly conducting grating already presents polarization effects. When the two causes both occur, which in common X-ray domain, grating efficiencies become difficult to predict with the aid of simple rules. Although the phenomenological formula (6.3) for RWF mounting still holds, the rules explained in Sect. 6.3.3 do not hold. No universal efficiency curve has been found. Even at very low λ/d ratio, the efficiency still strongly depends on incidence, groove depth, groove spacing, wavelength, and polarization. The only point that can be easily predicted is the optimum groove depth corresponding to a given blaze wavelength λ_B. For a sinusoidal profile, the near normal incidence formula $\lambda_B = 3.4$ h has simply to be turned into $\lambda_B = 3.4$ h cos (D/2), where D is the deviation between the incident beam and the desired spectral order. The same modification applies for ruled and lamellar gratings. Rigorous calculations have shown that such a formula predicts the right groove depth with only a few percent error till 80° incidence, and with 20% error at 88° [6.71]. In order to guide X-ray experimenters in the choice of a grating Fig.6.86 summarizes the results from many efficiency computations on a 1200 grooves/mm sinusoidal gold grating. It gives, on the ordinate axis, the optimal groove depth for a given wavelength and five typical grazing incidences. The figures near the curves indicate the efficiencies in natural light. The underlined ones represent, for a given wavelength, the maximum efficiencies found among the five incidences tested.

The effect of changing the groove frequency is smaller than the one predicted by scalar theories, from which the product resolution with luminosity is constant. The efficiency curves are improved when the groove spacing increases, but the improvement increases with the incidence. At 75°, the change in efficiency due to a multiplication by 4 of the groove spacing is quite small [6.71].

The optimization of grating geometry for ruled and lamellar gratings does not necessarily follow the same rules as under normal incidence. For ruled gratings, the apex angle has a negligible influence, due to the very small blaze angles used in this spectral domain. For lamellar gratings, the symmetrical profile ($c/d = 0.5$) no longer seems to be the best. Gratings with $c/d > 0.5$ behave better than those with $c/d < 0.5$. For given wavelength and incidence, the optimum c/d ratio seems to be close to the one for which the illuminated part of the bottom of the groove, determined by taking into account the shadow regions in both corners, has the same width as the plateau.

A final note has to be made concerning the influence of the groove shape. Let us recall first that for perfectly conducting gratings used in the scalar domain and near normal incidence, echelette gratings can reach 100% efficiency, sinusoidal gratings culminate at 34%, while lamellar gratings attain 40%. Things are

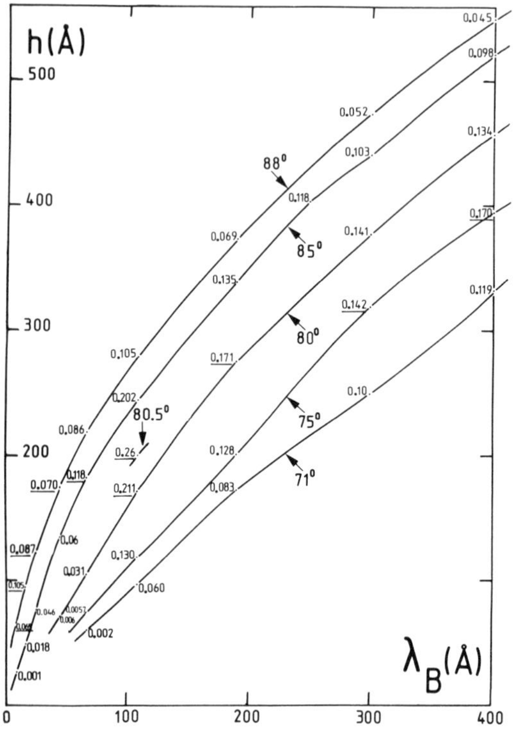

Fig. 6.86. Diagram showing optimal groove depth and corresponding efficiencies of a sinusoidal grating

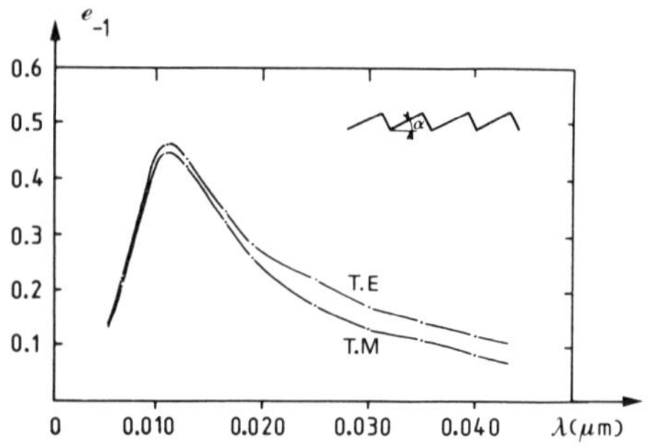

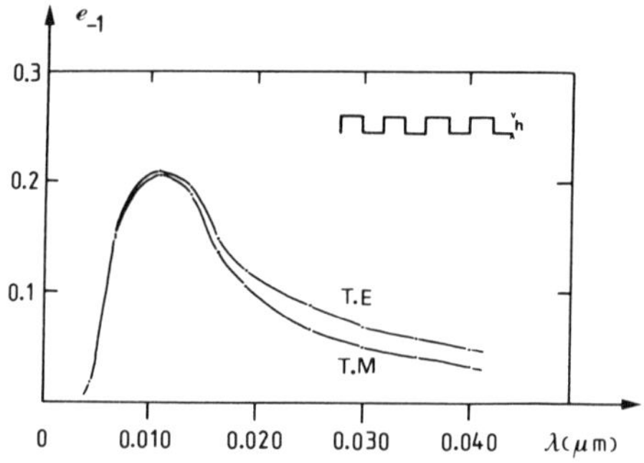

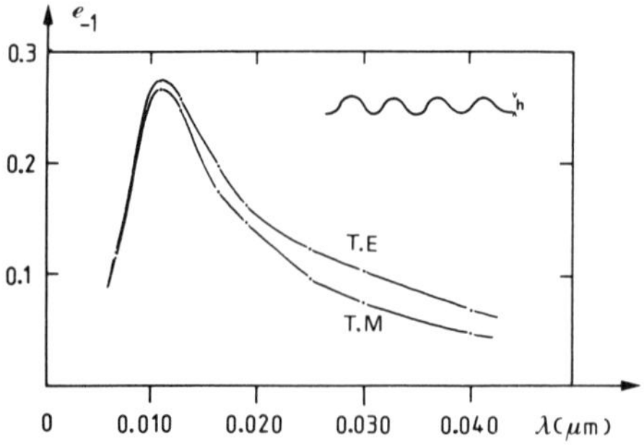

Fig. 6.87. Efficiency of a gold echelette grating in XUV domain, as a function of the wavelength: $\alpha = 1.624^{\circ}$

Fig. 6.88. Same as Fig.6.87 but for a sinusoidal grating: h = 302 Å

Fig. 6.89. Same as Fig.6.87, but for a symmetrical lamellar grating: h = 198 Å significantly different for gold gratings under grazing incidence. Figures 6.87-89 give, for a 600 groove/mm grating, the efficiency curves of an echelette, sinusoidal, and lamellar grating used under 85° incidence. All gratings have a groove depth optimized at λ = 109 Å. Sinusoidal and lamellar gratings seem to be close competitors, although this time a small advantage appears for the sinusoidal one. The echelette grating still wins the competition, but only takes a ratio of less than 2 on the sinusoidal grating. Thus grazing incidence improves the relative performances of sinusoidal gratings. A second difference is that, everything being kept constant, the nth order efficiency decreases much more slowly than for ruled and lamellar gratings when the number n is increased, a fact revealed both by electromagnetic computations and experiments [6.72].

6.3.5 Conical Diffraction Mountings

The reason for going to high angles of incidence is, of course, linked with the improvement of the reflectivity observed with a mirror. For a grating, however, an increase of the incidence concentrates more and more energy into the zero order, which is of no interest for spectroscopy. Consequently, the first-order efficiency decreases and at grazing incidence, culminates at very low values (see Fig. 6.86), despite the high reflectivity of the surface. The question which arises is whether a suitable mounting could concentrate most of the energy into a desired spectral order.

The mounting that we are going to describe differs from classical ones in that the incident beam propagates outside the cross-section plane of the grating; thus it is not perpendicular to the rulings. Under these conditions (see Sect.3.7), the diffracted beams lie on a cone whose axis is parallel to the ruling direction and whose half-angle is the angle between the incident beam and the ruling direction; hence the designation "conical diffraction". Among all the conical diffraction mountings, one of them, here called ITCD (invariance theorem in conical diffraction) mount, presents very interesting efficiency behavior. It is based on a theorem valid for perfectly conducting gratings that we briefly recall.

If the incident wave vector is varied in such a way that its projection on the cross-section plane of the grating is a fixed vector, them for a given n^{th} order:

- the projection on the cross-section plane of the diffracted wave vector also remains a fixed vector,
- the efficiency e_n is constant,

provided the incident wave polarization is fixed (see Sect. 3.7).

In order to keep its projection unchanged, the wave vector of the incident beam must move in the plane defined by the ruling direction and its initial position. Moreover, if ϕ is the angle between the incident beam and the cross-section plane, the wavelength λ has to be varied in such a way that k $\cos\phi$ be constant.

Let us suppose that we start from a particular in-plane mounting ($\phi = 0$), namely the -1 order Littrow mount, for which $\lambda = 2d \sin\theta$, i.e., $k = \pi/(d \sin\theta)$. For the ITCD mounting, we get

$$k \cos\phi = cte = \frac{\pi}{d \sin\theta}, \quad \text{or}$$

$$\lambda = 2d \sin\theta \cos\phi \ . \tag{6.4}$$

When ϕ and λ are varied in such a way as to keep this condition satisfied, the -1 order efficiency is kept constant and equal to the Littrow efficiency (which can reach 100% for an echelette grating if θ is equal to the blaze angle α !...). Of course, for real gratings, which are not perfectly conducting, the theorem does not hold rigorously. But experiments have shown that it holds almost perfectly in the infrared region, and that in the visible range, the efficiency drop is only in the ratio of metal reflectance. Thus, in these two domains, very high and constant efficiency conical diffraction mountings can be designed on this basis (see Sect. 6.4.2).

Recalling the low efficiencies that are obtained in X-ray and XUV regions, the question is to know whether the use of ITCD mountings can improve grating performances. Both theoretical studies [6.73-75] and experiments [6.76,42] allow us to answer in the affirmative. In order to prove its worth, let us first show that ITCD mountings are better than other conical diffraction mountings. Figure 6.90, taken from [6.74], shows theoretical efficiency curves as a function of the grazing angle $\psi = \pi/2 - \phi$, for an extremely off-plane mounting and two different wavelengths. The incident wave vector lies in a plane parallel to the grooves and perpendicular to the grating surface. Curves for zero, first and second orders are given, as well as the total energy reflected by the grating $\sum e_n$. From (6.4), the -1 order ITCD mounting condition which gives the highest efficiencies is satisfied if $\psi = 0.98^\circ$ when $\lambda = 0.83$ nm, and if $\psi = 1.57^\circ$ when $\lambda = 1.33$ nm. Both values perfectly agree with the location of the maxima given by electromagnetic computations. The same conclusion applies to the second order. Let us now demonstrate that ITCD mountings are also better that classical in-plane mountings. Figure 6.91, taken from [6.75], shows the comparison between the efficiency curve of a 3600 groove/mm, 5° blaze angle gold echelette grating used in the ITCD mount and the reflectivity[6] of the surface, both calculated for natural light. Contrary to what happens for in-plane mountings, the energy is almost all concentrated into the -1 order instead of the zero order. The same curve for a higher modulated grating, namely 14° blaze angle (Fig.6.92) shows a small reduction in efficiency compared to the reflectivity. In any case, efficiencies are still much higher than those found with

6 As given by the Fresnel formula for a plane parallel to the large facet and illuminated under the incidence ϕ.

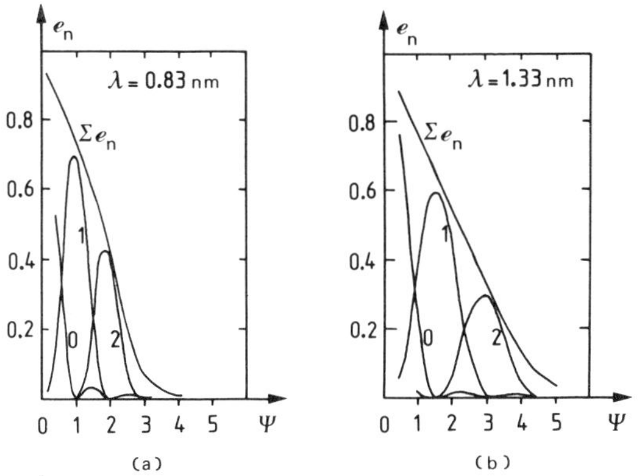

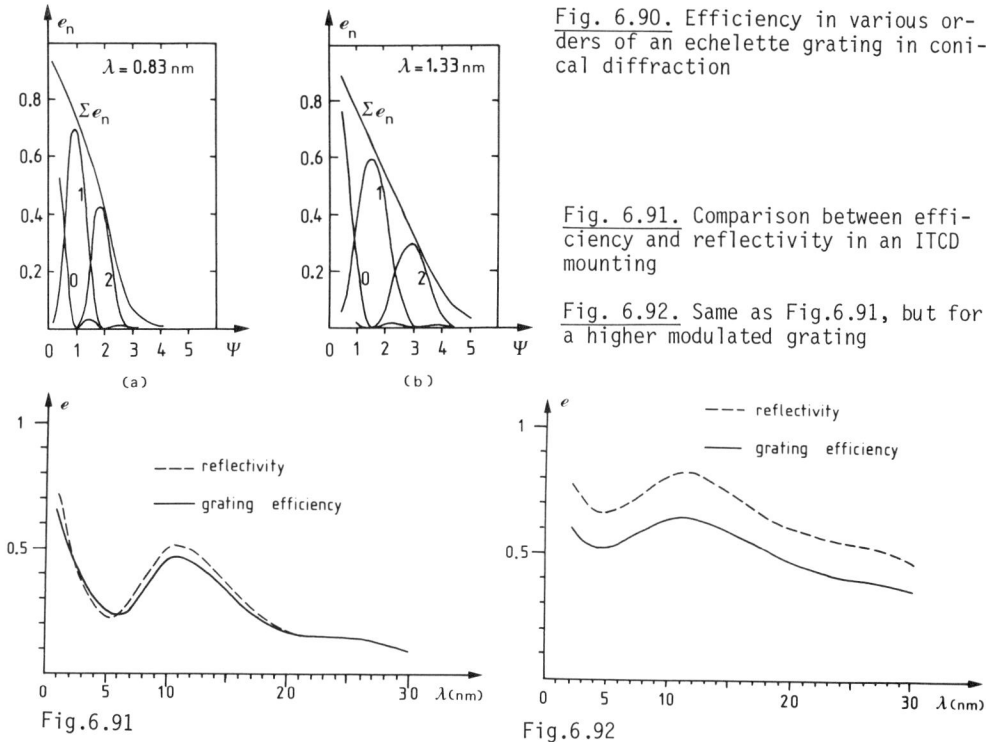

Fig. 6.90. Efficiency in various orders of an echelette grating in conical diffraction

Fig. 6.91. Comparison between efficiency and reflectivity in an ITCD mounting

Fig. 6.92. Same as Fig.6.91, but for a higher modulated grating

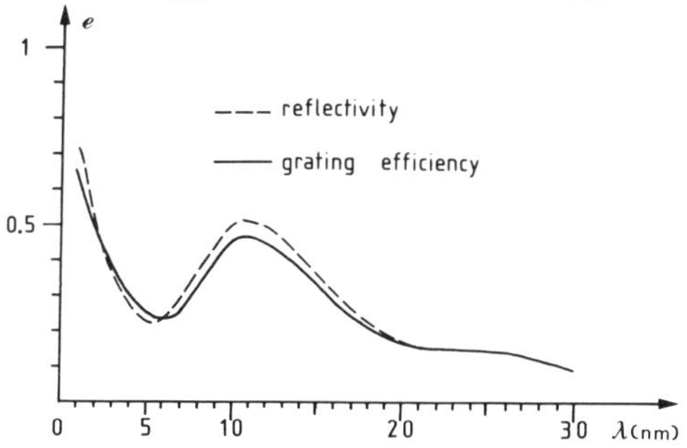

Fig.6.91

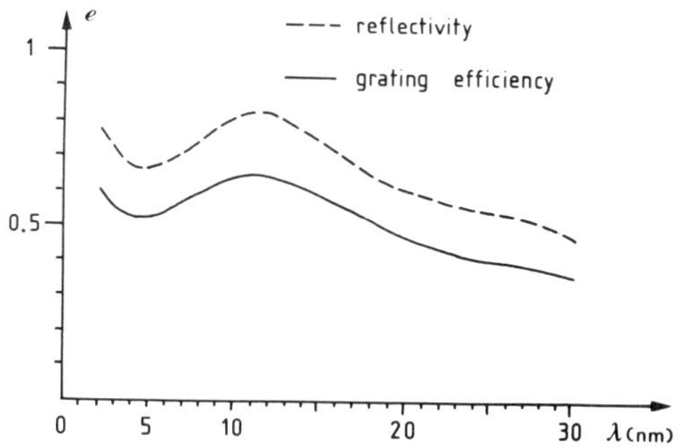

Fig.6.92

classical mountings. A more detailed study was published in [6.75]. As a simple rule of thumb, we can say that the efficieny curve for ITCD mounting can be derived from the reflectivity curve simply by multiplying by constant C, which is close to unity for low modulated gratings, and remains greater than 0.7 for high modulated ones.

The reason for the superiority of ITCD mountings can be easily explained from the phenomonological formula (6.3) previously established, which turns out to be still valid for off-plane mountings [6.76]. For in-plane RWF mounts, the ratio $\tau = \min[\cos\theta/\cos\theta', \cos\theta'/\cos\theta]$ becomes very low in grazing incidence, hence the reduction in efficiency. On the other hand, it is always equal to unity in ITCD mounting and the predicted efficiencies are equal to the corresponding reflectivities.

To conclude, let us emphasize that ITCD mountings are not restricted to ruled gratings, but apply to any kind of profile. Of course, for sinusoidal gratings, the incident and diffracted beams cannot lie in the plane of the small facet, since there is no facet. In order to realize a high efficiency ITCD mounting, one must first start from an in-plane Littrow mount and choose the incidence θ_B which gives the highest efficiency. This can easily be done looking at the corresponding efficiency curves. Then the incident wave vector $\underline{k}$ is moved in the plane parallel to the rulings and containing the initial wave vector $\underline{k}_0$ which makes the angle

θ_B with the grating normal, while the wavelength is varied according (6.4). At wavelength 2.36 nm, 27 % efficiency was obtained with a sinusoidal profile in this mounting, which has to be compared to the 42% found for a ruled grating and to the 24% found for the lamellar one [6.75]. In off-plane mountings too, grating incidence favors sinusoidal gratings, compared to the other types. However, for all types, efficiencies are much higher than for in-plane mountings.

6.4 Some Particular Applications

Thanks to the curves given in previous sections and using, if necessary, complementary data rapidly obtained from the computer, it is now possible to answer many questions in which experimenters are greatly interested. Here are a few examples.

6.4.1 Simultaneous Blazing in Both Polarizations

The possibility of obtaining a 100% efficiency in both polarization (TE and TM) has been pointed out by HESSEL et al. [6.77], at least, for perfectly conducting gratings. They suggested using a rectangular-groove grating placed in a -1 order Littrow mount. Of course, the parameters c/d and h/d which characterize such a grating (Fig.1.8) have to be adequately chosen. Additional calculated performance data, including the effect of finite conductivity, were provided by ROUMIGUIERES et al. [6.64]. Figures 6.93,94, taken from the latter paper, summarized the results, respectively, for TE and TM polarization. In these figures, each curve is associated with a given value of c/d. In fact, when c/d and θ are fixed, several depth h_B exist that give a 100% efficiency. Only the smallest has been plotted, except for c/d = 0.667, a value for which two curves are given in TE polarization (Fig.6.93). One has only to superimpose Figs.6.93,94 to determine the points of intersection of the two curves corresponding to the same value of the aspect ratio c/d. Such intersection points P_n have been plotted in Fig.6.94; their existence clearly establishes the feasibility of simultaneous perfect blazing. Each of them is associated with a grating G_n (completely defined by the ratios c/d and h/d) and with a value of the incidence θ. For each grating G_n. the efficiency curve (-1 order Littrow mount) can be found in the original publications [6.77,64]. Such curves, which have been confirmed by experiments in microwaves [6.78], can be used to optimize the profile with a special utilization in view: perfect blazing, filtering, mirror for tunable laser... However we must emphasize the fact that they have been obtained with the assumption of a perfectly conducting material. Consequently, they have to be used cautiously for spectroscopic purpose in the visible region, as shown by data published by ROUMIGUIERES et al. [6.64]. Further calculations seem to be necessary to know how a metallic grating G_n works when one takes into account its finite conductivity. Such calculations can be easily performed, it is

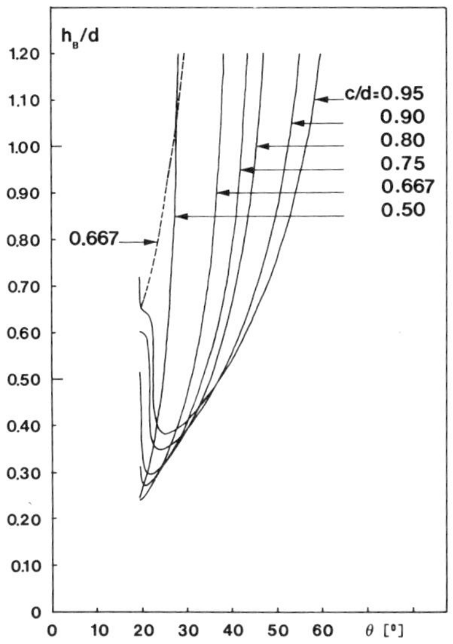

Fig. 6.93. Relative depth h_B/d of a 100% lamellar grating vs angle of incidence in a -1 Littrow mount and for the TE polarization. For c/d = 0.667, the solid line corresponds to the smallest depth, the dotted line to the other one

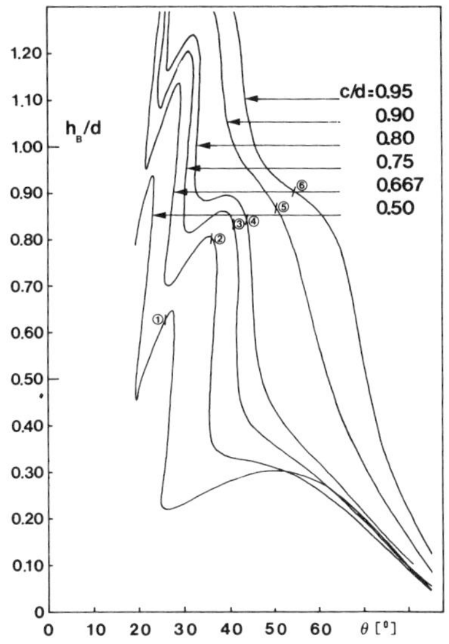

Fig. 6.94. Same as Fig.6.93, but for TM polarization. Only the smallest value of h_B/d has been plotted. An intersection point called P_n in the text is denoted by n

only a question of computation time. While waiting for such complementary data, the blazing properties of lamellar grating (which involve the vanishing of the 0 order efficiency) have already been considered with interest at ratio frequencies for which there is no difficulty with finite conductivity. Such surface corrugations have been proposed as a means of eliminating interfering specular reflection from metallic building surfaces such as airport hangars [6.78]. It only goes to show that electromagnetic theories of gratings are useful in numerous fields... including airport business.

The simultaneous and total blazing is not a prerogative of lamellar gratings. It can be also obtained with triangular-groove gratings, as shown by CHEO et al. [6.79]: several groove profiles can be used but they do not include the echelette grating for which the apex angle is 90°.

6.4.2 Spectrometers with Constant Efficiency

For spectroscopic purposes the efficiency curves must be smooth and free of anomalies. For many applications the ideal mountings would be those for which the efficiency is high and constant over as large as possible a range of wavelength. Such mountings do exist if we accept to work in conical diffraction (see Sect.1.2.13, 3.7 and 6.3.5) as explained by MAYSTRE and PETIT [6.80]. They have already been proposed as high efficiency mounts twenty years ago by WHITE in a US patent (N° 3069966, 1962) to which the interested experimenter can refer. For reasons of space, we can only give here the basic ideas.

It is well known that the efficiency in the -1 order is very close to unity for both fundamental cases of polarization provided that, in Littrow mounting, the incident wave vector $\underline{k}^i$ is perpendicular to the large facet of an echelette grating. The invariance theorem (see the end of Sect. 3.7) allows us to state the following proposition:

When, as assumed in Fig.6.95, the incident wave vector k $\hat{\underline{u}}_1$ remains parallel to the small facet and when moreover $\lambda/d = 2 \sin\alpha \cos\phi$, the -1 order diffracted wave can be deduced from the incident wave by a specular reflection on the large facet. Then, for this wavelength, the plane of the large facet works as a plane mirror, whose reflectance is very close to unity and does not depend on the wavelength.

We are now ready to propose the basic scheme of an ideal spectrometer (Fig. 6.96). Lenses L_1 and L_2 and opaque screen E are fixed. S is a polychromatic point source situated at the focus of lens L_1. Mirror M and grating R are fixed to the horizontal stand P and the large facet of R is vertical. When P rotates about the vertical axis, ϕ varies and the hole T is illuminated by different wavelengths. Some modifications are necessary [6.80] if one wishes to use a slit source rather than a point source because, in the present mounting, the image of the entrance slit through L_1, R, M and L_2 rotates when ϕ (i.e., λ) varies. It must be noticed that the value of the angle between the blank of R and the stand P is not critical; a small departure from the ideal value slightly changes the efficiency which remains constant when ϕ is varied. Strictly speaking, the invariance theorem is only valid for perfectly conducting grating. However MAYSTRE has reported having verified the invariance of the efficiency within six percent, using a commercial echelette grating (see table 6.1). Nevertheless, to our knowledge the type of mounting described in this section is not yet often used; regrettably because from many points of view its properties look indeed excellent.

6.4.3 Grating Bandpass Filter

It is known that gratings can be used as wavelength filters. This is clearly shown in Fig. 6.97, taken from [6.31], which gives for a fixed incidence ($\theta = 52^o$) the efficiency curves of a 38° blaze angle echelette grating illuminated by a plane wave in TM polarization. Looking at these curves, it seems possible to propose a grating arrangement [6.81] which, for polarized light, works as a bandpass filter whose transmission is very close to unity and whose bandwidth is tunable. The principle is explained in Fig.6.98 after the two curves of Fig.6.97 have been idealized, that is to say, replaced by the graphs of two step functions with discontinuity at $\lambda = \lambda_c$. Figure 6.98a gives the basic scheme. Figure 6.98b shows the efficiency of

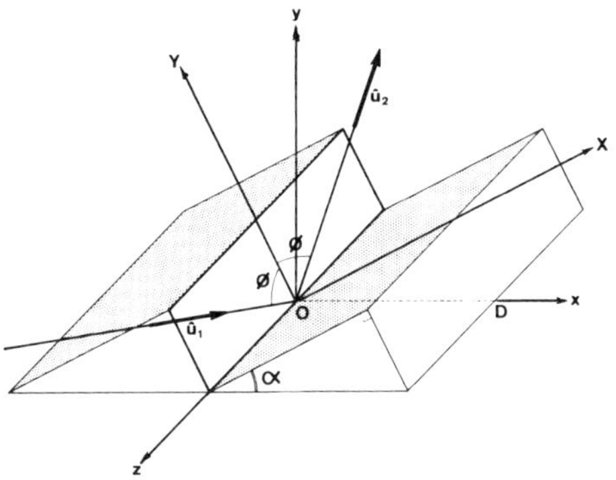

Fig. 6.95. An interesting ITCD configuration. u_1 and u_2 are, respectively, the unit vectors of the incident and -1 order diffracted directions. They lie in the zOY plane, the plane of the small facet

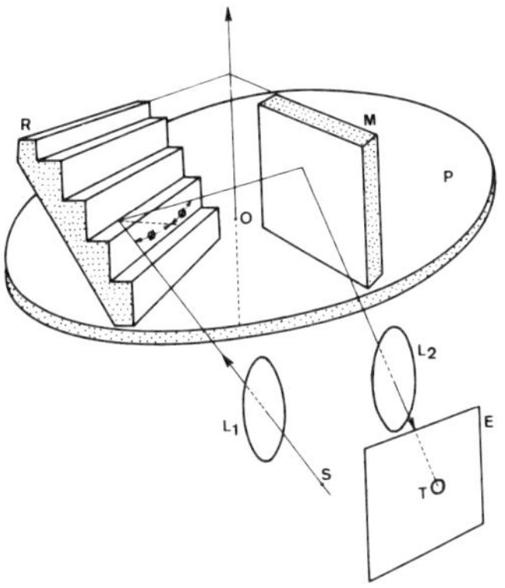

Fig. 6.96. Basic scheme of a spectrometer whose transmission does not depend on the wavelength

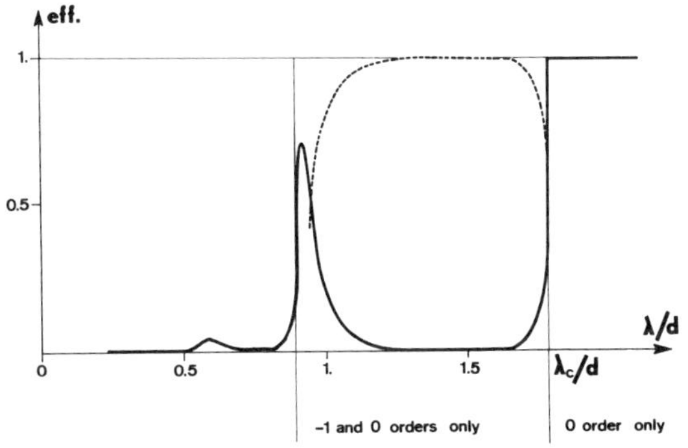

Fig. 6.97. Efficiencies for a fixed incidence and for TM polarization of a perfectly conducting echelette grating: $\lambda_c/d = 1 + \sin\theta$

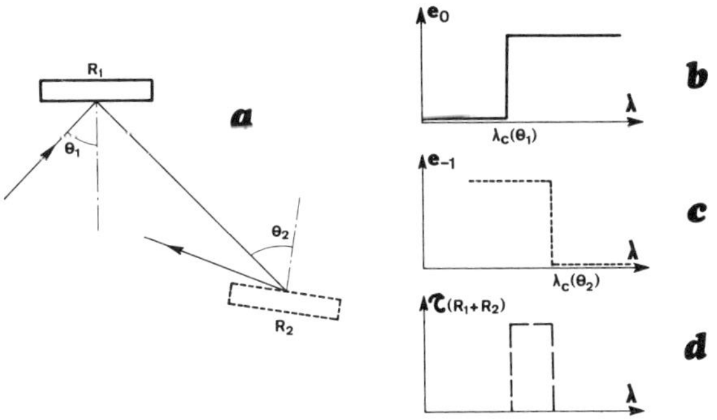

Fig. 6.98. Basic scheme of the bandpass filter

grating R_1 working in the zero order and Fig.6.98c the efficiency of R_2 in the -1 order. Finally the transmission of $R_1 + R_2$ is given in Fig.6.98d. Clearly, we control the bandwidth by choosing θ_1 and θ_2. Unfortunately the direction of the transmitted wave is depending on λ. This angular dispersion, which is of course a disadvantage, can be removed using a slightly more complicated mounting which needs four gratings [6.81]. Let us not forget that, again in this section, gratings have been supposed to be perfectly conducting, which is practically true in the infrared and a fortiori in the microwave region.

6.4.4 Reflection Grating Polarizer for the Infrared

The polarizing properties of the rectangular groove grating were reported ten years ago by WIRGIN and DELEUIL [6.13]. More recently, ROUMIGUIERES proposed, on the basis of theoretical computations, the use of such a grating as a polarizer at infrared wavelengths [6.82]. The experimental confirmation of his predictions has been given by KNOP [6.61]. A convenient grating structure is obtained by choosing the width c and depth h of the rectangular grooves such that $c/d = 0.5$ and $h/d = 0.275$. The zero-order efficiency is strongly polarization dependent when the grating is illuminated by a plane wave under the incidence $\theta = 50^\circ$, the incident wave vector being orthogonal to the groove (Fig.6.99). In the range of wavelengths λ, where $1.1 < \lambda/d < 1.7$, over 80% of the light polarized parallel to the grooves is reflected, and less than 5% for the other polarization. Moreover, the choice of parameters c, h, θ is not critical : deviations of about 10% from their optimum value can be tolerated [6.82]. For infrared wavelengths, the adequate grating profile can be easily realized using photoresist technology and "perfectly conducting metals" are available.

6.4.5 Transmission Gratings as Masks in Photolithography

Figure 6.100 represents a metallic transmission grating G (a grid made with rods of rectangular cross section) placed above a stack of two layers L_1 and L_2 lying on a substrate S. The grating itself rests on a glass plate L whose thickness is supposed to be infinite. Regions S, L_1 and L_2 are, respectively, filled with silicon, silicon oxide and photoresist. An air gap with thickness t separates the grating from the photoresist. The structure is illuminated by a plane wave (whose wave vector is represented by an arrow) in order to transfer the grating onto the silicon surface using photolithography techniques. The goal is to obtain in L_2 a highly

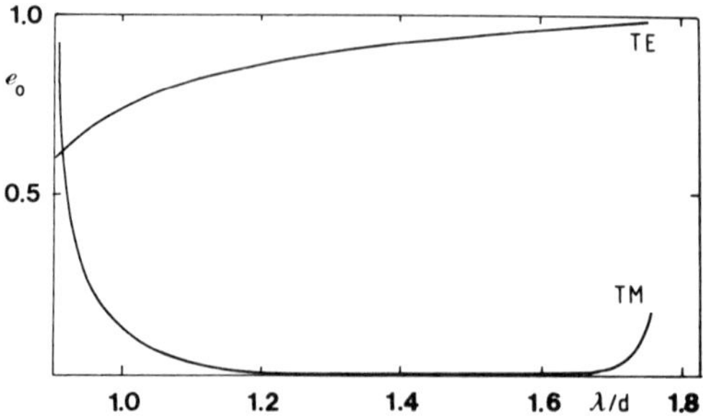

Fig. 6.99. Zero efficiency of a lamellar grating for a fixed incidence

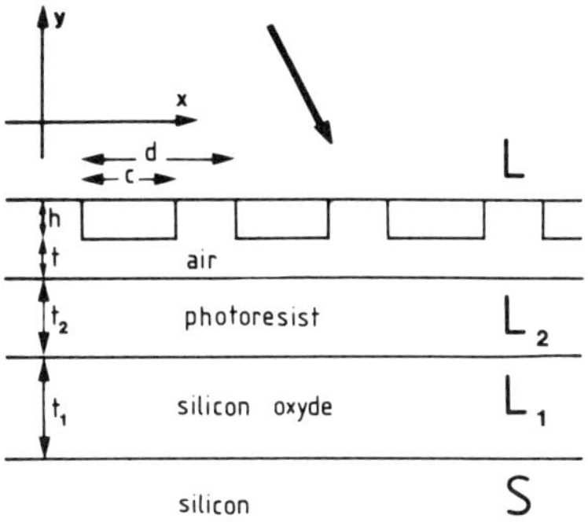

Fig. 6.100. Grating used as a mask in photolithography

contrasted distribution of intensity, with light under the holes (intervals between tow consecutive metallic rods) and shadow under the metal, as predicted by geometrical optics. This goal is not easily reached if, as is frequently the case nowadays due to circuit miniaturization, c, d and all the thicknesses are of the order of the wavelength. Then, neither geometrical optics nor physical optics are able to predict what happens under the grating. If only to take into account polarization effects which appear important, it is necessary to resort to Maxwell's equations. These equations have been solved by ROUMIGUIERES et al. [6.83,84] assuming the grating metal to be perfectly conducting and using modal methods described in Sect.1.2.12. Some numerical results are given in the quoted papers. Complementary data can be found in ROUMIGUIERES's thesis [6.85], where several electromagnetic field maps[7] are given in order to show the influence of incidence, polarization, wavelength, thickness of the layers and the air gap, aspect ratio c/d, rod thickness h... . In outline, the mask duplications is easy as long as the groove spacing d exceeds 1 μm. Theoretical problems linked with the diffraction phenomenon appear under this limit. The apparition of light under the metal may produce frequence doubling or even prevent the reproduction. Also the influence of a bad contact between the mask and the photoresist is investiqated by varying parameter t.

Let us emphasize that, in order to obtain a good duplication, the field maps in the photoresist must be independent of the ordinate y. Recent work has shown that this requirement can be satisfied for a wide class of structures using suitable illumination conditions (French patent N° 7922554).

By removing the substrate and the layers, the diffraction by the lamellar transmission grating (used as a grid) has been studied: influence of h on the well known frequency filtering effect in TE polarization, Talbot images, coupling effect between two consecutive "holes"... .

---

7 So called are the graph of the square modulus of the electric field as a function of x for a given ordinate y.

The theoretical studies mentioned in this section may be used as a guide in the manufacture of integrated circuits by contact masking techniques. The mask is then usually made with chromium whose low reflectivity (less than 50 %) allows us to question the use of the infinite conductivity model. Some work taking into account the finite conductivity of the mask [6.86] has led to field maps similar to those obtained by ROUMIGUIERES.

6.4.6 Gratings Used as Beam Sampling Mirrors for High Power Lasers

The use of high power gas dynamic lasers often requires continuous monitoring of wavefront phase and shape. Power levels are so high that no ordinary beam splitter can survive. The first alternative is to use high grade water cooled mirrors that reflect 98-99 % of the incoming light, but are modified as shallow diffraction gratings to diffract simultaneously 10^{-4} to 10^{-2} of that light toward suitable wavefront analyzers. The second solution is to put a stack of dielectric layers on top of a shallow grating, designed to minimize the energy absorption in the metal. On top of that, in many cases, the intensity information must be preserved, which requires minimal polarization effects.

Concerning the former alternative, the groove shape must be chosen in such a way as to be produced in solid metal. Lamellar gratings fulfill the requirements [6.87]. Since the wavelength is often 10.6 μm, a groove frequency of 80/mm is suitable to generate only two diffracted orders. The variables of interest are then limited

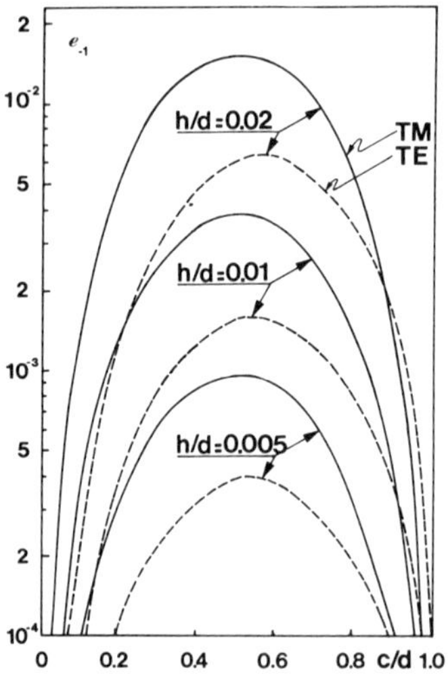

Fig. 6.101. Efficiency in the -1 order of a perfectly conducting lamellar grating as a function of c/d for various groove depths and both polarizations

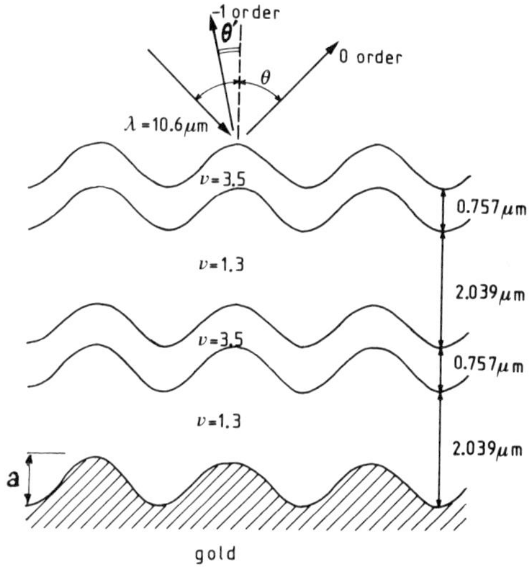

Fig. 6.102. Examples of dielectric coated gratings used as beam sampling mirrors. The bare gold grating can be covered by two or four dielectric films. The optical thickness of the dielectric films are equal to λ/4; d = 7.76 μm; λ = 10.6 μm

to two. The depth-to-period ratio h/d exercices primary control over efficiency; while c/d, the groove width-to-period ratio, controls efficiency as well as polarization. For an incidence of 45° and a λ/d ratio equal to 0.848, Fig.6.101 shows the TE and TM efficiencies as functions of c/d, for tree values of h/d. Since efficiency varies directly with $(h/d)^2$ for such shallow gratings, it is simple to extrapolate to other values. The solution for reducing polarization effects is to increase the c/d ratio to the vicinity of 0.9. The desired first-order efficiency gives the convenient h/d ratio.

Concerning the use of dielectric coated gratings, two different ways can be considered. As suggested by CHARLTON [6.88] the grating can be buried in a dielectric, on which a stack of plane players is superimposed. It is then possible to hope to simultaneously minimize energy absorption and polarization effects [6.87]. The other way, explored in our lab. [6.89], is to use a stack of two or four modulated dielectric layers, which follow the grating profile, and have alternately low and high refractive indices (Fig.6.102). Table 6.4 shows the energy E_d reflected by a metallic plane covered by the same stack of layers as the grating, the total energy E_t diffracted by the grating, and the grating -1 order efficencies e_{-1} for TE and TM polarizations.

For TM polarization, very interesting results are obtained. The grating absorbs the same amount as the metallic plane surface coated by dielectric films. The absorbed energy is reduced by a factor 10 by using four dielectric coatings. On the other hand, the introduction of dielectric films worsens the absorption for TE polarization. In one case, the computation could not be carried out because of the occurrence of an anomaly due to the propagation of guided waves in the dielectric films. The last columns show that the anomaly disappears if one modifies the angle of incidence θ, at least if two dielectric layers are used.

If the preceding results show how to minimize the energy absorption, they do not minimize polarization effects. A simultaneous reduction of both effects seems to necessitate the use of off-plane mountings, i.e., conical diffraction mountings.

The general conclusion of the use of dielectric overcoatings is that the best as well as the worst can be produced. Due to the introduction of new grating anom-

Table 6.4

	TM polarization $\theta = 60^{\circ}$, $\theta' = 30^{\circ}$, $a = 0.15$			TE polarization $\theta = 60^{\circ}$, $\theta' = 30^{\circ}$, $a = 0.5$			TE polarization $\theta = 50^{\circ}$, $\theta' = 36.9^{\circ}$, $a = 0.5$	
	E_d	E_t	e_{-1}	E_d	E_t	e_{-1}	E_t	e_{-1}
Bare grating	0.95	0.95	0.0089	0.987	0.987	0.0092	0.984	0.0108
Two films coated grating	0.987	0.987	0.0097	0.9988	0.923	0.022	0.9975	0.0088
Four films coated grating	0.9958	0.9953	0.0096	0.99987	unstable	unstable	0.78	0.21

alies, linked with the existence of guided waves in the films (see Chap.5), a complete destruction of the efficiency properties of a grating can occur. Outside such anomalies, the effect of the dielectric stack on top of a grating is about the same as on a mirror. Since no simple rule allows one to predict these anomalies, the study of dielectric coated gratings must necessarily be made with the electromagnetic theory.

6.4.7 Gratings as Wavelength Selectors in Tunable Lasers

A way to obtain a tunable laser is to place the grating at one of the ends of the laser cavity. This grating generally works in -1 order Littrow mounting and the selected wavelength is obtained by rotating the grating. The efficiency of such a grating must be as high as possible in a given range of wavelengths. The 35° blaze angle echelette grating used in TM polarization can reach this scope, as noted in our first systematic study of grating efficiency [6.90]: assuming a perfect conductivity, the efficiency exceeds 0.95 in a range of wavelength whose width is of the order of the groove spacing. But, for this type of application, the grating efficiency is of crucial importance. This is the reason why, in a recent collaboration with Jobin Yvon Inc., we have tried to improve this efficiency by superimposing stacks of dielectric layers on top of the gratings, as described in the preceding section. Figure 6.103, taken from [6.67], show the theoretical efficiency curves of a 3000 lines/mm, bare sinusoidal aluminum grating, having a groove depth of 0.12 μm in order to present a blaze effect at λ = 0.59 μm for which the efficiency reaches 87%. The second curve shows the efficiency of the same grating except a groove depth equal to 0.105 μm and two dielectric overcoatings also chosen to optimize e_{-1} for the same wavelength. The efficiency then culminates at 95% and its improvement extends from 0.52 μm to 0.63 μm. In order to put these predictions to the test, Jobin Yvon Inc. has produced a grating satisfying the abovementioned specifications. Due to problems linked with the high groove frequency, which resulted in a shift of the blaze wavelength, the agreement between theory and experiment was only qualitative. Nevertheless, the 77% efficiency of the bare grating at 0.63 μm was enhanced up to 84% with a stack of the two dielectric layers theoretically determined. It reached 87% with the superimposing of two identical dielectric stacks on the grating. Keeping in mind the experimental difficulties, this result can be considered very satisfactory.

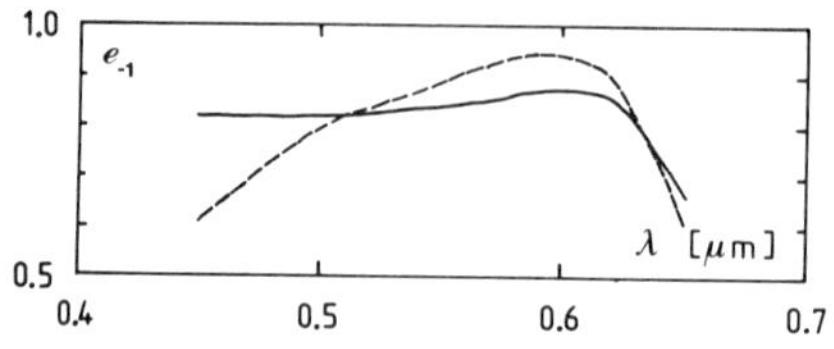

Fig. 6.103. Comparison between the efficiencies of a bare grating and those of a grating successively coated with a 0.105 μm layer of MgF_2 and a 0.07 μm layer of TiO_2: (——) bare grating; (----) coated grating;

In another configuration, the grating at the end of the cavity is fixed. It works at constant grazing incidence, in order to get a high dispersion. The -1 order is reflected towards the grating by a mirror which can be rotated to select the wavelength. The Littrow condition does not apply and the problem is to know whether it is possible to optimize the efficiencies by a correct choice of groove depth and groove frequencies. A numerical study conducted at the demand of Jobin Yvon Inc. has given 80 % efficiencies at $\lambda = 0.546$ μm under 75° of incidence for a 2400 lines/mm sinusoidal grating with h/d = 0.364.

6.4.8 Transmission Dielectric Gratings Used as Color Filters

As recently pointed out by KNOP [6.91] dielectric grating can be found with zero-order transmitted efficiency which are strongly wavelength dependent. Such gratings can replace an ordinary color filter and they have indeed been used in on-axis projection systems, which transmit the zero-order diffracted light only. They form the basis of the zero-order diffraction (ZOD) system [6.91,92] for recording color image as embossable surface relief structures in transparant plastic media. Beautiful ZOD images have been fabricated by R.C.A. laboratories in Zurich using techniques describided in [6.92]. Here we only want to deal with the associated grating problem. Because fine pitch gratings are required, with groove spacing comparable to the wavelength of light, the scalar diffraction theory used in the first reports on ZOD images [6.92] cannot provide us with quantitative results. Consequently a rigorous electromagnetic theory has been published by KNOP for the particular but useful case of deep lamellar gratings [6.93]. For this particular geometry, the numerical resolution of a differential system (see Sect.1.3.4) can be avoided because the matrix A does not depend on y. Let us emphasize that for *any* profile the differential method, as outline in Sect.1.3.4 and developed in Chap. 4, can be used, in principle at least. Only numerical difficulties linked with the great value of the groove depth to groove spacing ratio h/d may arise. Some computation trials, especially performed for this book, indicate that indeed our programs, initially devised for commercial gratings for which h/d does not exceed 0.5, seem at the limit of their possibilities. Nevertheless, they give data whose accuracy is probably about one percent and, no doubt, they could be adapted to the systematic study of gratings used in a ZOD system or in any recently proposed applications [6.94] ; it is essentially a question of FORTRAN programming. Figure 6.104 gives some efficiency curves for two dielectric gratings, having the same groove depth h = 1.56 μm, used in normal incidence. The optical index ν_2 is supposed to be 1.5 whatever the wavelength. The lamellar grating in Fig.6.104 is one of these already studied by KNOP; the associated data have to be compared with those given in Fig.3 of the quoted paper [6.93] and the agreement is excellent; the sinusoidal profile (Fig.6.104) do no show interesting properties for ZOD slides: The filtering effect is rather weak in the zero order and an important part of the diffracted energy is

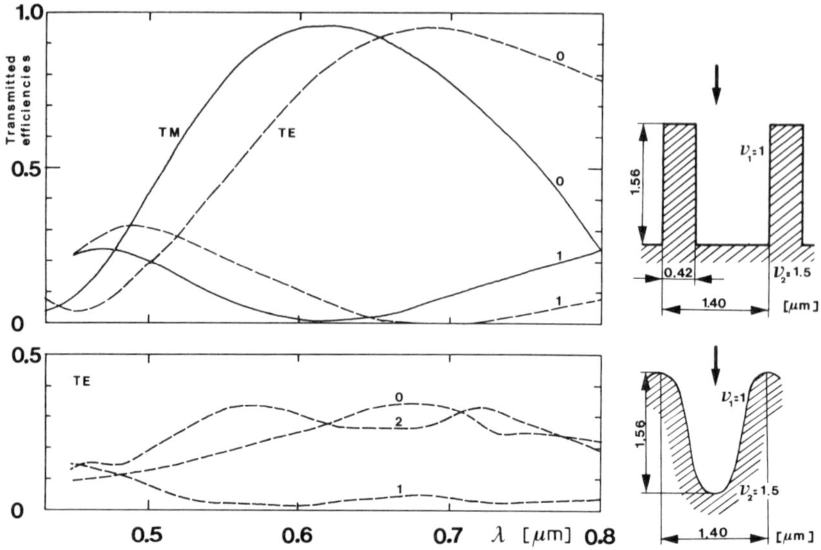

Fig. 6.104. Some efficiency curves for two deep dielectric gratings having a groove depth of 1.56 μm. The numbers refer to the spectral orders. (----) TE; (——) TM

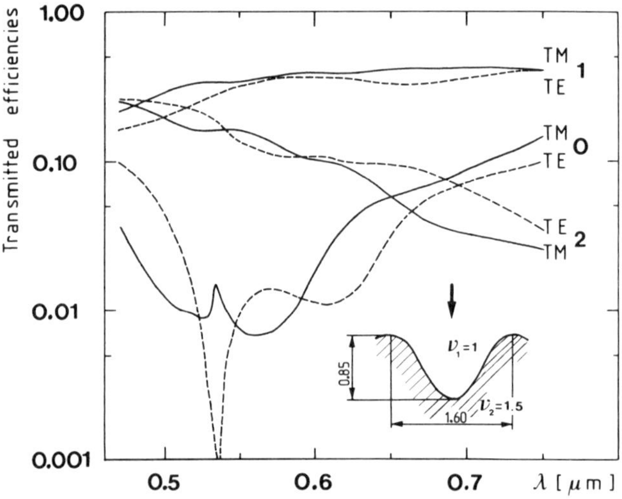

Fig. 6.105. Some efficiency curves for a sinusoidal grating in normal incidence. Note the use of a logarithmic scale and the Rayleigh anomaly for λ = 1.6/3

found in higher orders. Let us recall that, in ZOD techniques, sinusoidal gratings are used to produce black in zero order, i.e., to diffract substantially all visible light into the first and higher orders [6.92]. In fact, when the groove depth is 0.85 μm the simple scalar wave theory (Kirchhoff approximation) predicts a low efficiency in all the visible range. The electromagnetic theory leads to the same conclusion as shown in Fig.6.105. It is perhaps worth saying that, in Fig.6.104 as well as in Fig.6.105, the average computation time for one point (one wavelength value) is of the order of a second on the CDC 7600 computer.

Concluding Remarks

We hope that experimenters will find valuable data in this chapter and, maybe, a few additional ideas for the better use of gratings. All the practical problems, of course, have not been treated, but, with the computer codes available in our laboratory, most grating problems can probably be analyzed. We have been mainly concerned only with metal gratings, and dielectric gratings have been effectively neglected. This is not because the theoretical study of dielectric gratings presents specific difficulties, but because, until very recently, no electromagnetic study of them was required.

References

6.1 G.W. Stroke: "Diffraction gratings", in *Handbuch der physik*, Vol. 29 (Springer, Berlin, Heidelberg, New York 1967)
6.2 G.R. Harrison: Appl. Opt. 12, 2039-2049 (1973)
6.3 E.W. Palmer, M.C. Hutley, A. Franks, J.F. Verrill: Rep. Prog. Phys. 38, 975-1048 (1975)
6.4 R.P. Madden, J. Strong: *Concepts of Classical Optics* (Freeman, San Francisco 1958) p. 605
6.5 R. Petit: Rev. Opt. 45, 353-370 (1966)
6.6 W.C. Meecham, C.W. Peters: J. Appl. Phys. 28, 216 (1957)
6.7 J.H. Rohrbaugh, C. Pine, W. Zoellner, R.D. Hatcher: J. Opt. Soc. Am. 48, 710 (1958)
6.8 V.I. Aksenov: Radiotekhn. Electron. 5, 782 (1960)
6.9 T. Sueta: J. Inst. Electr. Comm. Engrs. Jpn. 42, 677 (1959)
6.10 G.W. Stroke: Rev. Opt. 39, 291 (1960)
6.11 C.H. Palmer, F.C. Evering, F.M. Nelson: Appl. Opt. 4, 1271 (1965)
6.12 R. Deleuil: Opt. Acta 16, 23 (1969)
6.13 A. Wirgin, R. Deleuil: J. Opt. Soc. Am. 59, 1348 (1969)
6.14 P. Bousquet, L. Capella, A. Fornier, J. Gonella: Appl. Opt. 8, 1229-1233 (1969)
6.15 J.M. Bennet: J. Phys. E 2, 816 (1969)
6.16 J.P. Chauvineau, L. Constanciel, A. Marraud, R. Petit: Rev. Opt. 46, 417-442 (1967)
6.17 R.G. Brandes, R.K. Curran: Appl. Opt. 10, 2101-2103 (1971)
6.18 D. Maystre: Thesis A.O. 9545, Université de Marseille, France (1974)
6.19 H. Volkmann: Optik 21, 385 (1964)
6.20 J.F. Verrill: J. Phys. E 6, 1199-1201 (1973)
6.21 M.C. Hutley: Opt. Acta 20, 607-624 (1973)
6.22 A. Roger, D. Maystre: J. Opt. Soc. Am. (to be published)
6.23 R.W. Hunter: Private communication
6.24 *American Institute of Physics Handbook*, 2nd ed. (McGraw-Hill, New York 1963) pp. 6-107
6.25 S. Robin: Private communication
6.26 A.P. Lukirskii, E.P. Savinov, O.A. Ershov, Yu.F. Shepelev: Opt. Spektrosk. 16, 310 (1964) [Opt. Sectrosc. 16, 168 (1964)]
6.27 G.B. Irani, T. Huen, F. Wooten: Appl. Opt. 61, 128 (1971)
6.28 E.G. Loewen, M. Nevière, D. Maystre: Appl. Opt. 16, 2711-2721 (1977)
6.29 E.G. Loewen, D. Maystre, R.C. McPhedran, I.J. Wilson: Jpn. J. Appl. Phys. Suppl. 14-1, 143-152 (1974)
6.30 M.C. Hutley, V.M. Bird: Opt. Acta 20, 771-782 (1973)

6.31 R. Petit, D. Maystre, M. Nevière: *Space Optics*, Proc. Ninth Intern. Cong. of the I.C.O., Santa Monica, Ca, USA, October 9-13 (1972) pp. 667-681
6.32 D. Maystre, R.C. McPhedran: Opt. Commun. 12, 164-167 (1974)
6.33 R.C. McPhedran, D. Maystre: J. Spectrosc. Soc. Jpn. 23, 13-20 (1974) Suppl. 1
6.34 R.C. McPhedran, D. Maystre: Opt. Acta 21, 413-421 (1974)
6.35 R.C. McPhedran, D. Maystre: Nouv. Rev. Opt. 5, 241-248 (1974)
6.36 M. Nevière, M. Cadilhac, R. Petit: Opt. Commun. 6, 34-37 (1972)
6.37 M.C. Hutley, J.P. Verrill, R.C. McPhedran: Opt. Commun. 11, 207 (1974)
6.38 M.C. Hutley, J.P. Verrill, R.C. McPhedran, M. Nevière, P. Vincent: Nouv. Rev. Opt. 6, 87-95 (1975)
6.39 M.C. Hutley, D. Maystre: Opt. Commun. 19, 431-436 (1976)
6.40 E.G. Loewen, M. Nevière: Appl. Opt. 16, 3009-3011 (1977)
6.41 M. Nevière, P. Vincent, D. Maystre: Appl. Opt. 17, 843-845 (1978)
6.42 W. Werner: Appl. Opt. 16, 2078 (1977)
6.43 I.J. Wilson, L.C. Botten, R.C. McPhedran: J. Opt. 8, 217 (1977)
6.44 D. Maystre: J. Opt. Soc. Am. 68, 490 (1978)
6.45 D. Maystre, R. Petit: Opt. Commun. 5, 90 (1972)
6.46 A. Marechal, G.W. Stroke: C. R. Ac. Sci. 249, 2042 (1959)
6.47 D. Maystre, R. Petit: Nouv. Rev. Opt. 2, 115 (1971)
6.48 M. Breidne, D. Maystre: Appl. Opt. 19, 1812 (1980)
6.49 J. Chandezon: Thèse d'état, Université de Clermont-Ferrand II, France (1978)
6.50 D. Maystre, R. Petit: Opt. Commun. 4, 25 (1971)
6.51 N.K. Sheridon: Appl. Phys. Lett. 12, 316 (1968)
5.52 S. Johansson, L.E. Nilsson, K. Biedermann, K. Kleveby: *Applications of Holography and Optical Data Processing*, ed. by E. Marom, A.A. Friesem, Wiener-Avnear (Pergamon Press, Oxford 1977) pp. 521-530
6.53 R.C. McPhedran: Ph. D. Thesis, University of Tasmania, Hobart, Australia (1973)
6.54 G. Pieuchard, J. Flamand: Jpn. J. Phys. 14, 153 (1975)
6.55 E.G. Loewen, M. Nevière, D. Maystre: J. Opt. Soc. Am. 68, 496 (1978)
6.56 A. Franks, V. Lindsay, J.M. Bennet, R.J. Speer, D. Turner, D.J. Hunt: Philos. Trans. R. Soc., Ser. A 277, 503 (1975)
6.57 E.G. Loewen, M. Nevière, D. Maystre: Appl. Opt. 15, 2937 (1976)
6.58 J.M. Elson: Appl. Opt. 16, 2872 (1977)
6.59 H.L. Garvin, E. Garmire, S. Somekh, H. Stoll, A. Yariv: Appl. Opt. 12, 455 (1973)
6.60 G. Schmahl, D. Rudolph: *Progress in Optics*, Vol. 14 (North-Holland, Amsterdam 1976) p. 195
6.61 K. Knop: Opt. Commun. 26, 281 (1978)
6.62 E.G. Loewen, M. Nevière, D. Maystre: Appl. Opt. 18, 2262 (1979)
6.63 C.F. Meyer: *The Diffraction of Light, X-Rays and Material Particles*, (U. Chicago Press, Chicago 1934) p. 135
6.64 J.L. Roumiguières, D. Maystre, R. Petit: J. Opt. Soc. Am. 66, 772-775 (1976)
6.65 E.G. Loewen, M. Nevière: Appl. Opt. 17, 1087-1092 (1978)
6.66 D. Maystre, R. Petit, M. Duban, J. Gilewicz: Nour. Rev. Opt. 5, 79-85 (1974)
6.67 D. Maystre, J.P. Laude, P. Gacoin, D. Lepere, J.P. Priou: Appl. Opt. (accepted for publication)
6.68a M. Nevière, W.R. Hunter: Appl. Opt. (submitted for publication)
6.68b W. Gudat, C. Kunz: "Instrumentation for Spectroscopy and Other Applications", in *Synchrotron Radiation*, ed. by C. Kunz, Topics in Current Physics, Vol. 10 (Springer, Berlin, Heidelberg, New York 1979) pp. 55-168
6.69 D. Maystre, R. Petit: J. Spectrosc. Soc. Jpn. 23, Suppl. 1, 61-65 (1974)
6.70 D. Maystre, R. Petit: Nouv. Rev. Opt. 7, 165-180 (1976)
6.71 M. Nevière, J. Flamand: Nucl. Instrum. Methods (to be published)
6.72 R.F. Johnson: Max Planck-Institut für Festkörperforschung, Stuttgart, Germany, private communication
6.73 P. Vincent, M. Nevière, D. Maystre: Nucl. Instrum. Meth. 152, 123-126 (1978)
6.74 M. Nevière, P. Vincent, D. Maystre: Appl. Opt. 17, 843-845 (1978)
6.75 P. Vincent, M. Nevière, D. Maystre: Appl. Opt. 18, 1780-1782 (1979)
6.76 M. Nevière, D. Maystre, W.R. Hunter: J. Opt. Soc. Am. 68, 1106-1113 (1978)
6.77 A. Hessel, J. Schmoys, D.Y. Tseng: J. Opt. Soc. Am. 65, 380 (1975)

6.78 E.V. Jull, G.R. Ebbeson: IEEE Trans. AP-25, 565 (1977)
6.79 L.S. Cheo, J. Schmoys, A. Hessel: J. Opt. Soc. Am. 67, 1686 (1977)
6.80 D. Maystre, R. Petit: Opt. Commun. 5, 35-38 (1972)
6.81 D. Maystre, R. Petit: Opt. Commun 4, 380-382 (1972)
6.82 J.L. Roumiguières: Opt. Commun. 19, 76-78 (1976)
6.83 J.L. Roumiguières, D. Maystre, R. Petit: Opt. Commun. 7, 402-405 (1973)
6.84 J.L. Roumiguières, D. Maystre, R. Petit: *Proc. Fifth Colloquium on Microwave Communication*, 3, ed. by Bognar (Akademiai kiado, Budapest 1974) ET 305-314
6.85 J.L. Roumiguières: Thèse de 3e cycle, Université d'Aix-Marseille III, Centre de St-Jérôme (1976)
6.86 M. Nevière: Thèse d'Etat AO 11556, Université d'Aix-Marseille III, Centre de St-Jérôme (1975)
6.87 J.M. Elson: Appl. Opt. 16, 2873 (1977)
6.88 J.B. Charlton: Kirtland Air Force Base, Albuquerque, New Mexico, USA, private communication
6.89 D. Maystre: Opt. Commun. 26, 127 (1978)
6.90 D. Maystre, R. Petit: Nouv. Rev. Opt. Appl. 2, 115-120 (1971)
6.91 K. Knop: Appl. Opt. 17, 3598-3603 (1978)
6.92 M.T. Gale, J. Kane, K.Knop: J. Appl. Photogr. Eng. 4, 41-47 (1978)
6.93 K. Knop: J. Opt. Soc. Am. 68, 1206-1210 (1978)
6.94 H. Dammann: Appl. Opt. 17, 2273-2278 (1978)

7. Theory of Crossed Gratings

R. C. McPhedran, G. H. Derrick, and L. C. Botten

With 20 Figures

In this chapter we will consider the theoretical analysis and the properties of doubly periodic diffraction gratings ("bigratings"). Our interest in these structures has been stimulated by two factors. Firstly, their theoretical study represents a natural extension of previous work on mono-periodic structures. Secondly, HORWITZ [7.1] suggested the use of metallic grids as selective filters (heat mirrors) in solar energy absorption. Of course such devices have application in far infrared spectroscopy as beam splitters [7.2], filters [7.3-5] and the reflecting elements in Fabry-Perot interferometers [7.5-10]. Similar applications exist in the field of microwave engineering [7.11.13].

7.1 Overview

The structures to be considered here divide themselves into a number of distinct categories. Firstly, there are inductive grids. These consist of a planar metallic sheet perforated in a doubly periodic fashion with apertures. Since the metal forms a connected region, surface current paths are unbroken and hence the structure has a high reflectance at long wavelengths. The complementary structure, the capacitive grid, is composed of a regular array of metallic plates surrounded by or supported upon an insulating material. Here the conducting material forms a disconnected region, inhibiting long wavelength surface current distributions and leading to poor reflectance at such wavelengths.

A further class of bigratings consists of a pair of spatially separated wire gratings. The properties of this structure depend fundamentally upon the angle (η) between the axes of the two sets of wires. When the gratings are orthogonal ($\eta=90^\circ$) the behavior resembles that of the inductive grid. For $\eta = 0^\circ$ the structure is now composed of two polarizing filters, which for $E_\parallel$ polarized light behaves as a Fabry-Perot interferometer.

The final category which we discuss is that of doubly periodic reflection gratings. These are gratings whose surface profile is modulated periodically in two directions. As we will see, the double modulation can profoundly influence the absorptance of the grating material in unpolarized light.

In keeping with the rest of this book the theoretical formulations discussed here are all entirely rigorous. This requirement limits the number of structures that can be usefully analyzed. The limitation in many cases arises more because of computational than theoretical difficulties. Indeed, if the standard integral equation techniques already discussed for classical gratings were applied to bigratings, the number of unknowns would be squared (because of the dual periodicity) and then multiplied by four (because of the necessity to solve for four field components rather than only one). Such techniques would require the inversion of complex matrices having more than a million elements.

Theoretical approaches to the problem of diffraction by biperiodic structures have followed two basic paths. *The earlier formalisms* (for example those of CHEN [7.11-13]) *rely upon the availability of a complete set of modal functions which can be superposed to specify the field distribution in the aperture region.* These modal functions must satisfy the wave equation and the boundary conditions on the aperture walls, and are only known analytically for a small number of perfectly conducting inductive geometries. The extension of modal formalisms to finitely conducting structures is difficult because the boundary conditions are not of the Dirichlet type or of the Neumann type, but rather refer to the continuity of the field and its derivatives across the boundary. For capacitive structures the difficulty in formulating satisfactory modes lies in reconciling the boundary and pseudoperiodicity conditions. In Sect.7.3 we outline the simplest of the modal formalisms, that for the inductive grid with rectangular apertures. We will also discuss the way in which modal approaches can be used for other aperture shapes and for spatially separated gratings (Sect.7.4).

More recently, attention has been turned to finitely conducting biperiodic reflection gratings. The theoretical treatments used are vastly different to those described above in that they represent generalizations of existing differential approaches for classical gratings discussed in earlier chapters. *The formalism to be outlined here is somewhat different, relying upon a transformation which flattens the bigrating profile.*

7.2 The Bigrating Equation and Rayleigh Expansions

Let us consider a plane wave incident upon any doubly periodic structure. Take the Oy axis of a rectangular, Cartesian coordinate system to be perpendicular to the plane of the grating, and the Ox axis to be aligned with one of the axes of periodicity. The second axis of periodicity makes an angle η with the Ox axis (see Fig. 7.1a). The two periodicities of the bigrating are defined by d, the period along Ox, and d', the projection of the second periodicity onto Oz.

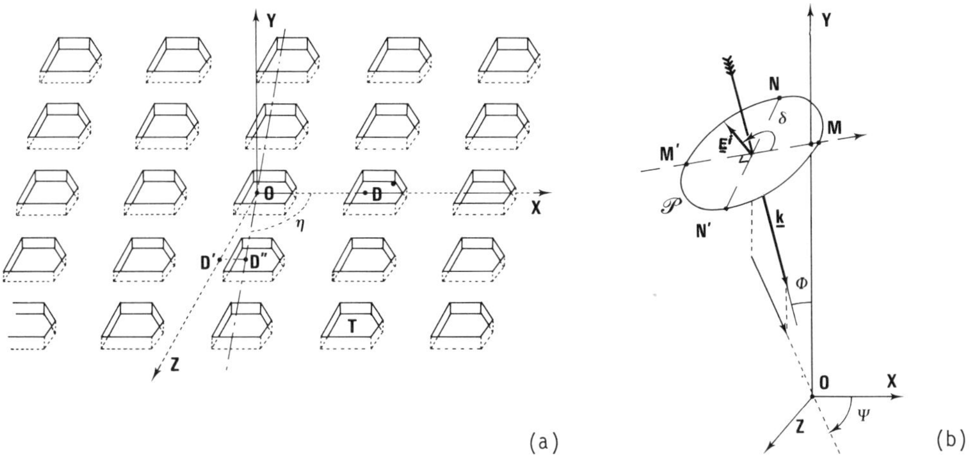

Fig. 7.1. (a) The general structure of inductive grids considered in this chapter. The apertures are spaced periodically along axes inclined at an angle η to each other, d is the period along the Ox axis, and d' is the projection of the other period onto the Oz axis [7.14]; (b) the specification of the incident field. The direction of propagation is determined by the polar angles ϕ and ψ and is represented by a vector $\underline{k}$ passing through O. Draw a circle $\mathscr{P}$ of unit radius centred at a point of $\underline{k}$ and perpendicular to it. The diameter MM' is that diameter of $\mathscr{P}$ which intersects the Oy axis. The polarization angle δ is then the angle between the electric field $\underline{E}^i$ and MM'. For $\delta = 0^0$ the electric field vector is parallel to MM' and lies in the plane of incidence, which is determined by $\underline{k}$ and the Oy axis. For $\delta = 90^0$ the electric vector is parallel to NN', i.e., perpendicular to the plane of incidence.

The direction of the incoming plane wave is specified by the two angles ϕ and ψ (Fig.7.1b) and its polarization by a third angle δ. ϕ is the angle between the incident wave vector $\underline{k}$ and the Oy axis, whereas ψ is the angle between the projection of $\underline{k}$ onto the xz plane and the Ox axis. Thus if $k = 2\pi/\lambda$ is the wave number of the incident beam then

$$\underline{k} = \alpha_0\hat{\underline{x}} - \beta_{00}\hat{\underline{y}} + \gamma_{00}\hat{\underline{z}} \tag{7.1}$$

where $\alpha_0 = k \sin\phi \cos\psi$, $\beta_{00} = k \cos\phi$ and $\gamma_{00} = k \sin\phi \sin\psi$.

In response to this incident field, the bigrating will give rise to a diffracted field in the region above and possibly below it. Away from the apertures or "maze" regions (our name for the two-dimensional analogue of groove) these diffracted fields may be written as superpositions of plane waves (i.e., Rayleigh expansions). Since the bigrating is doubly periodic these plane waves must propagate along discrete directions. We will now derive heuristically these possible directions of propagation.

In what follows we shall consistently suppress the temporal dependence of $\exp(-i\omega t)$ from all fields. The spatially dependent term describing the incident wave is then

$$\exp[i(\alpha_0 x - \beta_{00} y + \gamma_{00} z)] \quad .$$

Consider a diffracted plane wave whose spatial dependence takes the form $\exp[i(\alpha x+\beta y+\gamma z)]$. By moving the origin of coordinates through a distance d in the x direction, the diffraction problem is altered only in so far as the phase of the incident wave is modified; neither the boundary conditions nor the radiation conditions at $|y| = \infty$ are changed. Thus applying the uniqueness property of the solution, each plane wave element must be modified by the phase term $\exp(i\alpha_0 d)$. Thus,

$$\alpha = \alpha_p = \alpha_0 + pK \tag{7.2}$$

where p is an integer and $K = 2\pi/d$. We can also make a translation by one period along the line $z = x \tan\eta$. This means replacing x by $x+d' \cot\eta$ and z by $z+d'$. The resulting alteration to the incident wave is $\exp[i(\alpha_0 d'\cot\eta+\gamma_{00}d')]$. It then follows from the uniqueness condition that

$$\gamma = \gamma_{pq} = \gamma_{00} + qK' - pK \cot\eta \tag{7.3}$$

where q is also an integer and $K' = 2\pi/d'$. β is determined by requiring that the plane waves individually obey the Helmholtz equation. This leads to

$$\beta = \beta_{pq} = \sqrt{k^2-\alpha_p^2-\gamma_{pq}^2} \tag{7.4}$$

where the imaginary part of β_{pq} is, by definition, positive.

The orders of diffraction formed by a bigrating are specified by a pair of integers (p,q) *rather than just the single integer adequate in the case of classical gratings*. If β_{pq} is real then the order (p,q) is referred to as being a propagating order, since it can carry energy to infinity away from the grating surface. Should β_{pq} be imaginary then the order (p,q) is said to be evanescent and carries no energy away from the grating surface. Important physical phenomena arise in the vicinity of the transition point $\beta_{pq} = 0$. If the incidence parameters ϕ and ψ are fixed then β_{pq} vanishes at the Rayleigh wavelength

$$\lambda_R(p,q) = (-B+\sqrt{B^2+A \cos^2\phi})/A \tag{7.5}$$

where

$$A = \left(\frac{p}{d}\right)^2 + \left(\frac{q}{d'} - \frac{p}{d} \cot\eta\right)^2$$

and

$$B = \sin\phi\left[\frac{p}{d} \cos\psi+\left(\frac{q}{d'} - \frac{p}{d} \cot\eta\right)\sin\psi\right] .$$

We will now write down expressions for the electric and magnetic field vectors for the part of the reflected field region in which the Rayleigh expansion is valid. *Although in problems of diffraction by classical gratings it is the usual practice to employ Cartesian field components, it is advantageous in bigrating problems to use transverse electric (TE) and transverse magnetic (TM) vector fields.* These are defined in the same way as is habitual in the theory of waveguides, with the Oy axis being the preferred axis. (The reader unfamiliar with the TE/TM decomposition of vector fields is advised to refer to any of the standard texts on waveguides.) Thus "transverse" means the absence of an Oy component of the relevant field quantity. The TE and TM vector fields are linearly independent, have mutually orthogonal electric fields and can be normalized so that their inner product over a rectangle in the x-z plane with sides of length d,d' is unity. They then have the following resolved parts (resolutes) in the x-z plane:

$$\underline{RTE}_{pq}(x,y,z) = \frac{1}{\sqrt{dd'}\;\xi_{pq}}\left[\gamma_{pq}\hat{\underline{x}}-\alpha_p\hat{\underline{z}}\right]R_{pq}(x,y,z) \tag{7.6}$$

and

$$\underline{RTM}_{pq}(x,y,z) = \frac{1}{\sqrt{dd'}\;\xi_{pq}}\left[\alpha_p\hat{\underline{x}}+\gamma_{pq}\hat{\underline{z}}\right]R_{pq}(x,y,z) \quad , \tag{7.7}$$

where

$$R_{pq}(x,y,z) = \exp\left[i(\alpha_p x+\beta_{pq}y+\gamma_{pq}z)\right]$$

and

$$\xi_{pq} = \sqrt{\alpha_p^2+\gamma_{pq}^2} \quad .$$

The transverse resolute of the incident electric field is given by

$$\underline{E}_t^i = E_i\underline{RTE}_{00}(x,-y,z) + F_i\underline{RTM}_{00}(x,-y,z) \quad , \tag{7.8}$$

where $E_i = \cos\delta$ and $F_i = \frac{\beta}{k}\sin\delta$. From (7.8) and Maxwell's equations we deduce

$$\hat{\underline{y}} \times \underline{H}_t^i = \frac{1}{Z_0}\left[\frac{\beta_{00}}{k}\,E_i\underline{RTE}_{00}(x,-y,z)+\frac{k}{\beta_{00}}F_i\underline{RTM}_{00}(x,-y,z)\right] \quad , \tag{7.9}$$

from which the transverse resolute $\underline{H}_t^i$ is readily obtainable. Here Z_0 denotes the impedance of free space. Similarly the transverse resolutes of the reflected fields in the semi-infinite medium exterior to the maze region can be expressed by

$$\underline{E}_t^r = \sum_{pq}\left[E_{pq}\underline{RTE}_{pq}(x,y,z)+F_{pq}\underline{RTM}_{pq}(x,y,z)\right] \tag{7.10}$$

and

$$\hat{\underline{y}} \times \underline{H}_t^r = -\frac{1}{Z_0}\sum_{pq}\left[\frac{\beta_{pq}}{k} E_{pq}\underline{RTE}_{pq}(x,y,z)+\frac{k}{\beta_{pq}} F_{pq}\underline{RTM}_{pq}(x,y,z)\right] . \tag{7.11}$$

Here the notations "RTE" and "RTM", although at first sight clumsy, are introduced to enable the reader to keep track of the various field quantities. The first letter "R" refers to a Rayleigh or plane wave quantity. Later we shall introduce modal quantities which commence with the letter "M". For a problem involving a pure reflection grating the two sets of coefficients $\{E_{pq}\}$ and $\{F_{pq}\}$ form the fundamental unknowns of the diffraction problem. In the case of a transmission grating the field expansion coefficients $\{\hat{E}_{pq}\}$ and $\{\hat{F}_{pq}\}$ must also be determined in the region below the lower bigrating surface. In this latter case the transverse resolutes of the transmitted electric and magnetic fields are

$$\underline{E}_t^t = \sum_{pq}\left[\hat{E}_{pq}\underline{RTE}_{pq}(x,-y,z)+\hat{F}_{pq}\underline{RTM}_{pq}(x,-y,z)\right] \tag{7.12}$$

and

$$\hat{\underline{y}} \times \underline{H}_t^t = \frac{1}{Z_0}\sum_{pq}\left[\frac{\beta_{pq}}{k} \hat{E}_{pq}\underline{RTE}_{pq}(x,-y,z)+\frac{k}{\beta_{pq}} \hat{F}_{pq}\underline{RTM}_{pq}(x,-y,z)\right] . \tag{7.13}$$

Note that in what follows we will restrict our attention to the transverse resolutes of the electric and magnetic fields. From Maxwell's equations, when these resolutes satisfy the appropriate boundary and continuity conditions, the same properties must hold for the y components of the two fields.

The previous discussion is applicable to the diffraction problem for a general bigrating. In the following we will restrict ourselves to more particular geometries in order to be able to write down field expansions within the maze region. *These expansions will be in terms of TE and TM vector "modes", where by "mode" we mean a vector field not only satisfying analytically Maxwell's equations, but also the boundary conditions appropriate to the geometry.*

7.3 Inductive Grids

We will now consider theoretical approaches to the problem of diffraction by structures of the type shown in Fig.7.1a. The simplest of these is that in which the grid apertures are rectangular in cross section.

7.3.1 Grids with Rectangular Apertures

These grids were first studied by CHEN [7.11,13] and later by McPHEDRAN and MAYSTRE [7.14] and BOTTEN [7.15]. The grid apertures A are characterized by their dimensions c and c', measured parallel to the Ox and Oz axes, respectively, with their upper and lower faces being located at $y = \pm h/2$. On the perfectly conducting aperture walls parallel to the x axis

$$E_x = E_y = 0 \quad , \tag{7.14}$$

while on the metal walls parallel to the z axis

$$E_y = E_z = 0 \quad . \tag{7.15}$$

On the upper and lower faces of the grid, the tangential components of E (i.e., E_x and E_z) vanish for points not lying within an aperture.

The boundary conditions (7.14,15) are satisfied analytically by the following transverse electric and transverse magnetic modes

$$\underline{MTE}_{nm}(x,z) = g_{nm}\left[\frac{m\pi}{c'}\cos\left(\frac{n\pi x}{c}\right)\sin\left(\frac{m\pi z}{c'}\right)\hat{\underline{x}} - \frac{n\pi}{c}\sin\left(\frac{n\pi x}{c}\right)\cos\left(\frac{m\pi z}{c'}\right)\hat{\underline{z}}\right] \tag{7.16}$$

and

$$\underline{MTM}_{nm}(x,z) = g_{nm}\left[\frac{n\pi}{c}\cos\left(\frac{n\pi x}{c}\right)\sin\left(\frac{m\pi z}{c'}\right)\hat{\underline{x}} + \frac{m\pi}{c'}\sin\left(\frac{n\pi x}{c}\right)\cos\left(\frac{m\pi z}{c'}\right)\hat{\underline{z}}\right] \quad . \tag{7.17}$$

Here,

$$g_{nm} = \sqrt{\frac{(2-\delta_{n0})(2-\delta_{m0})}{cc'}}\left\{\left(\frac{n\pi}{c}\right)^2 + \left(\frac{m\pi}{c'}\right)^2\right\}^{-\frac{1}{2}} \tag{7.18}$$

is a normalization factor.
These modes have the following orthogonality properties

$$\iint_A \underline{MTF}_{nm} \cdot \underline{MTG}_{NM}\, dA = \delta_{nN}\delta_{mM}\delta_{FG}$$

where F and G are replaced by E and/or H, the other subscripts are integers.

Using these modal functions we can expand the transverse resolutes of the electric and magnetic fields as

$$\begin{aligned}\underline{E}_t = \sum_{nm} \Big\{ & \left[a_{nm}\sin(\mu_{nm}y)+b_{nm}\cos(\mu_{nm}y)\right]\underline{MTE}_{nm}(x,z) \\ & +\left[c_{nm}\sin(\mu_{nm}y)+d_{nm}\cos(\mu_{nm}y)\right]\underline{MTM}_{nm}(x,z)\Big\}\end{aligned} \tag{7.19}$$

and

$$\hat{\underline{y}} \times \underline{H}_t = \frac{i}{Z_0} \sum_{nm} \left\{ \frac{\mu_{nm}}{k} \left[a_{nm} \cos(\mu_{nm} y) - b_{nm} \sin(\mu_{nm} y) \right] \underline{MTE}_{nm}(x,z) \right.$$

$$\left. + \frac{k}{\mu_{nm}} \left[c_{nm} \cos(\mu_{nm} y) - d_{nm} \sin(\mu_{nm} y) \right] \underline{MTM}_{nm}(x,z) \right\} \quad . \tag{7.20}$$

Here

$$\mu_{nm} = \sqrt{k^2 - \left(\frac{n\pi}{c}\right)^2 - \left(\frac{m\pi}{c'}\right)^2} \quad .$$

We will now impose boundary and continuity conditions at the upper and lower faces of the grid. We exploit the up-down symmetry of the structure by adding and subtracting the constraints at $y = h/2$ and $y = -h/2$. This enables us to isolate y-symmetric and y-antisymmetric modal coefficients and so halves the dimensions of the matrices to be inverted. Let us define the following sums and differences of plane wave coefficients

$$MG^*_{pq} = (G_{pq} - \hat{G}_{pq}) \exp(i\beta_{pq} h/2) \tag{7.21}$$

$$PG^*_{pq} = (G_{pq} + \hat{G}_{pq}) \exp(i\beta_{pq} h/2) \tag{7.22}$$

where G may be replaced by either E or F. Modified modal coefficients are also useful

$$u^*_{nm} = 2u_{nm} \sin(\mu_{nm} h/2) \tag{7.23}$$

$$v^*_{nm} = 2v_{nm} \cos(\mu_{nm} h/2) \tag{7.24}$$

where u_{nm} and v_{nm} take, respectively, the values a_{nm} or c_{nm} and b_{nm} or d_{nm}.

The continuity of the transverse components of $\underline{E}$ over the apertures at $y = \pm h/2$ is combined with the boundary conditions on the metal surfaces there to give what are termed "reconstitution equations". These are derived by the method of moments, utilizing the orthogonality of the plane wave basis functions (7.6,7) when integrated over the unit period cell of the grid. They are

$$ME^*_{pq} = -E^*_i \delta_{p0} \delta_{q0} + \sum_{nm} \left(a^*_{nm} IEE^{pq}_{nm} + c^*_{nm} IEM^{pq}_{nm} \right) \tag{7.25}$$

$$MF^*_{pq} = -F^*_i \delta_{p0} \delta_{q0} + \sum_{nm} \left(a^*_{nm} IME^{pq}_{nm} + c^*_{nm} IMM^{pq}_{nm} \right) \quad , \tag{7.26}$$

where

$$IFG^{pq}_{nm} = \iint_A \underline{MTG}(x,z) \cdot \overline{\underline{RTF}}_{pq}(x,0,z) dA \tag{7.27}$$

in which F and G may once again be replaced by E and/or M. The modified incident plane wave amplitudes are defined by

$$E_i^* = E_i \exp(-i\beta_{00}h/2)$$

$$F_i^* = F_i \exp(-i\beta_{00}h/2) \quad .$$

Analytic expressions for the inner products (7.27) over the aperture A are easily derived. Equations for the quantities PE_{pq}^* and PF_{pq}^* may be obtained from (7.25,26), respectively, by replacing a_{nm}^* by b_{nm}^* and c_{nm}^* by d_{nm}^*.

The final step in this analysis is to apply the continuity of the transverse components of magnetic fields across the apertures at $y = \pm h/2$. The method of moments is again invoked, this time using the orthogonality properties of the vector modes (7.16,17) within the aperture A. This results in two pairs of coupled linear equations. The pair referring to the y-antisymmetric modal coefficients is

$$\sum_{nm} a_{nm}^* \left[\frac{i\mu_{NM}}{k} \cot(\mu_{NM}h/2)\delta_{mM}\delta_{nN} + \sum_{pq}\left(\frac{\beta_{pq}}{k} IEE_{nm}^{pq}\overline{IEE}_{NM}^{pq} + \frac{k}{\beta_{pq}} IME_{nm}^{pq}\overline{IME}_{nm}^{pq}\right)\right]$$

$$+ \sum_{nm} c_{nm}^* \left[\sum_{pq}\left(\frac{\beta_{pq}}{k} IEM_{nm}^{pq}\overline{IEE}_{NM}^{pq} + \frac{k}{\beta_{pq}} IMM_{nm}^{pq}\overline{IME}_{NM}^{pq}\right)\right]$$

$$= 2\left(\frac{\beta_{00}}{k} E_i^*\overline{IEE}_{NM}^{00} + \frac{k}{\beta_{00}} F_i^*\overline{IME}_{NM}^{00}\right) \quad , \qquad (7.28)$$

$$\sum_{nm} a_{nm}^* \left[\sum_{pq}\left(\frac{\beta_{pq}}{k} IEE_{nm}^{pq}\overline{IEM}_{NM}^{pq} + \frac{k}{\beta_{pq}} IME_{nm}^{pq}\overline{IMM}_{NM}^{pq}\right)\right]$$

$$+ \sum_{nm} c_{nm}^* \left[\frac{ik}{\mu_{NM}} \cot(\mu_{NM}h/2)\delta_{nN}\delta_{mM} + \sum_{pq}\left(\frac{\beta_{pq}}{k} IEM_{nm}^{pq}\overline{IEM}_{NM}^{pq} + \frac{k}{\beta_{pq}} IMM_{nm}^{pq}\overline{IMM}_{NM}^{pq}\right)\right]$$

$$= 2\left(\frac{\beta_{00}}{k} E_i^*\overline{IEM}_{NM}^{00} + \frac{k}{\beta_{00}} F_i^*\overline{IMM}_{NM}^{00}\right) \quad . \qquad (7.29)$$

The corresponding equations for the $\{b_{nm}^*\}$ and $\{d_{nm}^*\}$ can be obtained by replacing the impedance term "cot" by "-tan" in the above expressions. To solve these equations numerically it is necessary to truncate both the plane wave series (7.10,11) and modal expansions (7.19,20). The resulting systems of equations are then solved by standard elimination techniques and the plane wave amplitudes reconstructed from (7.25,26). The efficiencies of the various propagating orders can then be calculated simply from the expressions

$$\rho_{pq} = \frac{1}{\beta_{00}}\left(\beta_{pq}|E_{pq}|^2 + \frac{1}{\beta_{pq}}|F_{pq}|^2\right)$$

and

$$\hat{\rho}_{pq} = \frac{1}{\beta_{00}}\left(\beta_{pq}|\hat{E}_{pq}|^2 + \frac{1}{\beta_{pq}}|\hat{F}_{pq}|^2\right) \quad .$$

7.3.2 Numerical Tests and Applications

Integral and differential formalisms for classical grating problems can generally be tested numerically using the criterion of conservation of energy. This provides not only a test on the soundness of the computer implementation but also an estimate of the errors caused by the truncation of infinite series and the inaccuracies introduced by the use of numerical quadrature. However, as discussed by AMITAY and GALINDO [7.16] and McPHEDRAN and MAYSTRE [7.14] the truncation errors in modal formalisms of the above type do not cause corresponding errors in the energy balance between the incident and transmitted plus reflected fields. Quadrature errors do not arise because all integrations are performed analytically.

A number of other tests have accordingly been derived for crossed grating problems. We will discuss here the three most important of these. *As in the case of classical gratings the most fundamental of the theoretical tests is provided by the reciprocity theorem.* For crossed gratings this criterion has been discussed by McPHEDRAN and MAYSTRE [7.14] and DERRICK et al. [7.17]. We shall now outline its derivation. Let $\underline{E}$ and $\underline{H}$ denote the electric and magnetic field vectors corresponding to an initial diffraction problem. We define a second diffraction problem by choosing a propagating reflected order (p,q) of the first problem and returning an incident field of the same wavelength along the direction of this outgoing order. The choice of the polarization for this second problem is quite arbitrary.

Using a superscripted prime to denote quantities associated with the second problem we form the field quantity

$$\underline{F} = \underline{E} \times \underline{H}' - \underline{E}' \times \underline{H} \quad . \tag{7.30}$$

The divergence of $\underline{F}$ may be shown to be zero using Maxwell's curl equations. The divergence theorem is then used to show that

$$\iint_S \hat{\underline{n}} \cdot \underline{F} ds = 0 \tag{7.31}$$

where $\hat{\underline{n}}$ denotes the outward pointing unit normal vector to the closed surface S. We take S to be the union of the three surfaces enclosing the regions:

A - a rectangular parallelepiped whose cross section is that of the unit period cell in the region $y \geqq h/2$ and whose height is arbitrary.

B - the aperture volume $(-h/2 < y < h/2)$.

C - the analogue of A in the region $y \leqq -h/2$.

The contribution of the integral (7.31) from the vertical walls of A and C is zero because of pseudoperiodicity conditions whereas that from the metallic walls of B vanishes because of the boundary conditions. Thus (7.31) may be interpreted as a statement that the integral of $\hat{\underline{y}} \cdot \underline{F}$ over any horizontal unit period cell is y invariant. In fact this invariant quantity is zero as can be seen by considering its value in the region $y \leq -h/2$. In this way we derive the following expression for the reciprocity theorem as it applies to the return of a reflected order

$$\beta_{00} \cos(\delta)E'_{pq} + k \sin(\delta)F'_{pq} = \beta_{pq} \cos(\delta')E_{pq} + k \sin(\delta')F_{pq} \tag{7.32}$$

where δ and δ' define the polarization angles of the two problems.

In an analogous manner we can derive the following relation pertaining to the return of a transmitted order

$$\beta_{00} \cos(\delta)\hat{E}'_{pq} - k \sin(\delta)\hat{F}'_{pq} = \beta_{pq} \cos(\delta')\hat{E}_{pq} - k \sin(\delta')\hat{F}_{pq} \quad . \tag{7.33}$$

The relationships (7.32,33) are in general not analytically satisfied, and can be used to estimate the size of truncation errors.

It will be noted that we have not given an equation relating the efficiencies in the first and second problems. Because of the free choice of the polarization angle δ' in relation to δ, the efficiency ρ'_{pq} is in general not equal to ρ_{pq}. However, if we consider the efficiencies in unpolarized light, formed by integrating over all angles δ and δ', these are equal in the two cases [7.17]. (Of course, the efficiency in unpolarized light is equal to the mean of the efficiencies for any two values of δ, δ' differing by 90°.)

The second theoretical test applies to square-symmetric crossed gratings used in the Littrow mount in a particular order (f,g). By now operating the grating in a (-f,-g) order Littrow configuration (i.e., by returning the zeroth order) and superposing these two fields we can derive the so-called Littrow constraint from the principle of conservation of energy

$$\mathrm{Re}\left\{\sum_{p,q}\left[\beta_{pq}\left(E_{pq}\bar{E}_{f-p,g-q}+\hat{E}_{pq}\bar{\hat{E}}_{f-p,g-q}\right)+\frac{1}{\beta_{pq}}\left(F_{pq}\bar{F}_{f-p,g-q}+\hat{F}_{pq}\bar{\hat{F}}_{f-p,g-q}\right)\right]\right\} = 0 \ . \tag{7.34}$$

The ultimate test of any theoretical formulation is to compare its results with experimental measurements. For inductive grids with square apertures, the theoretical predictions have been compared [7.14] with a number of measurements made in the far-infrared region. The theory and experiment were seen to be in good agreement, even though it was often necessary to assume the value of a parameter such as grid thickness in order to make the calculations. We will give an example of a comparison between theory and experiment later, for the case of inductive grids with circular apertures.

Let us now look at a typical example of an efficiency curve for an inductive grid. In Fig.7.2 the energy transmitted through a grid having square apertures is shown as a function of normalized wavelength, for normally incident radiation. The curve may be divided into three regions. For $\lambda/d \gg 1$, the grid behaves as a filter, with the sharpness of the cutoff increasing with thickness (h) and decreasing with aperture size (c). As the normalized wavelength diminishes, the transmittance increases,

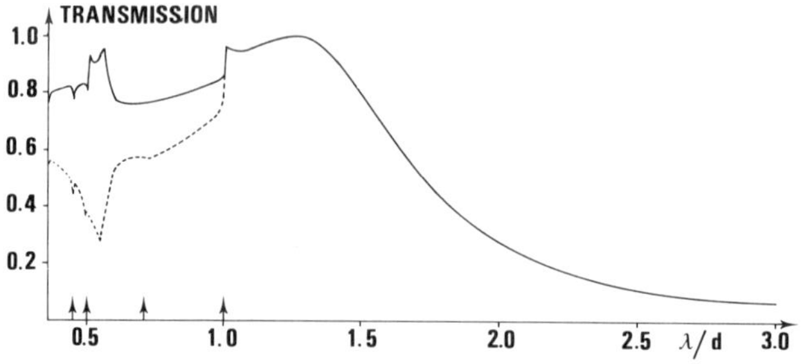

Fig. 7.2. The energy transmitted by a square grid is shown as a function of normalized wavelength (λ/d). The grid has apertures of side length $c = 0.9$ d, and thickness $h = 0.4$ d, and is used with normally incident radiation [7.14]

approaching unity in the resonance region. The value of λ/d at which the transmittance is unity typically lies between 1.05 and 1.30, and increases with h but decreases with c. Below the resonance region, the curve passes through a cusp-like Wood anomaly at $\lambda/d = 1.0$, associated with the appearance of the four orders $(\pm1,0)$ and $(0,\pm1)$. In the transmission region ($\lambda/d < 1.0$), both the total energy passing through the grid and that carried away normal to its surface ($\hat{\eta}_{00}$) are shown. The former value undergoes local variations due to higher order diffraction anomalies, but in general lies close to a mean value given by the area ratio c^2/d^2: the geometrical optics transmission. This mean value is largely unaffected by grid thickness. It is quite astounding how rapidly the curve approaches the geometrical limit which one usually takes as being applicable only when the dimensions d and c are of the order of tens of wavelengths, but which holds here when $\lambda/d \leqq 0.95$.

The difference between the zeroth order and total transmittance curves gives the energy carried by the dispersed orders. As soon as the nonzero orders start propagating (just below $\lambda/d=1$) they carry a considerable fraction of the incident energy even though they are travelling almost parallel to the grid surface.

Square symmetric grids have the behavior required of a solar selective heat mirror, when used with normally incident light. If placed in front of an absorbing body, they can transmit onto it a high fraction of the incident solar flux (if $c/d \geqq 0.9$) and can largely block the infrared re-emission from the absorber (provided that $h/d \gtrsim 0.4$). Grids not possessing square symmetry (i.e., $\eta \neq 90^\circ$, $c \neq c'$ or $d \neq d'$) tend to polarize the transmitted flux to the detriment of the selective properties for the unpolarized solar radiation.

As soon as the incident flux no longer strikes the grid normally, its transmission properties become polarization dependent. In Fig.7.3 we show transmission spectra for a grid having square symmetry used with an angle of incidence (ϕ) of 45°. The curves for the three polarization angles $\delta = 90^\circ$, 45°, 0° are strikingly dissimilar. The increase in the angle of incidence has moved the resonance and filtering regions to much longer wavelengths and has greatly reduced the performance in the transmission region. It has been shown [7.14] that the integrated solar transmittance for incident

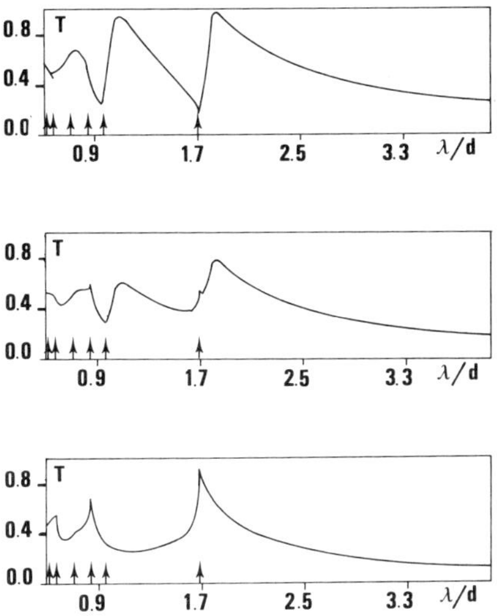

Fig. 7.3. The graphs show the energy carried by the transmitted order (0,0) as a function of normalized wavelength. In order to facilitate comparison with experimental results, [7.14] they have been obtained by averaging the calculated results for two grids, one having c/d = 0.78, h/d = 0.03, and the second having c/d = 0.91, h/d = 0.07. The curves all correspond to a constant direction of incidence specified by the angles $\phi = 45^o$ and $\psi = 0^o$. The respective values of the polarization angle are $\delta = 90^o$, 45°, 0° (proceeding from top to bottom)

radiation distributed over the hemisphere of all possible directions is almost precisely half that for normally incident radiation. This means that if grids were to be used as solar-selective heat mirrors, it would be in systems used in direct illumination and tracking its source.

7.3.3 Inductive Grids with Circular Apertures

Theoretical studies of such grids have been made by CHEN [7.12], McPHEDRAN and BOTTEN [7.18] and McPHEDRAN et al. [7.19]. The formalism is slightly more complicated than that for grids with rectangular apertures because the modal expansions are more involved. Using standard polar coordinates r and θ, with θ being measured in an anticlockwise sense from the z axis, we write down four basic modal functions

$$\underline{MTE}_{nm1}(r,\theta) = g_{nm}\left[\frac{na}{\chi'_{nm}r}J_n\left(\frac{\chi'_{nm}r}{a}\right)\cos(n\theta)\hat{\underline{r}}-J'_n\left(\frac{\chi'_{nm}r}{a}\right)\sin(n\theta)\hat{\underline{\theta}}\right] \quad , \tag{7.35}$$

$$\underline{MTE}_{nm2}(r,\theta) = g_{nm}\left[\frac{na}{\chi'_{nm}r}J_n\left(\frac{\chi'_{nm}r}{a}\right)\sin(n\theta)\hat{\underline{r}}+J'_n\left(\frac{\chi'_{nm}r}{a}\right)\cos(n\theta)\hat{\underline{\theta}}\right] \quad , \tag{7.36}$$

$$\underline{MTM}_{nm1}(r,\theta) = h_{nm}\left[J'_n\left(\frac{\chi_{nm}r}{a}\right)\cos(n\theta)\hat{\underline{r}}-\frac{na}{\chi_{nm}r}J_n\left(\frac{\chi_{nm}r}{a}\right)\sin(n\theta)\hat{\underline{\theta}}\right] \quad , \tag{7.37}$$

and

$$\underline{MTM}_{nm2}(r,\theta) = h_{nm}\left[J'_n\left(\frac{\chi_{nm}r}{a}\right)\sin(n\theta)\hat{\underline{r}}+\frac{na}{\chi_{nm}r}J_n\left(\frac{\chi_{nm}r}{a}\right)\cos(n\theta)\hat{\underline{\theta}}\right] \quad . \tag{7.38}$$

Here χ_{nm} and χ'_{nm} denote the mth nontrivial zeros of the Bessel functions J_n and J'_n. The $\{g_{nm}\}$ and $\{h_{nm}\}$ are normalization factors and are given by

$$g_{nm} = \sqrt{\frac{\varepsilon_n}{\pi}} \frac{\chi'_{nm}}{aJ_n(\chi'_{nm})\sqrt{\chi'^2_{nm}-n^2}} \tag{7.39}$$

and

$$h_{nm} = \sqrt{\frac{\varepsilon_n}{\pi}} \frac{1}{aJ_{n-1}(\chi_{nm})} \quad . \tag{7.40}$$

Also

$$\nu'_{nm} = \sqrt{1-\left(\frac{\chi'_{nm}}{k_2 a}\right)^2} \quad , \quad \nu_{nm} = \sqrt{1-\left(\frac{\chi_{nm}}{k_2 a}\right)^2} \quad , \quad \text{and} \quad k_2 = kr_2 \quad .$$

In the above expressions r_2 is the refractive index of the medium filling the apertures, $\varepsilon_n = 2-\delta_{n0}$, and a is the aperture radius.

Using these modal functions we can expand the transverse resolutes of the electric and magnetic fields as

$$\underline{E}_t = \sum_{n=0}^{\infty}\sum_{m=0}^{\infty}\sum_{\ell=1}^{2}\Big\{\big[a_{nm\ell}\sin(k_2\nu'_{nm}y)+b_{nm\ell}\cos(k_2\nu'_{nm}y)\big]\underline{MTE}_{nm\ell}(r,\theta)$$
$$+\big[c_{nm\ell}\sin(k_2\nu_{nm}y)+d_{nm\ell}\cos(k_2\nu_{nm}y)\big]\underline{MTM}_{nm\ell}(r,\theta)\Big\} \tag{7.41}$$

and

$$Z_0\hat{\underline{y}}\times\underline{H}_t = ir_2\sum_{n=0}^{\infty}\sum_{m=0}^{\infty}\sum_{\ell=1}^{2}\Big\{\nu'_{nm}\big[a_{nm\ell}\cos(k_2\nu'_{nm}y)-b_{nm\ell}\sin(k_2\nu'_{nm}y)\big]\underline{MTE}_{nm\ell}(r,\theta)$$
$$+\frac{1}{\nu_{nm}}\big[c_{nm\ell}\cos(k_2\nu_{nm}y)-d_{nm\ell}\sin(k_2\nu_{nm}y)\big]\underline{MTM}_{nm\ell}(r,\theta)\Big\} \quad . \tag{7.42}$$

Equations (7.41,42) are separable expansions in which terms are grouped according to

(a) their y symmetry or antisymmetry;
(b) the nature of the polarization (TE or TM) and
(c) their orientation (ℓ=1,2 running over "horizontal" and "vertical" states).

The remainder of the theoretical treatment follows the same lines as that used for rectangular apertures. It is to be noted that analytic expressions are available for the inner product of the above modal functions with the plane wave vector modes over the circular apertures [7.20].

Should one choose to envelope the grid in a symmetric dielectric sandwich of arbitrary thickness then this results in little additional complication of the analysis. This, in the main, is due to the use of TE/TM decompositions of fields. All that is necessary is to define an effective incident field at the upper surface of the grid and to link this field with the actual incident field by means of plane wave admittances.

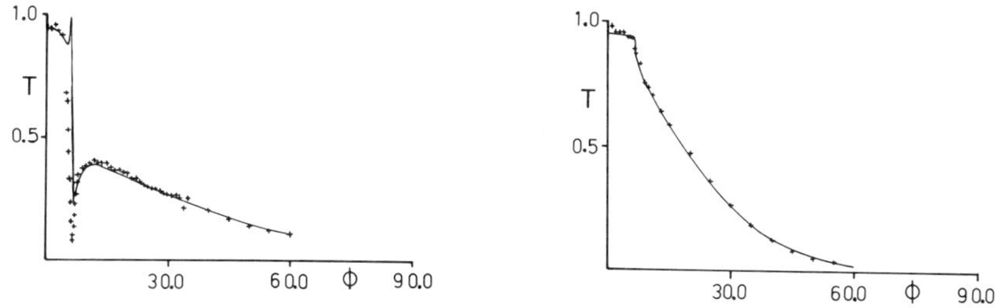

Fig. 7.4. The energy transmitted through a round-holed grid (with aperture radius $a = 0.45$ d and thickness $h = 0.40$ d) is shown as a function of the angle of incidence ϕ (with ψ=90°) for a fixed wavelength (λ=1.12 d). For the upper and lower curves, the electric vector of the incident wave is, respectively, parallel and perpendicular to the plane of incidence. The continuous line gives the theoretical values, while the points are experimental measurements made using microwaves [7.19]

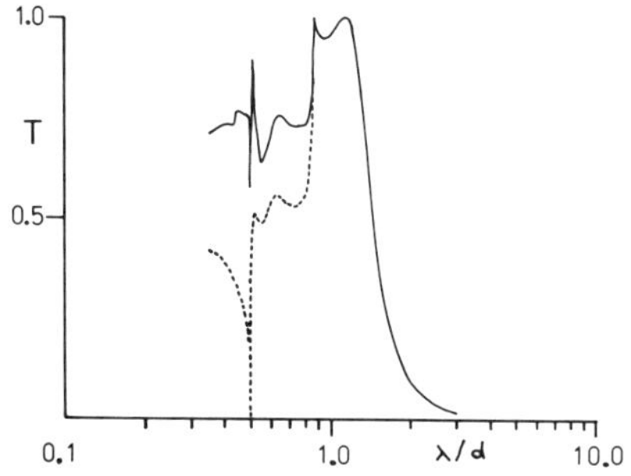

Fig. 7.5. The total energy transmittance (full line) and that in the order (0,0) (broken line) are shown as a function of normalized wavelength for normally incident radiation on a grid having the parameters $d' = 0.866$ d, $a = 0.45$ d, $h = 0.40$ d and $\eta = 60°$

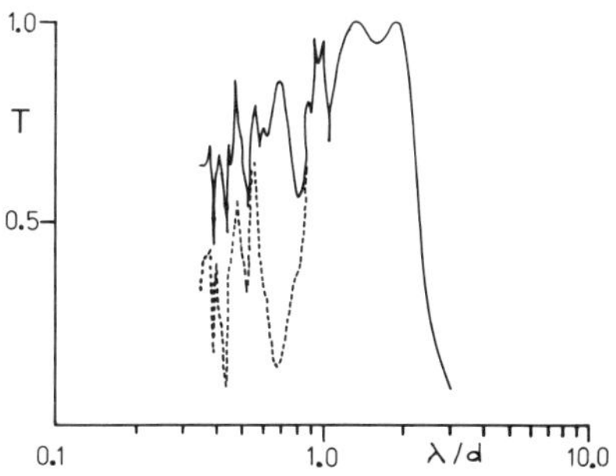

Fig. 7.6. The total and zeroth order transmittances are shown for the grid of Fig.7.5, now with its apertures filled with a dielectric of refractive index 1.5 and surrounded with a symmetric sandwich of the same material having a total thickness of 0.24 d

A careful comparison [7.19] has been made between theory and experiment for round-holed grids. We show in Fig.7.4 curves for the energy transmitted in the zeroth order as a function of the angle of incidence (ϕ) at a fixed normalized wavelength of $\lambda/d = 1.12$, for the two principal cases of polarization, $E_{\parallel}$(δ=90°) and $E_{\perp}$(δ=0°). (For crossed gratings it is no longer possible to specify principal polarizations with respect to an axis of periodicity, and so the terms $E_{\parallel}$ and $E_{\perp}$ refer to the plane of incidence.) The points are measurements made at millimetric wavelengths and generally are in remarkably good agreement with the theory.

From the curves we see that the polarization independence of the grid at normal incidence is very quickly lost as ϕ departs from zero. This rapid increase in polarization is caused by the presence of the Wood anomaly associated with the order (0,-1) at $\phi = 6.5°$. In $E_{\parallel}$ polarization, this anomaly causes an abrupt loss of energy from the transmitted spectrum, this energy plunging by 80% in 2°! For $E_{\perp}$ polarization, the anomaly is evident only as a change in the gradient of the curve.

The solar selective behavior of this class of grid has also been investigated [7.18]. In contrast with the case of grids with rectangular apertures, optimal transmission occurs here with $\eta = 60^{\circ}$ rather than with $\eta = 90^{\circ}$. The former choice permits a closer packing of the circular apertures. The transmission curve shown in Fig.7.5 corresponds to a grid having aperture diameters equal to 90% of the fundamental period, and even so the optimized integrated transmittance is only 74%. By introducing dielectric plugs in the apertures and surrounding the grid with a symmetric dielectric film we have managed to increase the transmittance to 84%, for the structure of Fig.7.6. In order to obtain this transmittance however, much finer grid periods need to be used (0.5 μm rather than 1.0 μm). The improvements in transmittance are due to the better impedance matching between the grid and free space.

7.4 Capacitive and Other Grid Geometries

In the previous section, we have considered two inductive structures, for which modal expansions in the aperture regions were analytically known. Here, we will consider other geometries, for which analytic modal expansions do not exist. By this, we mean that analytic expressions cannot be written down for the modes which will satisfy all physical constraints in the aperture regions, and which only need to be matched to plane wave expansions in free space to give a complete solution of the diffraction problem. For example, in the case of a capacitive grid of perfectly conducting discs, it is a simple matter to satisfy analytically the boundary conditions on the curved surface of the discs using cylindrical waveguide modes. However, in order to satisfy the pseudoperiodicity conditions which we need to be enforced at the edge surfaces of the unit period cell for capacitive structures, an infinite set of such modes is necessary. The modes must in fact be determined numerically, for example, by a least squares technique.

Rigorous theoretical studies of the grid geometries to be described here have yet to be made. Thus, in this section we shall be giving a list of problems worthy of future attention by grating theorists. To guide them in their task, we will describe the properties of the structures, as determined experimentally in the far-infrared and quasi-optical regions.

The grids described below all function in the wavelength or frequency region in which only the zeroth orders in reflection and transmission propagate, and in which absorption losses into the metallic regions making them up are negligible. Thus, they can all be said to act as diplexers, dividing the energy of the incident wave between a single reflected channel and a single transmitted channel. At shorter wavelengths or higher frequencies, the grids act as multiplexers. While grid behavior can often be conveniently summarized by an equivalent circuit representation in the diplexer region, no such representation is possible in the multiplexer region.

In our description, we will group structures according to their diplexer function, with the terminology referring to the zeroth-order transmittance and its variation as a function of frequency. Only structures having square symmetry will be considered, so that the transmittance is polarization independent for normally incident radiation.

7.4.1 High-Pass Filters

These may be made using inductive grids, having apertures of square, circular or other form. As we have seen, the transmittance in the passband region is generally close to the geometrical-optics value (the hole-area fraction). The steepness of the filter cutoff is controlled by the grid thickness and by the hole-area fraction (with small, deep apertures giving the best filtering).

As noted by ULRICH [7.5], improved filtering action may be obtained by using two or more grids in series. The structure is then of the spatially separated type to be considered in Sect.7.5.

A high pass filter can also be made using two spatially separated wire gratings with orthogonal directions of periodicity. The behavior of the crossed lamellar grating (for which the wires are rectangular in section) will be considered below.

7.4.2 Low-Pass Filters

Babinet's principle [7.21,22] may be used to derive the properties of a grid of infinitesimal thickness from that of the complementary structure, in which apertures are replaced by perfectly conducting metal and vice versa. In the case of a grid whose transmittance is unaffected when the polarization of the incident wave is rotated through 90°, Babinet's principle states that not only the power transmittances but also the complex amplitude transmittances for the complementary structures add up to unity [7.4].

As an application of the above, we note that a thin capacitive grid must act as a low pass filter. Its filtering action will improve as the gaps between metallic discs become larger, just as the filtering action of inductive grids improves with increasing wire width. We cannot comment on the effect of the thickness of the capacitive grid on its filtering action.

We are aware of no rigorous theoretical studies of capacitive grids of nonzero thickness. EGGIMANN and COLLIN [7.23] have considered the case of a planar array of circular discs of vanishingly small thickness, while OTT et al. [7.24] have considered an array of narrow plates.

7.4.3 Bandpass Filters

As we have seen, the transmission curve for an inductive grid may be divided into three regions, with the transmittance falling away on both sides of the resonance region. As pointed out by SAKAI and YOSHIDA [7.25], such a grid acts as a bandpass filter. However, its rejection characteristics on the high-frequency side of the passband are very different from those on the low-frequency side, since in the former region the total transmittance falls quickly to the geometrical optics value and varies little about this level, while in the latter region the transmittance falls off as the square of frequency [7.3]. In order to have good rejection on the high-frequency side, inductive grids with low hole-area fractions should be used. Of course, in this frequency region the total transmitted energy is shared between at least five propagating orders, so that the grid is not acting as a diplexer.

Other geometries capable of providing passband action have been discussed by ULRICH [7.5] and by ARNAUD and PELOW [7.26]. As well as using multigrid systems for this purpose, they have considered the properties of so-called "resonant grids". Of course, any grid will undergo resonances, but these may be heightened by incorporating extra capacitive features into an inductive geometry, for example. If the square apertures of an inductive grid are replaced by crosses, this brings the opposite sides of the holes closer together, and so heightens their capacitive action [7.5]. Apertures of quite complicated shape have been constructed and employed at quasi-optical frequencies [7.26].

7.4.4 Bandstop Filters

From Babinet's principle, we see that a grid composed of capacitive crosses can provide a satisfactory bandstop filter. The shape and width of the stop band can be adjusted by altering the aperture geometry, or by employing two or more spatially separated grids [7.5,26].

7.5 Spatially Separated Grids or Gratings

As we have seen above, the intrinsic properties of a single grid may sometimes be enhanced by constructing a system of several such elements separated in space. It is our purpose now to indicate how the properties of such structures may be analyzed rigorously. Two approaches will be discussed. In the first, the system is considered as an entity. Modal expansions applicable to the apertures of each element are related to the plane expansions existing between elements by the method of moments. The second treatment utilizes a multiple-scattering approach. The scattering matrix of the system viewed as a whole is derived from the scattering matrices of the individual components and transfer matrices which relate the field quantities between successive elements.

7.5.1 The Crossed Lamellar Transmission Grating

Here we consider the structure illustrated in Fig.7.7. This consists of a pair of nonidentical perfectly conducting lamellar transmission gratings separated in space and oriented with orthogonal periodicity axes. As has been discussed previously, the fields outside the grooves are expanded in Rayleigh series (7.6,7). It will be noted that four infinite sets of plane waves are necessary to specify field behavior above, between and below the gratings.

The structure of Fig.7.7 does not possess up-down symmetry. For this, and other reasons, the theoretical analysis developed for it by ADAMS et al. [7.27] is considerably less elegant than the formalism described in Sect.7.3. Because of this we choose not to present any details of the formulation here, other than a description of the form of the modal expansions.

Let us restrict our attention to the mode structure of the upper array. On the aperture walls $x = \pm c/2$ the relevant boundary conditions are

$$E_y = E_z = 0 \quad , \quad H_x = 0 \quad .$$

These three components can thus be built up from separable modes having an x dependence of the form $\sin\left[\frac{m\pi}{c}\left(x+\frac{c}{2}\right)\right]$, where m is a positive integer. No boundary conditions apply to the other three Cartesian field components of the electric and magnetic fields on the metal walls and so the appropriate form for their x dependence is $\cos\left[\frac{m\pi}{c}\left(x+\frac{c}{2}\right)\right]$, where m is a nonnegative integer.

The grooves of the upper lamellar grating are open ended in the z direction and so the modes depend on this variable in the same way as they would in free space. Thus the field components E_y, E_z and H_x are represented within the grooves of the upper grating as sums of modes of the form

$$\sin\left[\frac{m\pi}{c}\left(x+\frac{c}{2}\right)\right]\exp(i\gamma_q z)\begin{matrix}\sin\\ \cos\end{matrix}\left[\mu_m^q(y-s)\right] \quad . \tag{7.43}$$

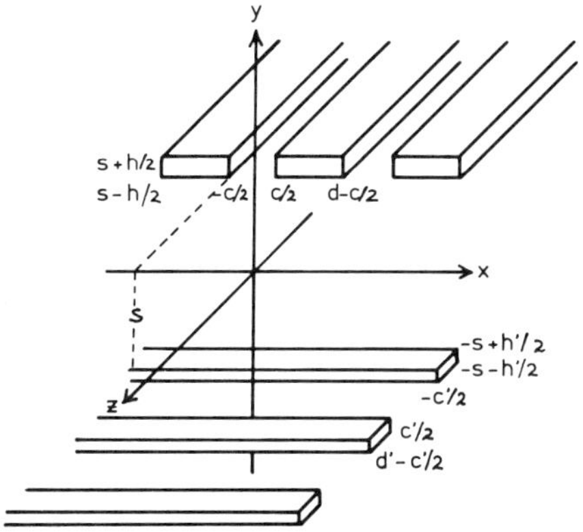

Fig. 7.7. The structure of the general crossed lamellar grating having orthogonal periodicity axes

In (7.43) the sine or cosine y dependence gives, respectively, antisymmetry or symmetry about the line $y=s$. The parameter γ_q is obtained from (7.3) (putting $\eta=90^\circ$), while the value of μ_m^q follows from the Helmholtz equation

$$\mu_m^q = \sqrt{k^2-\left(\frac{m\pi}{c}\right)^2-\gamma_q^2} \quad . \tag{7.44}$$

Modes for the other three components are obtained by replacing the sinusoidal x dependence by a cosinusoidal form.

In the lower grating the expression analogous to (7.43) applies to the field components E_x, E_y and H_z, and is

$$\exp(i\alpha_p x)\,\sin\left[\frac{n\pi}{c'}\left(z+\frac{c'}{2}\right)\right]\begin{matrix}\sin\\ \cos\end{matrix}\left[\nu_n^p(y+s)\right] \quad , \tag{7.45}$$

where

$$\nu_n^p = \sqrt{k^2-\alpha_p^2-\left(\frac{n\pi}{c'}\right)^2} \quad . \tag{7.46}$$

The diffraction problem is treated in terms of four Cartesian components (E_x,E_z, H_x,H_z), with Rayleigh expansions in three separated regions linked to one another by modal expansions in the upper and lower groove regions. This specification of the problem involves 32 infinite sets of coefficients — 16 sets of modal and 16 sets of Rayleigh coefficients. Maxwell's curl equations enable the elimination of all magnetic field coefficients, while the continuity conditions at the surfaces $y=s\pm(h/2)$ and $y=-s\pm(h'/2)$ can be used to remove the remaining eight sets of Rayleigh coefficients. Because of the absence of reflection symmetry about the plane $y=0$, no further reduction is possible: eight coupled sets of linear equations must be solved in order to obtain the eight infinite sets of modal coefficients. It is to be noted that one modal index runs over positive integral values, and may safely be truncated with a small upper limit, while the second index runs over the diffraction orders, with far more of these being necessary to achieve good accuracy. A typical problem involves the inversion of a complex matrix of dimension 120.

A complete description of this crossed grating formulation may be found in [7.28], together with details of its numerical verification and of a study of the use of the structure as a solar-selective element. This study showed that the crossed lamellar transmission grating and the inductive grid are equally well adapted to this application. A typical wavelength spectrum is shown in Fig.7.8. This curve possesses the same filtering, resonance and transmission regions already discussed for inductive grids. Once again, optimal transmission and filtering characteristics are obtained with square-symmetric structures, made up of narrow and deep rectangular wires. The additional parameter present for the spatially separated structure is the array separation S. As can be seen from Fig.7.9, this parameter controls the position of a broad capacitive-grid resonance on the shoulder of the main inductive-grid resonance.

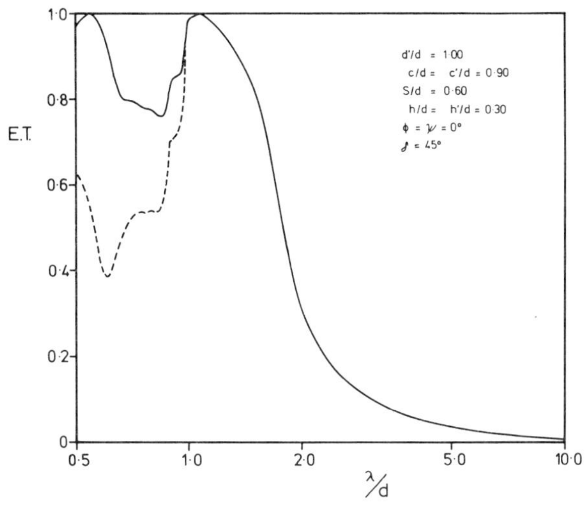

Fig. 7.8. A typical spectrum of a crossed lamellar grating. The solid curve gives the total transmittance, while the broken curve gives that of the zeroth order [7.27]

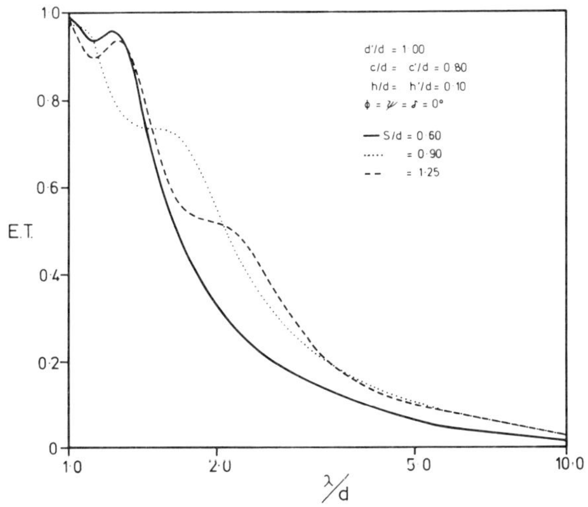

Fig. 7.9. The energy transmittance as a function of normalized wavelength is shown for three array separations of the crossed lamellar grating [7.27]

A judicious choice of the separation leads to a broadening of the transmission region, without adversely influencing the filtering action.

The above theory refers to the case of gratings arranged with orthogonal periodicity axes ($\eta=90^{\circ}$). The method has been generalized in current work by ADAMS at the University of Tasmania to deal with arbitrary values of η.

7.5.2 The Double Grating

If the inclination angle η *of the crossed lamellar grating approaches zero, we obtain a new structure, the double grating.* This is shown in Fig.7.10. It has been the subject of several experimental and theoretical investigations. CASEY and LEWIS [7.6]

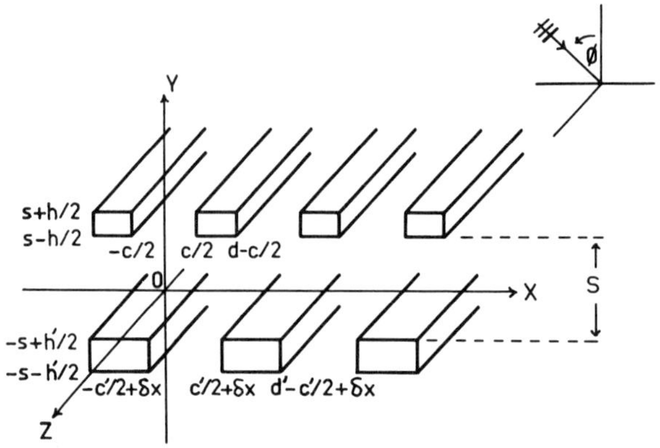

Fig. 7.10. The structure of the general double grating [7.30]

showed that a double grating (operated in P polarized light) could successfully replace the highly conducting thin films which were conventionally used as the reflecting elements in long wavelength Fabry-Perot interferometers. They demonstrated that the transmission properties of the grating interferometer were far superior to those of the conventional design. BLOK and MUR [7.29] have given a rigorous integral theory for double gratings in which the laminae composing each array were assumed to be perfectly conducting and of infinitesimal thickness. (In the discussion of double gratings, we shall call radiation P polarized if its electric vector is parallel to the wires, and S polarized if its magnetic vector is parallel to them.)

In this section we will briefly describe a rigorous modal formalism for double gratings composed of rectangular perfectly conducting wires. Fuller details may be found elsewhere [7.30].

The theory is in essence a much simplified version of the formalism for crossed lamellar gratings. The modal expansions used in the upper and lower gratings take exactly the same form as those already encountered for the single lamellar grating. As in the case of the crossed lamellar grating plane wave expansions in three separated regions are necessary. The plane wave expansions take the same form as those for a classical grating with period d_T, where d_T is the lowest common integral multiple of the periods of the upper and lower arrays, d and d'. A feature of the double grating having some theoretical interest is that if the ratio of d' to d is irrational the overall period of the structure becomes infinite and it gives rise to a continuous rather than a discrete spectrum of plane waves.

In general a knowledge of the field in one aperture of each grating is not sufficient to enable a solution of the diffraction problem. Instead we need to specify the behavior of the field in each of the (d_T/d) apertures of the upper grating lying within the overall period, and also within each of (d_T/d') apertures of the lower grating. The use of Maxwell's equations and the method of moments eliminates any reference to the plane wave coefficients and reduces the diffraction problem to one of solving four coupled sets of simultaneous equations in the four infinite sets of

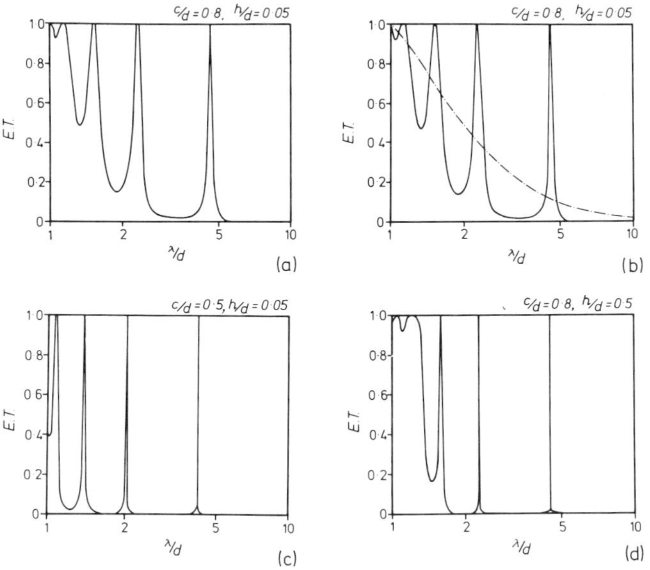

Fig. 7.11. P polarization normal incidence wavelength spectra for a double grating composed of a pair of identical lamellar arrays separated by $S = 2.0$ d and aligned such that $\delta x=0$. The solid curve (b) is the reconstruction of curve (a) using the multiple scattering method based on the transmission data for a single grating shown in the broken curve of (b) [7.30]

modal coefficients. These four sets of coefficients describe the y symmetric and y antisymmetric behavior of the fields in the upper and lower arrays.

In Fig.7.11a we show a typical interference spectrum for a double grating operated in P polarized light, computed using the theory described above. *The features of this curve may be readily understood using a multiple scattering treatment*. The crucial assumption in this form of analysis is that the zeroth order provides the only means of interaction between the two gratings. As has been shown recently [7.30] the evanescent orders can provide no significant coupling between gratings provided that their separation exceeds half of their period.

Let R_0 and T_0, respectively, denote the zeroth order amplitude reflection and transmission coefficients of each of two identical lamellar gratings. With the above assumption, the overall transmitted amplitude of the double grating is given by the geometric series

$$\tau = T_0^2 \sum_{\ell=0}^{\infty} (R_0\rho)^{2\ell} = \frac{T_0^2}{1-R_0^2\rho^2} , \qquad (7.47)$$

where $\rho = \exp(i\delta)$ and $\delta = (2\pi/\lambda)(S+h)\cos\phi$, with ϕ denoting the angle of incidence. The transmitted efficiency is

$$|\tau|^2 = \frac{1}{1+F\sin^2\zeta} \qquad (7.48)$$

where $\zeta = \delta + \arg(R_0)$ and F is the finesse of the interferometer

$$F = \frac{4|R_0|^2}{\left(1-|R_0|^2\right)^2} \quad .$$

Using a diffraction theory for a single lamellar grating, R_0 may be calculated, and used in (7.48) to find the transmission of the double grating for a given separation S. This technique was used to construct the curve of Fig.7.11b. The exact agreement between the curves of Figs.7.11a,b indicates the accuracy of the multiple scattering treatment.

The resolution of the interferometer increases with the finesse F and thus with $|R_0|$. Just as for the inductive grid, the filtering action of the lamellar grating for P polarized, long-wavelength radiation increases with the thickness parameter (h), and with the width of the rectangular wires (d-c). In other words, at a fixed wavelength in the filtering region, $|R_0|$ will increase with these parameters, as will the interferometer resolution. These properties are evident from the four curves of Fig.7.11.

The dependence of the interferometer transmission upon such parameters as c, h, λ and ϕ has been investigated [7.30] using a monomodal approach, of the type used by CHEN [7.11] and by McPHEDRAN and MAYSTRE [7.14] in their studies on inductive grids. In this approach, the modal expansion is reduced to a single, dominant term, and the resulting equations are so simplified as to be soluble in closed form. Monomodal expressions have been shown [7.14] to be quite accurate even in the wavelength region below the first Wood anomaly.

Thus far we have not considered the effects of the array phasing parameter δx. The multiple scattering treatment predicts no dependence of array transmission upon δx, and for large array separations $S/d \gtrsim 0.5$ this is in agreement with the numerical results obtained from the rigorous theory. However, for $S/d \lesssim 0.5$ the evanescent coupling becomes quite noticeable and a marked dependence upon δx is observed. This is illustrated in Fig.7.12 which shows the effects of array phasing on the transmission properties of a double grating with an array separation of $S/d = 0.02$.

One of the interesting features evident in this figure is the total blocking of P polarized radiation for $\delta x/d = 0.5$, but the significant "leakage" of S polarized radiation for the same structure. The blocking action for P polarization is readily understood: the bottom grating is here placed so as to fill in the apertures of the upper grating, and because surface currents run along the grating wires there is no mechanism which can lead to significant coupling of the incident field to the transmitted field region. However, for S polarization, induced surface currents run round the wires, enabling re-radiation from the lower surface of each grating into the transmitted field region.

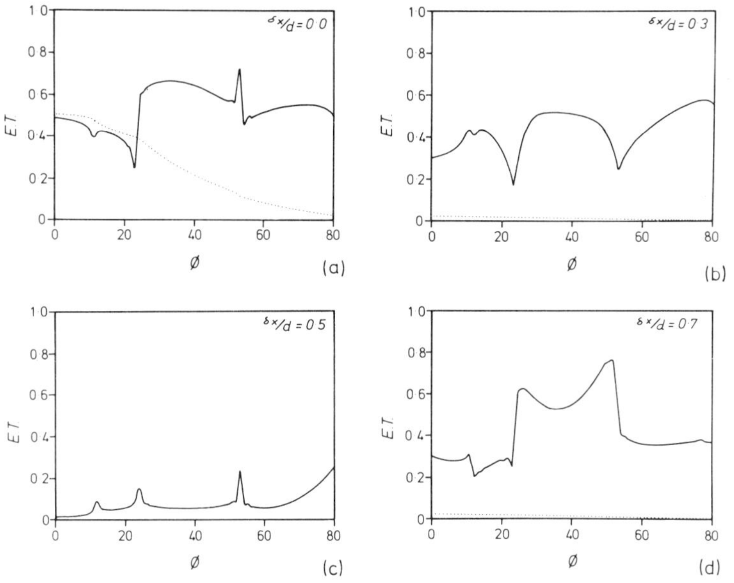

Fig. 7.12. The effect of the array displacement parameter δx. The structure has $d' = d$, $S = 0.02$ d, $c = 0.5$ d, $h = 0.4$ d and is used with radiation of wavelength $\lambda = 0.6$ d. P and S polarization spectra are shown in the broken and solid curves, respectively [7.30]

7.5.3 Symmetry Properties of Double Gratings

We will now consider two of a general class of properties which may be derived for lossless diffracting structures which possess some degree of structural symmetry. As shown elsewhere [7.31] such conservation relations may be established using the concept of time reversibility. GRIVET [7.32] has given a lucid discussion of this concept. In our context the following principle is sufficient: *the field distribution in the lossless systems is unmodified if the directions of all wave vectors are reversed and the conjugates of all complex amplitudes are taken.*

In the following analysis we will consider a plane wave I incident upon the lossless up-down and left-right symmetric structure shown in Fig.7.13. We will choose the wavelength (λ) and the angle of incidence (ϕ) such that there are only two plane wave output energy channels, R_0 and T_0 (state 1). Upon application of time reversibility, there are now two input channels, $\overline{R}_0$ and $\overline{T}_0$, and a single output channel $\overline{I}$ (state 2).

Let us now discuss the far field behavior. State 2 of Fig.7.13 may be regarded as the superposition of states 2A and 2B of that same figure. In these diagrams each plane wave is represented by a vector, associated with which is its amplitude. The bracketed quantity adjacent to the amplitude indicates the symmetry property required to obtain that value, with LR and UD denoting, respectively, left-right and up-down symmetries. Upon identifying states 1 and 2 we obtain the following constraints

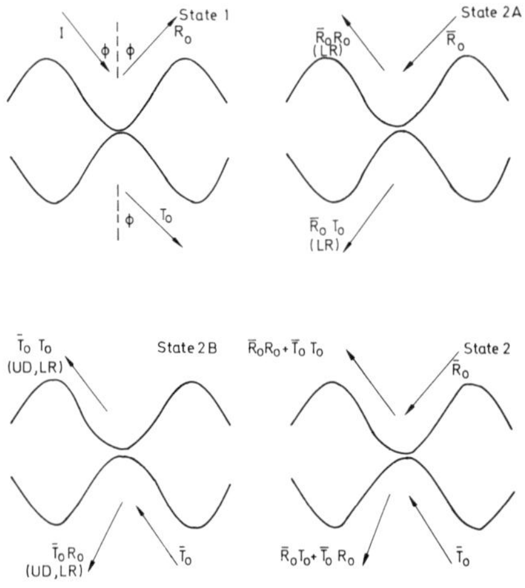

Fig. 7.13. The long-wavelength problem is shown in state 1 for a lossless double grating having up-down and left-right symmetries. In state 2A, the zeroth order in reflection is returned, while in state 2B the zeroth order in transmission is returned. By superposition of states 2A and 2B we obtain state 2, which may also be obtained by time reversal from state 1

$$|R_0|^2 + |T_0|^2 = 1 \quad , \tag{7.49}$$

$$\mathrm{Re}\{R_0\overline{T}_0\} = 0 \quad . \tag{7.50}$$

Equation (7.49) is simply a statement of conservation of energy while (7.50) indicates that

$$\arg\{T_0\} = \arg\{R_0\} \pm \pi/2 \quad . \tag{7.51}$$

This property of the reflected and transmitted zeroth orders of the double grating also holds true for inductive and capacitive grids [7.3,14].

The uniqueness condition of diffraction theory [7.33] is now invoked in order to extend these far-field results to all space. Here the assertion is that if two fields diffracted by the same structure exhibit the same far-field behavior, then their difference has zero far field and is zero everywhere in space. In the upper and lower free space regions D_u and D_ℓ, field distributions $E_u(x,y)$ and $E_\ell(x,y)$ are produced in response to the incident plane wave I. Time reversal then implies that the fields in D_u and D_ℓ excited by the incident fields $\overline{R}_0$ and $\overline{T}_0$ of state 2 are given by $\overline{E}_u(x,y)$ and $\overline{E}_\ell(x,y)$, respectively.

In D_u, an incident field $\overline{R}_0$ gives rise to a wave field $\overline{R}_0E_u(-x,y)$ because of LR symmetry. In order for us to be able to relate fields corresponding to an incident wave in D_ℓ to those applying in the initial situation, the additional assumption of UD symmetry must be invoked. Thus the form of this field corresponding to an incident wave of amplitude $\overline{T}_0$ in D_ℓ is $\overline{T}_0E_\ell(-x,-y)$. Hence for all UD, LR symmetric structures,

$$\overline{E}_u(x,y) = \overline{R}_0E_u(-x,y) + \overline{T}_0E_\ell(-x,-y) \quad . \tag{7.52}$$

Similarly, for fields in D_ℓ

$$\overline{E}_\ell(x,y) = \overline{T}_0 E_u(-x,-y) + \overline{R}_0 E_\ell(-x,y) \quad . \tag{7.53}$$

Equations (7.52,53) are the most general form of the constraints on the lossless symmetric system.

From (7.52,53) we can derive constraints on the amplitudes R_p, T_p of the pth orders (assumed to be evanescent) above and below the grating. On putting

$$A_p = R_p + T_p$$

and

$$B_p = R_p - T_p \quad ,$$

we may express these constraints in the form

$$2\arg\{A_p\} = \arg\{A_0\}$$

$$2\arg\{B_p\} = \arg\{B_0\} \quad .$$

The importance of the general field constraints (7.52,53) to the modal expansions of the double grating will now be considered. For structures of this type it is possible to expand the groove fields in terms of separable modes which are classified as follows:

a) LR symmetric and antisymmetric modes:

$$\mathcal{E}_m^x(x) = \mathcal{E}_m^x(-x) \quad , \quad \mathcal{O}_m^x(x) = -\mathcal{O}_m^x(-x)$$

and

b) UD symmetric and antisymmetric modes:

$$\mathcal{E}_m^y(y) = \mathcal{E}_m^y(-y) \quad , \quad \mathcal{O}_m^y(y) = -\mathcal{O}_m^y(-y) \quad .$$

Naturally, in these definitions the axes of the coordinate system must coincide with the axes of symmetry of the physical system. The modal functions are taken to be real and orthogonal, so that when the flux of energy across any surface $y = $ constant within a groove region is calculated, no coupling terms arise between different modes. Hence we may refer unambiguously to the "energy flux carried by a mode".

With these definitions, we may now write the field in the upper grooves as

$$E_u(x,y) = \sum_{m=1}^{\infty}\left[a_m^e \mathcal{O}_m^y(y') + b_m^e \mathcal{E}_m^y(y')\right]\mathcal{E}_m^x(x) + \sum_{m=1}^{\infty}\left[a_m^o \mathcal{O}_m^y(y') + b_m^o \mathcal{O}_m^y(y')\right]\mathcal{O}_m^x(x) \tag{7.54}$$

where

$$y' = y - s \quad .$$

In a similar manner the field in the lower set of grooves is specified by

$$E_\ell(x,y) = \sum_{m=1}^{\infty}\left[\hat{a}_m^e \mathcal{O}_m^y(y'')+\hat{b}_m^e \mathcal{E}_m^y(y'')\right]\mathcal{E}_m^x(x) + \sum_{m=1}^{\infty}\left[\hat{a}_m^o \mathcal{O}_m^y(y'')+\hat{b}_m^o \mathcal{O}_m^y(y'')\right]\mathcal{O}_m^x(x) \tag{7.55}$$

where

$$y'' = y + s \quad .$$

On substituting (7.54,55) into the general constraints (7.52,53) the following relations are obtained:

$$\overline{A}_m^e = \overline{M}\, A_m^e \tag{7.56}$$

$$\overline{B}_m^e = \overline{M}\, B_m^e \tag{7.57}$$

$$\overline{A}_m^o = -\overline{M}\, A_m^o \tag{7.58}$$

$$\overline{B}_m^o = -\overline{M}\, B_m^o \tag{7.59}$$

where

$$A_m^{e/o} = \begin{bmatrix} a_m \\ -\hat{a}_m \end{bmatrix}^{e/o} , \tag{7.60}$$

$$B_m^{e/o} = \begin{bmatrix} b_m \\ \hat{b}_m \end{bmatrix}^{e/o} \tag{7.61}$$

and

$$M = \begin{bmatrix} R_0 & T_0 \\ T_0 & R_0 \end{bmatrix} \quad . \tag{7.62}$$

Manipulation of (7.56-59) enables modal phase constraints such as

$$2\arg\left\{a_m^e+\hat{a}_m^e\right\} = \arg\{B_0\} \tag{7.63}$$

and

$$2\arg\left\{a_m^e-\hat{a}_m^e\right\} = \arg\{A_0\} \tag{7.64}$$

to be derived.

Equations (7.56-59) *enable us to establish a rather startling result which we refer to as the detailed balance of the mode structure in the upper and lower arrays of grooves*. Let us define this result before indicating its proof. In each of the upper and lower arrays of grooves it is possible to calculate a set of modal energy fluxes $\{F_m^{e/o}\}$ and $\{\hat{F}_m^{e/o}\}$, respectively. Because of the LR and UD symmetries, it is possible to show that the flux of each mode is array invariant, i.e.,

$$F_m^{e/o} = \hat{F}_m^{e/o} \quad \text{for all } m \quad . \tag{7.65}$$

This flux invariance is equivalent to

$$\mathrm{Im}\{a_m\bar{b}_m\}|^{e/o} = \mathrm{Im}\{\hat{a}_m\bar{\hat{b}}_m\}|^{e/o} \quad . \tag{7.66}$$

To demonstrate the detailed balance property (7.66) we note that the matrix M is unitary, as a consequence of (7.49,50). In deriving (7.66) we consider the product $\tilde{A}\bar{B}$, where A and B refer, respectively, to $A_m^{e/o}$ and $B_m^{e/o}$. Here a superscripted ~ denotes matrix transposition. Now,

$$\tilde{A}\bar{B} = \bar{\tilde{A}}\tilde{M}\bar{M}\bar{B} = \bar{\tilde{A}}B \quad , \quad \text{i.e.,} \quad \mathrm{Im}\{a_m\bar{b}_m - \hat{a}_m\bar{\hat{b}}_m\}|^{e/o} = 0 \quad .$$

Numerical examples of these properties may be found in [7.31].

7.5.4 Multielement Grating Interference Filters

In this section we will consider the problem of finding the transmission characteristics of an interference filter composed of (n+1) identical gratings or grids. *These will be used so that only the zeroth orders in reflection and transmission propagate and with a center to center separation t chosen such that these orders provide the only coupling between successive elements*. This problem has been considered by PRADHAN [7.10] and by PRADHAN and GARG [7.34,35], but the treatment given here relies upon more recent work [7.36].

As before, R_0 and T_0 denote the amplitude reflection and transmission coefficients of a single element, and satisfy (7.49,50). Also let R_n and T_n be the corresponding quantities for the (n+1) element structure, relative to a phase origin taken at the vertical center of the uppermost grid.

We may regard this (n+1) grating stack as a single element placed above an n element stack. The analogue of (7.47) for this system is

$$R_n = R_0 + R_{n-1}T_0^2\rho^2 \sum_{k=0}^{\infty} \left(R_0R_{n-1}\rho^2\right)^k = R_0 + \frac{R_{n-1}T_0^2\rho^2}{1-R_0R_{n-1}\rho^2} \quad , \tag{7.67}$$

where $\rho = \exp(i\,\frac{2\pi}{\lambda}\,t\cos\phi)$. Similarly,

$$T_n = \frac{T_{n-1}T_0}{1-R_0R_{n-1}\rho^2} \quad . \tag{7.68}$$

Equation (7.67) may be reduced to the simpler form

$$R_n = \frac{R_0 - R_{n-1}\rho^2\xi^2}{1 - R_0 R_{n-1}\rho^2} , \tag{7.69}$$

where $\xi = \exp[i \arg(R_0)]$. Putting

$$A = \frac{1}{R_0\rho^2} , \quad B = \frac{\xi^2}{R_0} , \quad C = \frac{1}{\rho^2}$$

and making the substitution

$$R_n = \frac{U_{n+1}}{U_n} + A \tag{7.70}$$

we can transform (7.69) into a second-order, homogeneous, linear difference equation

$$U_{n+1} - (B-A)U_n + (C-AB)U_{n-1} = 0 . \tag{7.71}$$

Such an equation has a solution of the form

$$U_n = V\alpha^n + W\beta^n , \tag{7.72}$$

where

$$\alpha = \frac{1}{2}[(B-A)-\sqrt{D}] , \quad \beta = \frac{1}{2}[(B-A)+\sqrt{D}] , \quad D = (A+B)^2 - 4C .$$

The constants V and W are specified by selecting the value of U_0 arbitrarily to be unity and by fixing U_1 in accordance with (7.70). Thus

$$V = \frac{\beta-(R_0-A)}{\beta-\alpha} , \quad W = \frac{(R_0-A)-\alpha}{\beta-\alpha} .$$

The final expression for R_n is then

$$R_n = \frac{(A+\alpha)V\alpha^n+(A+\beta)W\beta^n}{V\alpha^n+W\beta^n} . \tag{7.73}$$

From this, we can show that the positions at which transmission maxima occur are given by

$$\frac{2\pi}{\lambda} t \cos\phi + \arg(R_0) = \ell\pi \pm \frac{1}{2}\cos^{-1}\left[|R_0|^2 - (1-|R_0|^2)\cos\left(\frac{2\pi m}{n+1}\right)\right] \tag{7.74}$$

where ℓ is a nonnegative integer and m is an integer such that $1 \leq m \leq n$. The integer ℓ is called the order. From (7.74), each transmission peak for a Fabry-Perot interferometer (n=1) is split up into multiple peaks for an interferometer having three or more elements. The integer m runs over these multiple peaks, of which there are just one less than the number of elements in the system.

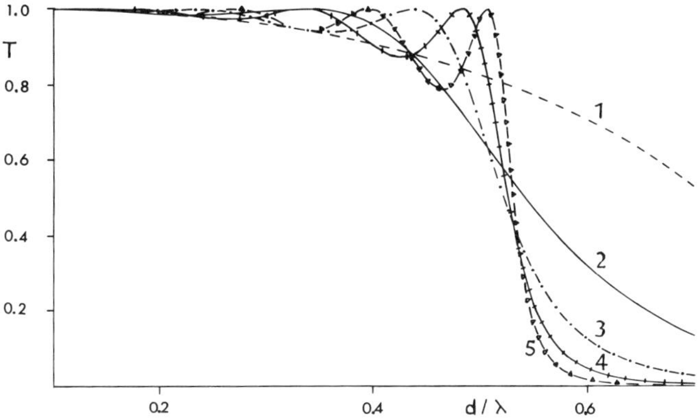

Fig. 7.14. The zeroth order transmittance is shown as a function of inverse wavelength, for normally incident radiation. Curve 1 represents the behavior of a single element, while curves are also given for a system composed of 2, 3, 4 and 5 such capacitive elements

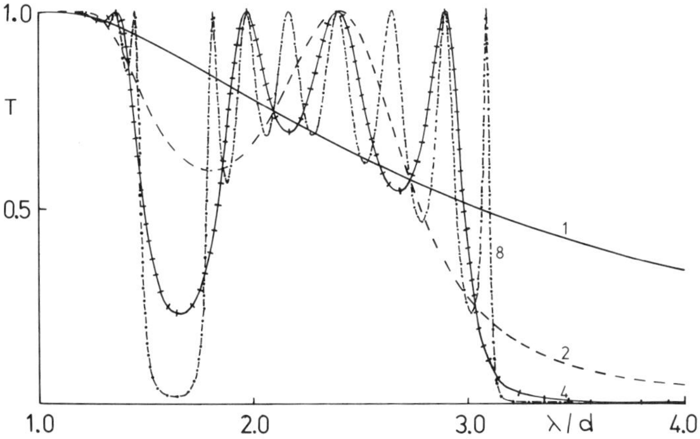

Fig. 7.15. The zeroth order transmittance is shown as a function of normalized wavelength for one inductive element, and for systems containing 2, 4 and 8 such elements

In Figs.7.14,15 (taken from [7.37]) we illustrate typical behavior of multielement systems composed, respectively, of capacitive and inductive grids. (If the system were to be used in linearly polarized rather than unpolarized radiation, then gratings rather than grids could be used, with the magnetic and electric field vectors of the incident wave being aligned along the grating wires in the two cases.) In Fig.7.14, we see that a very good low pass filter can be made up using a capacitive system of high multiplicity. In Fig.7.15 we illustrate the way in which an inductive system of high multiplicity can be used as a passband filter. A common feature of the two multiple systems is that the finesse increases with the number of elements. In Figs.7.14,15 this is manifest in the ever steepening edges of the passbands with increasing multiplicity.

7.6 Finitely Conducting Bigratings

In Sect.7.6 we shall consider the diffraction of a plane wave by a surface which is periodic along two orthogonal axes and which separates free space from a medium of complex refractive index ν. The method we will adopt to solve this problem is rather different from methods habitually used in the field of diffraction theory. For this reason we will preface our mathematical description of it with an explanatory passage in which we emphasize basic ideas worth knowing. After reading this introduction, the reader not interested in the mathematical details may wish to turn to Sect.7.6.5 in which we commence our discussion of the results obtained using the formalism.

7.6.1 A Short Description of the Method

We shall adopt a convention for labelling the coordinates axes which differs from that of Sect.7.2 in order to facilitate the use of tensor analysis. Set up a rectangular coordinate system $Ox_1'x_2'x_3'$ with the axis Ox_3' perpendicular to the plane of the grating and axes Ox_1' and Ox_2' aligned with the two axes of periodicity (Fig.7.16). The two periods, along Ox_1' and Ox_2', respectively, are denoted d_1 and d_2. Let the equation of the grating surface referred to these Cartesian coordinates be

$$x_3' = f(x_1',x_2') \quad . \tag{7.75}$$

As in Sect.7.2 the direction of the incoming plane wave may be specified by the two angles ϕ and ψ and its polarization by a third angle δ. For each ϕ, ψ and δ, it is required to calculate the efficiency for each propagating reflected order, and if the refractive index ν be real, then also the efficiency for each propagating transmitted order.

As we have seen, a variety of methods are available for the solution of the diffraction problem with a singly periodic surface. When an extra direction of period-

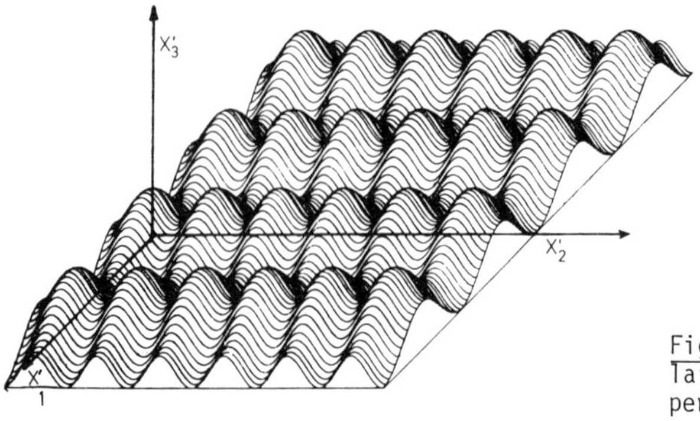

Fig. 7.16. A general doubly modulated surface having orthogonal periodicity axes [7.17]

icity is added, the number of plane waves with a propagation constant smaller than a given tolerance is approximately squared. This results in too great a number of unknowns if the diffraction problem is formulated using an integral equation approach culminating in a matrix inversion. Differential formulations have been developed for doubly periodic dielectric and conducting surfaces [7.38,39], but these give rise to intractable numerical problems when dealing with refractive indices of large modulus.

The formulation we will deal with here is one of a class which has received sporadic attention in the past [7.40-42]. *Their common factor is the use of a coordinate transformation which replaces the Cartesian coordinates* (x_1',x_2',x_3') *by a new set* (x_1,x_2,x_3) *chosen so the grating surface becomes the coordinate surface* $x_3 = 0$. The effect of such a coordinate transformation, already used for single periodic gratings by ELSON [7.41] and CHANDEZON et al. [7.42], is to simplify the boundary conditions which now apply at $x_3 = 0$. Conversely, the expression of Maxwell's equations is more complicated. After the field has been expanded in terms of a double Fourier series (with respect to x_1 and x_2) we are led to a differential system which can be deduced from the one linked with the physical coordinates (x_1',x_2',x_3') by introducing convenient "source functions". Using a Green's function technique this system can be written in an integral form. Because the "source functions" actually depend on the unknown functions, we obtain a set of coupled integral equations which is solved by an iterative technique to avoid huge matrix inversions.

7.6.2 The Coordinate Transformation

A suitable transformation is one of the form

$$
\begin{aligned}
x_1' &= x_1 \\
x_2' &= x_2 \\
x_3' &= x_3 + F(x_1,x_2,x_3) \quad .
\end{aligned}
\tag{7.76}
$$

Here F is arbitrary apart from the constraints

$$F(x_1,x_2,0) = f(x_1,x_2) \tag{7.77}$$

$$F(x_1,x_2,\infty) = g_1 \tag{7.78}$$

$$F(x_1,x_2,-\infty) = g_2 \tag{7.79}$$

$$1 + \partial F/\partial x_3 > 0 \quad , \tag{7.80}$$

where g_1 and g_2 are arbitrary constants. Equation (7.77) ensures that $x_3' = f(x_1',x_2')$ transforms into $x_3 = 0$, while (7.78,79) guarantee that far from the grating surface the physical and transformed far fields coincide, apart from irrelevant phase factors. Thus the outgoing wave condition is not affected by the coordinate transformation. Equation (7.80) is the Jacobian condition which ensures that the transformation be one-one. In the case of metallic gratings, no fields penetrate deeply so that(7.79) may be waived.

A simple choice of F which satisfies all the constraints (7.77-80) is

$$F(x_1,x_2,x_3) = K_1(x_3)\left[f(x_1,x_2)-g_1\right] + g_1 \qquad x_3 > 0 \ ,$$
$$= K_2(x_3)\left[f(x_1,x_2)-g_2\right] + g_2 \qquad x_3 < 0 \ ,$$

where

$$K_1(0) = 1 \ , \quad K_1(\infty) = 0 \ ,$$
$$K_2(0) = 1 \ , \quad K_2(-\infty) = 0 \ ,$$
$$dK_1/dx_3 \leqq 0 \quad x_3 > 0 \ ,$$
$$dK_2/dx_3 \geqq 0 \quad x_3 < 0 \ ,$$

and g_1 and g_2 are upper and lower bounds to $f(x_1,x_2)$.

After removing the time-dependent factor $\exp(-i\omega t)$, with $\omega/c = k = 2\pi/\lambda$, Maxwell's equations in Cartesian coordinates take the form

$$\mathrm{curl}'(\underline{E}') - ikZ_0\underline{H}' = 0$$
$$\mathrm{curl}'(Z_0\underline{H}') + ikn^2(x_3)\underline{E}' = 0 \ , \qquad (7.81)$$

where

$$n(x_3) = 1 \quad \text{for} \quad x_3 > 0 \ , \quad \text{i.e.,} \quad x_3' > f(x_1',x_2')$$
$$= \nu \quad \text{for} \quad x_3 < 0 \ , \quad \text{i.e.,} \quad x_3' < f(x_1',x_2') \qquad (7.82)$$

and Z_0 is the impedance of free space. Throughout Sect.7.6 a prime on a variable indicates that its components are taken with respect to the rectangular system $Ox_1'x_2'x_3'$, while unprimed variables refer to the coordinates x_1,x_2,x_3.

We seek solutions of (7.81) subject to the following boundary conditions:

a) There is an input plane wave in free space of spatial dependence (for large positive x_3) $\exp[i(k_1x_1+k_2x_2-k_3x_3)]$ with TE amplitude $\cos\delta$ and TM amplitude $\sin\delta$ (these being normalized to make the electric vector of unit magnitude). In the notation of Sect.7.2

$$k_1 = \gamma_{00} = k\sin\phi\sin\psi \ , \quad k_2 = \alpha_0 = k\sin\phi\cos\psi \ , \quad k_3 = \beta_{00} = k\cos\phi \ . \qquad (7.83)$$

b) All diffracted fields are purely outgoing.

c) The tangential components of $\underline{E}'$ and $\underline{H}'$ are continuous at the surface $x_3' = f(x_1', x_2')$.

Let us define covariant field vector components in the $x_1x_2x_3$ coordinate system in the standard way (see, for example, KORN and KORN [7.43])

$$E_i = E_j'(\partial x_j'/\partial x_i) \quad , \quad \text{i.e.,}$$

$$E_1 = E_1' + F_1E_3' \quad , \quad E_2 = E_2' + F_2E_3' \quad , \quad E_3 = (1+F_3)E_3' \tag{7.84}$$

where a repeated index implies summation and $F_i = \partial F/\partial x_i$. Similar relations pertain to the magnetic field components H_i. To transform (7.81) into the new variables we note that the left-hand member of each equation of the pair is a contravariant vector density of weight unity [7.43]. This enables us to write down the transformed equation immediately in terms of $E_1, E_2, E_3, H_1, H_2, H_3$ and the components of the metric $g^{ij} = \delta_{\ell m}(\partial x_i/\partial x_\ell')(\partial x_j/\partial x_m')$. For details see [7.17,44]. Writing $\underline{E} = (E_1, E_2, E_3)$ and $\underline{H} = (H_1, H_2, H_3)$ the transformed equations are

$$\mathrm{curl}\,(\underline{E}) - ikZ_0(\underline{H}+\underline{\sigma}) = 0$$

$$\mathrm{curl}(Z_0\underline{H}) + ikn^2(x_3)(\underline{E}+\underline{\tau}) = 0 \quad . \tag{7.85}$$

In (7.85), "curl" is to be interpreted in the same way as in Cartesian coordinates – i.e., $[\mathrm{curl}(\underline{E})]_1 = (\partial E_3/\partial x_2) - (\partial E_2/\partial x_3)$, etc. These equations differ from Maxwell's equations by short-range source terms $\underline{\sigma} = (\sigma_1, \sigma_2, \sigma_3)$ and $\underline{\tau} = (\tau_1, \tau_2, \tau_3)$ given by[1]

$$\sigma_1 = F_3H_1 - F_1H_3$$

$$\sigma_2 = F_3H_2 - F_2H_3$$

$$\sigma_3 = -F_1(\sigma_1+H_1) - F_2(\sigma_2+H_2) - F_3(\mathrm{curl}\,\underline{E})_3/(ik) \quad . \tag{7.86}$$

$$\tau_1 = F_3E_1 - F_1E_3$$

$$\tau_2 = F_3E_2 - F_2E_3$$

$$\tau_3 = -F_1(\tau_1+E_1) - F_2(\tau_2+E_2) + F_3(\mathrm{curl}\,\underline{H})_3/\left[ikn^2(x_3)\right] \quad . \tag{7.87}$$

The continuity conditions become very simple in the new variables: E_1, E_2, H_1, H_2 are continuous across the plane $x_3 = 0$. This is readily proved by noting that these field components are simple linear combinations of the parallel components of $\underline{E}'$ and $\underline{H}'$ at the grating surface.

---

1 Note that henceforth we will incorporata a factor Z_0 in the magnetic field components H_1, H_2, H_3.

7.6.3 Integral Equation Form

We can write the solution of (7.85) with boundary conditions as integrals over products of the source functions $\underline{\sigma}$, $\underline{\tau}$ with suitable kernels of the Green's function type. *Since the source terms are themselves functions of the unknown fields we obtain thereby a set of coupled integral equations, whose superiority over the differential system lies in their explicit incorporation of the boundary conditions.*

First let us exploit the pseudoperiodicity of the fields with respect to X_1 and x_2 and expand the dependence on thse two variables in terms of the complete set related to the input wave vector of (7.83)

$$\chi_{rs}(x_1,x_2) = (d_1d_2)^{-\frac{1}{2}} \exp[i(k_{1r}x_1+k_{2s}x_2)] \quad ,$$

where

$$k_{1r} = k_1 + 2\pi r/d_1 \quad , \quad k_{2s} = k_2 + 2\pi s/d_2 \tag{7.88}$$

for all positive, zero or negative integral values of r,s. Each component of $\underline{E}$, $\underline{H}$, $\underline{\sigma}$, $\underline{\tau}$ is written as a linear combination of the above complete set with Fourier coefficients of order r,s denoted by a subscript rs. For example, we write

$$E_1(x_1,x_2,x_3) = \sum_{r,s=-\infty}^{\infty} E_{1rs}(x_3)\chi_{rs}(x_1,x_2)$$

where

$$E_{1rs}(x_3) = \int_0^{d_1} dx_1 \int_0^{d_2} dx_2 E_1(x_1,x_2,x_3)\bar{\chi}_{rs}(x_1,x_2) \quad .$$

Substituting such expansions into (7.85) then leads to the following system:

$$\frac{d}{dx_3} \mathcal{G}_{rs}(x_3) - \mathcal{L}_{rs}\mathcal{G}_{rs}(x_3) = \rho_{rs}(x_3) \tag{7.89}$$

$$E_{3rs}(x_3) = -\tau_{3rs}(x_3) - \left[k_{1r}H_{2rs}(x_3)-k_{2s}H_{1rs}(x_3)\right]\Big/\left[n^2(x_3)k\right] \tag{7.90}$$

$$H_{3rs}(x_3) = -\sigma_{3rs}(x_3) + \left[k_{1r}E_{2rs}(x_3)-k_{2s}E_{1rs}(x_3)\right]/k \quad . \tag{7.91}$$

Here $\mathcal{G}_{rs}(x_3)$ is the column vector formed from the four components $E_{1rs}(x_3)$, $E_{2rs}(x_3)$, $H_{1rs}(x_3)$, $H_{2rs}(x_3)$, and $\rho_{rs}(x_3)$ is the column vector

$$\rho_{rs}(x_3) = i\begin{bmatrix} -k_{1rs}\tau_{3rs}(x_3)+k\sigma_{2rs}(x_3) \\ -k_{2s}\tau_{3rs}(x_3)-k\sigma_{1rs}(x_3) \\ -n^2(x_3)k\tau_{2rs}(x_3)-k_{1r}\sigma_{3rs}(x_3) \\ n^2(x_3)k\tau_{1rs}(x_3)-k_{2s}\sigma_{3rs}(x_3) \end{bmatrix} \quad . \tag{7.92}$$

$\mathcal{L}_{rs}$ is a 4×4 matrix which is spatially constant apart from the change associated with the transition in refractive index $n(x_3)$ from 1 to ν at $x_3=0$

$$\mathcal{L}_{rs} = (i/k)\begin{bmatrix} 0 & \ell_{rs}/n^2(x_3) \\ -\ell_{rs} & 0 \end{bmatrix}$$

with

$$\ell_{rs} = \begin{bmatrix} k_{1r}k_{2s} & k^2-k_{1r}^2 \\ -k^2+k_{2s}^2 & -k_{1r}k_{2s} \end{bmatrix} .$$

Regarding $\rho_{rs}(x_3)$ as a source function (7.89) has the solution

$$\mathcal{G}_{rs}(x_3) = [\exp(\mathcal{L}_{rs}x_3)]\mathcal{G}_{rs}(0) + \int_0^{x_3} dt \exp[\mathcal{L}_{rs}(x_3-t)]\rho_{rs}(t) \tag{7.93}$$

where no ambiguity is attached to the symbol $\mathcal{G}_{rs}(0)$ since all its components are continuous across the boundary $x_3=0$.

In order to evaluate the matrix exponentials in (7.93) it is useful to consider the eigenvalues and eigenvectors of the matrix $\mathcal{L}_{rs}$ [Ref.7.43, pp.403-424,pp.457-463]. The four eigenvalues occur in degenerate pairs and take the values $\pm ik_{3rs}$ where

$$k_{3rs} = \left[n^2(x_3)k^2-k_{1r}^2-k_{2s}^2\right]^{\frac{1}{2}} .$$

The square root is chosen to give k_{3rs} a positive imaginary part for evanescent orders or a positive real part for the propagating orders. For the degenerate eigenvalue pair ik_{3rs} the right eigenvectors may be taken as

$$u_{1rs} = \left(k_{1r}^2+k_{2s}^2\right)^{-\frac{1}{2}}\begin{bmatrix} -k_{2s} \\ k_{1r} \\ -k_{1r}k_{3rs}/k \\ -k_{2s}k_{3rs}/k \end{bmatrix}$$

$$u_{2rs} = \left(k_{1r}^2+k_{2s}^2\right)^{-\frac{1}{2}}\begin{bmatrix} -k_{1r}k_{3rs}/k \\ -k_{2s}k_{3rs}/k \\ n^2(x_3)k_{2s} \\ -n^2(x_3)k_{1r} \end{bmatrix} \tag{7.94}$$

with corresponding left eigenvectors

$$v_{1rs} = \frac{1}{2}\left(k_{1r}^2+k_{2s}^2\right)^{-\frac{1}{2}}\left[-k_{2s},k_{1r},-k_{1r}k/k_{3rs},-k_{2s}k/k_{3rs}\right]$$

$$v_{2rs} = \frac{1}{2}\left(k_{1r}^2+k_{2s}^2\right)^{-\frac{1}{2}}\left[-k_{1r}k/k_{3rs},-k_{2s}k/k_{3rs},k_{2s}/n^2(x_3),-k_{1r}/n^2(x_3)\right] . \tag{7.95}$$

The right eigenvectors u_{3rs}, u_{4rs} and left eigenvectors v_{3rs}, v_{4rs} belonging to the degenerate eigenvalue pair $-ik_{3rs}$ are obtained by replacing k_{3rs} by $-k_{3rs}$ in (7.94,95). Note that the value of refractive index appropriate to each region must be inserted, viz., 1 or ν, giving a different set of eigenvalues and eigenvectors for $x_3 > 0$ and $x_3 < 0$. When necessary a superscript +,- will be affixed to eigenvariables to indicate to which region they belong. The eigenvectors have been normalized to give the scalar products $v_{irs}u_{jrs} = \delta_{ij}$.

Now $u_{1rs}\exp(ik_{3rs}x_3)$, $u_{2rs}\exp(ik_{3rs}x_3)$, $u_{3rs}\exp(-ik_{3rs}x_3)$ and $u_{4rs}\exp(-ik_{3rs}x_3)$ are four linearly independent solutions of the homogeneous equation obtained from (7.89) by setting the source term zero, i.e., of the flat-space Maxwell equations. *They correspond in the vacuum region* $x_3 > 0$, *respectively, to E-outgoing, TM-outgoing, TE-incoming and TM-incoming plane waves.* For $x_3 < 0$ the roles of outgoing and incoming waves are reversed.

We now determine $\mathcal{G}_{rs}(0)$ from the boundary conditions at infinity. As $x_3 \to +\infty$

$$v_{3rs}\mathcal{G}_{rs}(x_3) \to \delta_{r0}\delta_{s0}\cos\delta\, \exp(-ik_3x_3)$$

$$v_{4rs}\mathcal{G}_{rs}(x_3) \to \delta_{r0}\delta_{s0}\sin\delta\, \exp(-ik_3x_3)$$

and as $x_3 \to -\infty$

$$v_{1rs}\mathcal{G}_{rs}(x_3) \to 0$$

$$v_{2rs}\mathcal{G}_{rs}(x_3) \to 0 \quad .$$

Substituting the above conditions into (7.93) yields for each r,s four linear algebraic equations which may be solved for the four components of $\mathcal{G}_{rs}(0)$. The amplitudes V^+_{1rs} and V^+_{2rs} for TE-outgoing and TM-outgoing waves in the vacuum can now be obtained from (7.93) as the limiting values of $v_{1rs}\mathcal{G}_{rs}(x_3)\exp(-ik_{3rs}x_3)$ and $v_{2rs}\mathcal{G}_{rs}(x_3)\exp(-ik_{3rs}x_3)$, respectively, as $x_3 \to \infty$, while the TE-outgoing and TM-outgoing amplitudes in the grating material, V^-_{3rs} and V^-_{4rs}, are the limits of $v_{3rs}\mathcal{G}_{rs}(x_3)\exp(ik_{3rs}x_3)$ and $v_{4rs}\mathcal{G}_{rs}(x_3)\exp(ik_{3rs}x_3)$, respectively, for $x_3 \to -\infty$. The explicit solution is

$$V^+_{1rs} = \delta_{r0}\delta_{s0}\cos\delta\, R_{Ers} + (1-R_{Ers})\int_{-\infty}^{0} dt\, v_{1rs}(t)\rho_{rs}(t)$$
$$+ \int_0^\infty dt\left[v_{1rs}(t)\rho_{rs}(t) - R_{Ers}v_{3rs}(t)\rho_{rs}(t)\right]$$

$$V^+_{2rs} = \delta_{r0}\delta_{s0}\sin\delta\, R_{Mrs} + \nu^2(1-R_{Mrs})\int_{-\infty}^{0} dt\, v_{2rs}(t)\rho_{rs}(t)$$
$$+ \int_0^\infty dt\left[v_{2rs}(t)\rho_{rs}(t) - R_{Mrs}v_{4rs}(t)\rho_{rs}(t)\right]$$

$$V_{3rs}^{-} = \delta_{r0}\delta_{s0}\cos\delta(1+R_{Ers}) - (1+R_{Ers})\int_{0}^{\infty} dt\ v_{3rs}(t)\rho_{rs}(t)$$
$$-\int_{-\infty}^{0} dt\left[R_{Ers}v_{1rs}(t)\rho_{rs}(t)+v_{3rs}(t)\rho_{rs}(t)\right]$$
$$V_{4rs}^{-} = \delta_{r0}\delta_{s0}\sin\delta(1+R_{Mrs})/\nu^2 - (1+R_{Mrs})(1/\nu^2)\int_{0}^{\infty} dt\ v_{4rs}(t)\rho_{rs}(t)$$
$$-\int_{-\infty}^{0} dt\left[R_{Mrs}v_{2rs}(t)\rho_{rs}(t)+v_{4rs}(t)\rho_{rs}(t)\right] . \tag{7.96}$$

We have introduced above the reflection coefficients

$$R_{Ers} = \left(k_{3rs}^{+}-k_{3rs}^{-}\right)/\left(k_{3rs}^{+}+k_{3rs}^{-}\right)$$
$$R_{Mrs} = \left(\nu^2 k_{3rs}^{+}-k_{3rs}^{-}\right)/\left(\nu^2 k_{3rs}^{+}+k_{3rs}^{-}\right)$$

and defined

$$v_{1rs}(t) = v_{1rs}\exp(-ik_{3rs}t) \quad , \quad v_{2rs}(t) = v_{2rs}\exp(-ik_{3rs}t) \quad ,$$
$$v_{3rs}(t) = v_{3rs}\exp(ik_{3rs}t) \quad , \quad v_{3rs}(t) = v_{4rs}\exp(ik_{3rs}t) \quad .$$

In the expressions for the reflection coefficients above and in what follows a superscript +,- indicates that a quantity is evaluated using the refractive index values appropriate to the vacuum or grating, respectively, i.e., 1, ν. In the integrations of (7.96) a similar convention applies in the choice of the refractive index used in the evaluation of $v_{irs}(t)$.

The fields at the grating surface are

$$\mathcal{G}_{rs}(0) = u_{1rs}^{-}\int_{-\infty}^{0} dt\ v_{1rs}(t)\rho_{rs}(t) + u_{2rs}^{-}\int_{-\infty}^{0} dt\ v_{2rs}(t)\rho_{rs}(t)$$
$$+ u_{3rs}^{-}\left[V_{3rs}^{-}+\int_{-\infty}^{0} dt\ v_{3rs}(t)\rho_{rs}(t)\right]$$
$$+ u_{4rs}^{-}\left[V_{4rs}^{-}+\int_{-\infty}^{0} dt\ v_{4rs}(t)\rho_{rs}(t)\right] . \tag{7.97}$$

We have now completed the specification of the integral equations. Equations (7.90-97) express the fields $\underline{E}_{rs}(x_3)$ and $\underline{H}_{rs}(x_3)$ in terms of integrals involving the short-range sources $\underline{\sigma}_{rs}(x_3)$ and $\underline{\tau}_{rs}(x_3)$, with no coupling terms between different values of r,s. The sources in turn depend on $\underline{E}_{rs}(x_3)$ and $\underline{H}_{rs}(x_3)$ through (7.86,87), which couple together different rs orders through convolutions of the fields with the Fourier transform of the transformation function $F(x_1,x_2,x_3)$. The continuity conditions at $x_3=0$ and the boundary conditions at $x_3=\pm\infty$ have been explicitly incorporated into the integral equation system.

Having obtained a solution of these equations by some means, such as the iterative technique described in the next section, the efficiencies for propagating orders follow immediately from the amplitudes given in (7.96). For an order in free space the efficiency is

$$\eta_{rs} = \left(k^{+}_{3rs}/k_3\right)\left(|V^{+}_{1rs}|^2+|V^{+}_{2rs}|^2\right) \quad , \tag{7.98}$$

while if ν is real the corresponding efficiency transmitted into the grating is

$$\hat{\eta}_{rs} = \left(k^{-}_{3rs}/k_3\right)\left(|V^{-}_{3rs}|^2+\nu^2|V^{-}_{4rs}|^2\right) \quad . \tag{7.99}$$

For ν complex the rate of energy dissipation may be expressed as the integral of the normal component of the Poynting vector over the surface of the grating. In our coordinate system $x_1x_2x_3$ this integral is trivial to compute, and yields for the normalized rate of energy dissipation

$$\eta_D = \sum_{rs} \eta_{Drs}$$

with

$$\eta_{Drs} = -(k/k_3)\mathrm{Re}\left\{E_{1rs}(0)\bar{H}_{2rs}(0)-E_{2rs}(0)\bar{H}_{1rs}(0)\right\} \tag{7.100}$$

The fields at $x_3 = 0$ are given by (7.97). Thus we have a useful check on the accuracy of the calculations in any departure from zero of the energy defect

$$\mathrm{E.D.} = 1 - \eta_D - \sum \eta_{rs} - \sum \hat{\eta}_{rs} \quad , \tag{7.101}$$

where the summations are restricted to propagating orders.

7.6.4 Iterative Solution of the Integral Equations

Direct solution of the integral equations of the previous section is impracticable on account of the huge matrices involved. However, the system lends itself to solution by an iterative technique. We start from some value of the sources $\underline{\sigma}_{rs}(x_3)$, $\underline{\tau}_{rs}(x_3)$, or equivalently $\rho_{rs}(x_3)$, $\sigma_{3rs}(x_3)$, $\tau_{3rs}(x_3)$, and compute $\underline{E}_{rs}(x_3)$, $\underline{H}_{rs}(x_3)$ by evaluating the integrals of (7.93-97) and applying (7.90,91). During this stage there is no coupling between different orders or polarizations in that the calculation of the integrals and their linear combinations proceeds separately for each r,s and polarization (TE or TM). The second phase of the iteration is the recreation of the sources $\underline{\sigma}_{rs}(x_3)$, $\underline{\tau}_{rs}(x_3)$ using (7.86,87). This involves convolution of the field vectors with the Fourier transform (with respect to x_1, x_2) of the transformation function $F(x_1,x_2,x_3)$. This operation introduces coupling between the different orders and polarizations but proceeds separately for each value of x_3. During each cycle of the iteration one obtains the current estimate of the efficiencies and energy dissipation from (7.98,99). The process is stopped when these quantities become stable and the energy defect becomes acceptably small.

The conditions of convergence may be deduced by a matrix representation of the iterative scheme. Let us denote by G the column vector formed from the values of $\underline{E}_{rs}(x_3)$, $\underline{H}_{rs}(x_3)$ as r,s,x_3 range over all allowed values. For computing purposes the latter must be a finite subset of the infinite domains of these variables. Typically, r and s might each range from -3 to +3 and x_3 take 60 discrete values, giving $6 \times 7 \times 7 \times 60 = 17640$ components to G. Similarly let us collect together into a single column vector ρ all values of $\rho_{rs}(x_3)$, $\sigma_{3rs}(x_3)$, $\tau_{3rs}(x_3)$. Then the iterative scheme is of the form

$$G_n = G_f + N\,\rho_n \tag{7.102}$$

$$\rho_{n+1} = M\,G_n \quad , \tag{7.103}$$

where G_n and ρ_n denote the estimates of G and ρ after n iterations. G_f is the Fresnel field appropriate to a plane interface and N and M are constant square matrices associated with the integration and convolution processes, respectively. Combining (7.102,103) we obtain

$$G_{n+1} = G_f + NM\,G_n = \sum_{\ell=0}^{n-1} (NM)^{\ell} G_f + (NM)^n G_1 \quad . \tag{7.104}$$

Clearly the process will converge to $(I-NM)^{-1}G_f$ if the modulus of the largest eigenvalue of NM is less than 1 and diverge otherwise [Ref.7.43, p.415]. The convolution matrix M vanishes for zero groove depth, suggesting that the iterative method will succeed for small groove depths and fail for deep grooves. Numerical experience indicates convergence for groove depths up to about 30-40% of the period, depending on wavelength. However convergent results can sometimes be obtained for depths up to 100% of the period by using the modified iterative scheme

$$G_n = G_f + N\,\rho_n$$

$$\rho_{n+1} = (1-p)M\,G_n + p\rho_n$$

where p is a complex number, and by projecting out the largest eigenvalues of NM [7.44].

7.6.5 Total Absorption of Unpolarized Monochromatic Light

As we saw in Chaps.5 and 6, one of the most surprising predictions of classical grating theory was that the use of a shallow surface modulation can transform a highly

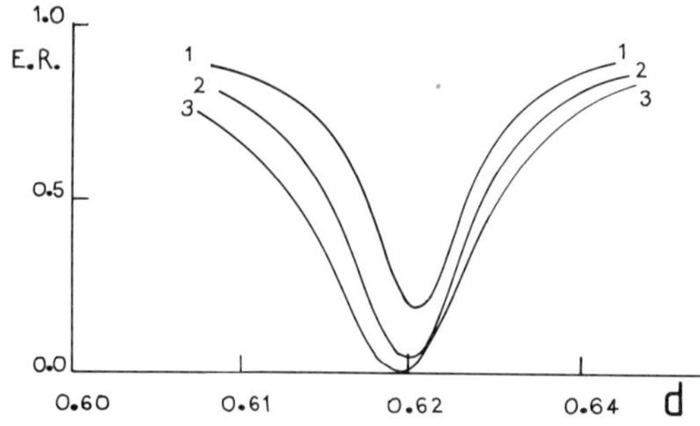

Fig. 7.17. The energy reflected by a crossed grating in gold used with normally incident light of wavelength $\lambda = 0.65$ μm. The profile is a square-symmetric sum of sinusoids with groove depth 1. $H = 0.0.40$ μm; 2. $H = 0.055$ μm; 3. $H = 0.070$ μm [7.17]

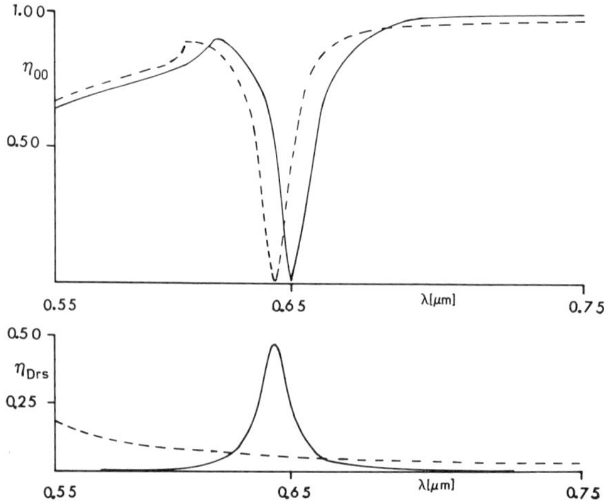

Fig. 7.18. In the upper curves, theoretical (broken line) and experimental (solid line) curves are given for the zeroth-order efficiency, for a crossed grating in gold having a groove depth $H = 0.080$ μm and a period 0.60 μm, used with normally incident light. In the lower curves, the solid line gives the energy dissipated in the order (1,0), while the broken line gives the same quantity for order (0,0)

reflecting metallic surface into one capable of totally absorbing incident light having a particular polarization and wavelength. *A crossed grating with a profile exhibiting square symmetry has a total reflected energy which is independent of polarization angle* (δ) *for normally incident light. An interesting question is whether this reflectance can be annulled, given a suitable choice of the grating profile parameters.*

In order to answer this question, a classical grating theory was first used to obtain a totally absorbing surface for normally incident light of $H_{\parallel}$ polarization and wavelength 0.65 μm. This was achieved with a sinusoidal grating in gold having a period $d = 0.62$ μm and a groove depth $h = 0.033$ μm. The crossed-grating formalism was then used to study the zeroth-order reflectance as a function of groove depth H, for a doubly modulated surface having a profile given by

$$f(x_1',x_2') = \frac{H}{4}\left(\sin\frac{2\pi x_1'}{d} + \sin\frac{2\pi x_2'}{d}\right) \quad .$$

In Fig.7.17 we show curves giving the zeroth-order reflectance as a function of the period d, for three values of the groove depth. It is interesting that in Fig.7.17 the minimum of η_{00} occurs for a value of the period d again close to 0.62 μm, but for a groove depth H very different from 0.033 μm. In fact, further calculations have shown that η_{00} falls to zero (to within numerical accuracy) for a groove depth H equal to 0.067 ± 0.002.

An experimental confirmation [7.45] of the theoretical prediction concerning the total absorption of unpolarized light has been undertaken. A crossed grating having a period of 0.60 μm and a total depth of 0.08 μm was made by recording interference fringes in a photoresist, and was then covered with gold. In Fig.7.18 we compare the measured reflectance curve with one calculated using the coordinate-transformation formalism. The only discrepancy between the two curves is a small wavelength shift (of about 0.005 μm). This we attribute either to a small period difference between the theoretical and experimental gratings, or to a slight inaccuracy in the experimental wavelength scale. (It will be noted that not only the reflectance minimum, but also the Wood anomaly at a wavelength equal to the grating period has been displaced in the two curves.)

The second set of curves in Fig.7.18 shows the energy dissipated in the gold grating, for the orders (0,0) and (1,0). It is a feature of the transformation formalism that in the unprimed coordinates the contributions made by the various spatial harmonics (r,s) to the energy dissipation in the metal are not coupled [see (7.100)]. In other words, it is possible to evaluate numerically the energy dissipated by each order. From Fig.7.18 we see that the energy dissipation in the order (1,0) undergoes a resonance in the wavelength region near 0.634 μm. The energy dissipation in the order (-1,0) is exactly equal to that in the order (1,0), as a consequence of the choice of the polarization of the incident wave. The energies dissipated by other orders in the metal do not resonate at all near $\lambda = 0.634$ μm, with only that due to the order (0,0) being significant. It is a feature of gratings with shallow grooves that orders can undergo resonant behavior without affecting adjacent orders. For gratings with deep grooves, all evanescent orders tend to resonate simultaneously.

7.6.6 Reduction of Metallic Reflectivity: Plasmons and Moth-Eyes

We have seen that the modulation of the surface of a metal may lead to a substantial diminution of its reflectance over a narrow wavelength band. This effect is associated with a resonance in the energy dissipated by a diffracted order which has just become evanescent, and is said to be due to the excitation of a surface plasmon. *It is an interesting question whether this effect could be exploited in making a roughened metal surface solar-selective, i.e., highly absorbing throughout the visible region, and highly reflecting in the near infrared.*

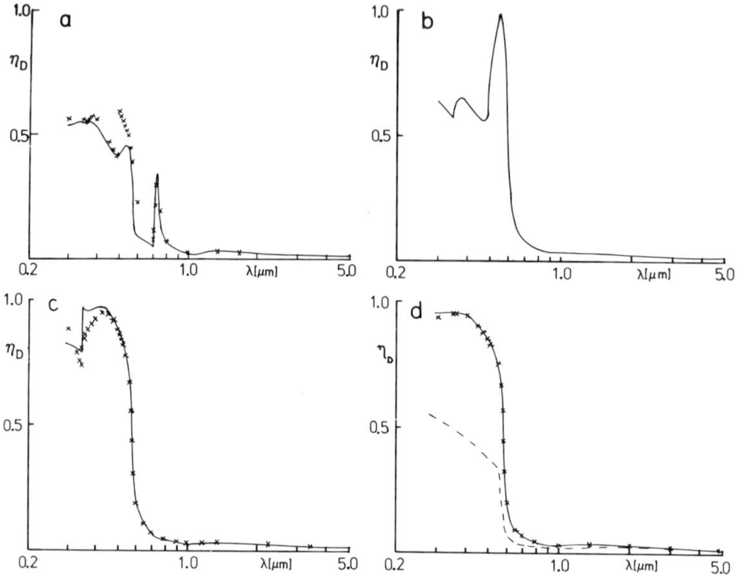

Fig. 7.19. The energy absorbed by crossed gratings in copper, having a groove depth of 0.20 μm, as a function of wavelength for normally incident light. The solid line represents the theoretical curve obtained using the transformation formalism, while the points have been obtained using the equivalence formula. (a) d = 0.7071 μm; (b) d = 0.50 μm; (c) d = 0.35 μm; (d) d = 0.20 μm

In Fig.7.19 we show a series of curves which represent the first results of a numerical study of this question. The energy absorbed by a copper surface sinusoidally modulated in two orthogonal directions is shown as a function of wavelength, for normally incident radiation. The curves for bigrating periods of 0.707 μm and 0.50 μm have quite localized plasmon absorption peaks, rather too narrow to significantly affect the absorptance integrated over the solar spectrum. However, the curves for finer bigrating periods show a greatly enhanced absorption throughout the region where the Fresnel reflectance of copper is smaller than 0.6.

This effect occurs over far too wide a wavelength band to be attributable to surface plasmon resonances, and in fact has a different origin. A hint as to its nature may be found by examining the cornea of certain nocturnal insects, which is covered with a fine array of conical protuberances, typically of about 0.2 μm height and spacing [7.46]. CLAPHAM and HUTLEY [7.47] have made such an array in photoresist, and have interpreted its properties in terms of the protuberances providing a gradual transition between the refractive index of air and that of the substrate. THORNTON [7.48] has used the same explanation of the "moth-eye effect", and has shown how it may be optimized over a wide bandwidth.

Our experience with diffraction gratings has led us to doubt the validity of the model of a corrugated surface in which it is replaced by a graded refractive index layer, if the wavelength of the incident radiation is comparable with the period.

Numerical studies of classical gratings by BOTTEN and RITCHIE [7.49] have shown a clear difference between the effects introduced by modulating a surface and those consequent upon a grading of refractive index.

We interpret the moth-eye effect as being due to the weakening of the zeroth order caused by the surface modulation. Thinking in terms of a photon model, the corrugation imparts a transverse momentum which diminishes the probability of re-radiation into the zeroth order (for the case of near-normal incidence). If the grating period is such that only the zeroth order propagates, any weakening of this order in reflection must lead to an enhancement of the energy absorbed, when region 2 is dissipative. Of course, when the grating reflectivity becomes sufficiently large, it is difficult to achieve significant absorption of energy. With the shallow grooves so far studied, the region of enhanced absorption for bigratings in copper has been confined to wavelengths smaller than 0.6 μm. From the point of view of solar-energy applications, it would be desirable to move the transition wavelength to 1.0 μm or 1.5 μm. This would require groove depths far in excess of those which can be handled at present using our numerical implementation of the transformation formalism.

7.6.7 Equivalence Formulae Linking Crossed and Classical Gratings

We saw in Sect.7.6.5 that a classical grating in gold having a period of 0.62 μm and a groove depth of 0.033 μm will totally absorb normally incident light of $H_{\parallel}$ polarization and wavelength 0.65 μm. We also saw that if a modulated surface is constructed in gold which represents the superposition of the classical grating with an identical grating orthogonal to the first, then this will totally absorb incident light of wavelength 0.65 μm, irrespective of its polarization state. This remarkable fact suggested that perhaps one could decompose an incident wave of arbitrary polarization on the bigrating into principal polarizations, each of which interacted only with the classical grating component of the surface aligned with axis of invariance parallel to its magnetic vector. *In short, it prompted a search for simple formulae linking bigrating and classical grating behavior.*

The following discussion will be restricted to the case of radiation incident normally on a crossed grating whose profile is a sum of two functions, one of groove depth h_1 and the second of groove depth h_2

$$f(x_1,x_2) = \hat{f}(x_1) + \tilde{f}(x_2) \quad . \tag{7.105}$$

We will suppose that $\hat{f}$ and $\tilde{f}$ are up-down symmetric functions, so that replacing h_1 by $(-h_1)$ or h_2 by $(-h_2)$ does not change the diffracting structure. It then follows that the bigrating efficiencies will be functions of h_1^2 and h_2^2 only. Let us restrict our attention to the zeroth order in reflection

$$\eta_{00} = R + a_{00}h_1^2 + b_{00}h_2^2 + c_{00}h_1^4 + d_{00}h_2^4 + e_{00}h_1^2h_2^2 + O(h^6) \quad , \tag{7.106}$$

for some set of coefficients R, a_{00}, b_{00}, c_{00}, d_{00} and e_{00}. Equation (7.106) is a Taylor series, valid for sufficiently small h_1 and h_2.

If we put $h_1 = 0$ and $h_2 = 0$, we see that R is the Fresnel reflectance for the unmodulated surface. Putting $h_2 = 0$, we revert to the diffraction problem for the first grating

$$\hat{\eta}_0 = R + a_{00}h_1^2 + c_{00}h_1^4 + O(h_1^6) \quad . \tag{7.107}$$

Putting $h_1 = 0$, we get a classical diffraction problem for the second grating

$$\tilde{\eta}_0 = R + b_{00}h_2^2 + d_{00}h_2^4 + O(h_2^6) \quad . \tag{7.108}$$

Using (7.107,108),

$$\eta_{00} = \hat{\eta}_0 + \tilde{\eta}_0 - R + e_{00}h_1^2h_2^2 + O(h^6) \quad , \tag{7.109}$$

and

$$\eta_{00} = \frac{\hat{\eta}_0\tilde{\eta}_0}{R} + \left(e_{00} - \frac{a_{00}b_{00}}{R}\right)h_1^2h_2^2 + O(h^6) \quad . \tag{7.110}$$

The two expressions $(\hat{\eta}_0+\tilde{\eta}_0-R)$ and $(\hat{\eta}_0\tilde{\eta}_0/R)$ can therefore be used as approximations to η_{00} for sufficiently shallow bigratings.

We have investigated numerically both these rival equivalence formulae, and have found the product form to be the superior of the two. One physical reason for the superiority of the product formula can be given. Numerical and experimental experience with classical metal gratings and metallic bigratings has shown that

$$\tilde{\eta}_0 < R \quad , \quad \hat{\eta}_0 < R \quad , \quad \eta_{00} < R \quad .$$

The first two constraints limit $(\hat{\eta}_0+\tilde{\eta}_0-R)$ to lie between -R and R, with this expression often becoming negative for quite small groove depths. On the other hand, $(\hat{\eta}_0\tilde{\eta}_0/R)$ always lies between the physical limits 0 and R.

Consider next the case of an order (r,0). This will have zero efficiency if $h_1 = 0$, and so each term in the expansion for η_{r0} must contain at least one factor of h_1^2

$$\eta_{r0} = a_{r0}h_1^2 + c_{r0}h_1^4 + e_{r0}h_1^2h_2^2 + O(h^6) \quad . \tag{7.111}$$

Now,

$$\hat{\eta}_r = a_{r0}h_1^2 + c_{r0}h_1^4 + O(h_1^6) \quad , \tag{7.112}$$

and so

$$\eta_{r0} = \frac{\hat{\eta}_r \tilde{\eta}_0}{R} + \left(e_{r0} - \frac{a_{r0} b_{00}}{R}\right) h_1^2 h_2^2 + O(h^6) \quad . \tag{7.113}$$

We would expect quite accurate results from the product formula for orders (r,0), since it gives the correct term of second order, and one of the two fourth order terms. Similar remarks apply to orders (0,s).

In the case of an order (r,s) with $r \neq 0$, $s \neq 0$, this can only be formed in the presence of both component gratings, and hence η_{rs} can only contain product terms of the form $h_1^{2n} h_2^{2m}$ (for integral n,m). The error in the product formula is in the leading term, and so this formula cannot be expected to hold good for bigratings with h_1 and h_2 sufficiently large to ensure the presence of orders (r,s) with significant efficiencies.

In order to obtain from the equivalence formula the energy absorbed by a finitely conducting bigrating, we subtract from one the efficiencies of all propagating orders, as obtained from products of classical grating efficiencies divided by R. This method has been used to give the results plotted as points in Fig. 7.19.

From Fig.7.19 we see that the product formula is very accurate in predicting the energy absorbed by crossed gratings in the region where only the zeroth reflected order propagates. The accuracy declines notably when the wavelength falls close to the Rayleigh wavelength for the order (±1,±1). It is interesting that the most important parameter affecting accuracy appears to be λ/d rather than h/d. If the quality of the agreement for curve (d) persists at deeper groove depths, then the equivalence will permit very interesting studies of solar-selective absorbers based on the moth-eye effect.

7.6.8 Coated Bigratings

The performance of a metallic bigrating as a solar-selective absorber may be improved by coating it with a suitable layer of another material. We then have a system of three media, requiring two profile functions and two complex refractive indices for its specification. We shall designate the three media from top to bottom by A, B, C and their corresponding refractive indices by $\nu_A = 1$, ν_B, ν_C. The spatial domains expressed in Cartesian coordinates are

Upper layer A = air: $f_A(x_1', x_2') < x_3' < \infty$,

Middle layer B: $f_C(x_1', x_2') < x_3' < f_A(x_1', x_2')$,

Bottom layer C: $-\infty < x_3' < f_C(x_1', x_2')$.

The AB interface profile function f_A and the BC interface profile function f_C are each periodic functions of x_1' and x_2'. The periods of f_A and f_C for each direction must be equal or in the ratio of integers in order that the overall structure be biperiodic.

The coordinate transformation technique of Sect.7.6.2 is readily adapted to treat biperiodic multilayer stacks. The object is to choose the transformation function $F(x_1,x_2,x_3)$ of (7.76) to map every interface into a plane surface $x_3 = \text{constant}$. For our structure containing two interfaces a suitable transformation function is

$$\begin{aligned} F(x_1,x_2,x_3) &= K_A(x_3)[f_A(x_1,x_2)-g_1] + g_1 \qquad x_u < x_3 < \infty \ , \\ &= K_B(x_3)f_A(x_1,x_2) + [1-K_B(x_3)]f_C(x_1,x_2) \qquad x_\ell < x_3 < x_u \ , \\ &= K_C(x_3)[f_C(x_1,x_2)-g_2] + g_2 \qquad -\infty < x_3 < x_\ell \end{aligned} \tag{7.114}$$

where x_u, x_ℓ, g_1, g_2 are constants.

The transformation specified by (7.76,114) maps the AB interface $x_3' = f_A(x_1',x_2')$ onto the plane $x_3 = x_u$ and the BC interface $x_3' = f_C(x_1',x_2')$ onto the plane $x_3 = x_\ell$, provided we impose the following constraints:

$$\begin{aligned} &K_A(\infty) = 0 \ , \quad K_A(x_u) = 1 \ , \\ &dK_A/dx_3 \leqq 0 \ , \quad x_u < x_3 < \infty \ , \\ &K_B(x_u) = 1 \ , \quad K_B(x_\ell) = 0 \ , \\ &K_C(-\infty) = 0 \ , \quad K_C(x_\ell) = 1 \ , \\ &dK_C/dx_3 \geqq 0 \ , \quad -\infty < x_3 < x_\ell \ , \\ &g_2 < f_C(x_1,x_2) < f_A(x_1,x_2) < g_1 \ . \end{aligned} \tag{7.115}$$

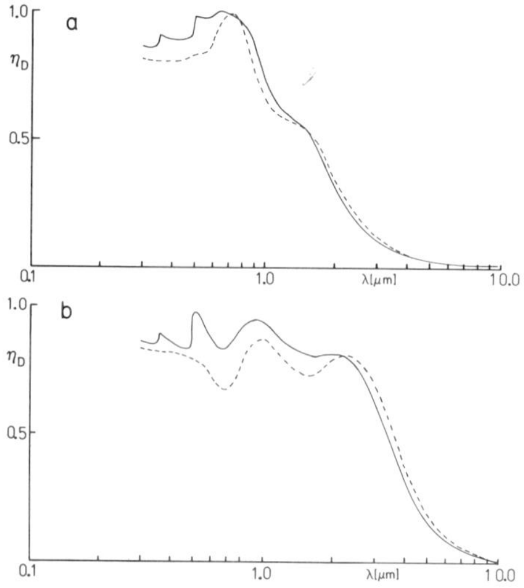

Fig. 7.20. The absorption of a sinusoidal copper bigrating conformally coated with arc evaporated carbon (full curve) compared with that of a plane copper surface coated with the same thickness of carbon (broken curve). The periods are both 0.50 μm and the depth of the modulation of the copper is 0.20 μm. (a) Carbon thickness 0.055 μm; (b) Carbon thickness 0.12 μm

We now proceed as in Sect.7.6.3 to construct an integral equation equivalent to (7.85-87) which incorporates the boundary conditions of continuity of E_1, E_2, H_1, H_2 at the plane interface $x_3 = x_u$, $x_3 = x_\ell$ and the outgoing-waves-only condition at infinity. We again are able to find explicit expressions for the TE and TM amplitudes, which are analogous to (7.96). These expressions involve linear combinations of integrals of the type $\int v_{nrs}(t)\rho_{rs}(t)dt$ with factors formed from the various reflection and transmission coefficients relating to a plane layer of medium B of thickness $(x_u - x_\ell)$ sandwiched between air and medium C. The formulae are too lengthy to reproduce here but will be published elsewhere [7.50].

The integral equations may now be solved by the technique of Sect.7.6.4. Figure 7.20 shows theoretical curves obtained using the transformation formalism for the absorptance of a copper grating conformally coated with carbon. A comparison of Fig.7.20 with Fig.7.19b shows how much the performance as a solar selective absorber has been enhanced by the addition of the carbon layer.

References

7.1 C.M. Horwitz: Opt. Commun. *11*, 210 (1974)
7.2 E.E. Russell, E.E. Bell: Infrared Phys. *6*, 75 (1966)
7.3 R. Ulrich: Infrared Phys. *7*, 37 (1967)
7.4 R. Ulrich: Infrared Phys. *7*, 65 (1967)
7.5 R. Ulrich: Appl. Opt. *7*, 1987 (1968)
7.6 J.P. Casey, E.A. Lewis: J. Opt. Soc. Am. *42*, 971 (1952)
7.7 G.D. Holah, J.P. Auton: Infrared Phys. *14*, 217 (1974)
7.8 R.D. Rawcliffe, C.M. Randall: Appl. Opt. *6*, 1353 (1967)
7.9 J.C. Lecullier: Nouv. Rev. Opt. *5*, 313 (1974)
7.10 M.M. Pradhan: Infrared Phys. *11*, 241 (1971)
7.11 C-C Chen: IEEE Trans. MTT-*18*, 627 (1970)
7.12 C-C Chen: IEEE Trans. MTT-*19*, 475 (1971)
7.13 C-C Chen: IEEE Trans. MTT-*21*, 1 (1973)
7.14 R.C. McPhedran, D. Maystre: Appl. Phys. *14*, 1 (1977)
7.15 L.C. Botten: "Theories of Singly and Doubly Periodic Diffraction Gratings"; Ph.D. Thesis, University of Tasmania (1978)
7.16 N. Amitay, V. Galindo: IEEE Trans. AP-*17*, 747 (1969)
7.17 G.H. Derrick, R.C. McPhedran, D. Maystre, M. Nevière: Appl. Phys. *18*, 39 (1979)
7.18 R.C. McPhedran, L.C. Botten: "Inductive Grids with Circular Apertures"; Report SP 77/5, University of Sydney (1977)
7.19 R.C. McPhedran, L.C. Botten, P. Bliek, R. Deleuil, D. Maystre: 8th European Microwave Conf., Paris (1978)
7.20 N. Amitay, V. Galindo: IEEE Trans. MTT-*16*, 265 (1968)
7.21 M. Born, E. Wolf: *Principles of Optics*, 3rd ed. (Pergamon, Oxford 1965) pp.559-560
7.22 D.S. Jones: *The Theory of Electromagnetism*, 1st ed. (Pergamon, Oxford 1964) pp.569-572
7.23 W.H. Eggimann, R.E. Collin: IRE Trans. MTT-*10*, 528 (1962)
7.24 R.H. Ott, R.G. Kouyoumjian, L. Peters: Radio Sci. *2*, 1347 (1967)
7.25 K. Sakai, T. Yoshida: Infrared Phys. *18*, 137 (1978)
7.26 J.A. Arnaud, F.A. Pelow: Bell Syst. Tech. J. *54*, 263 (1975)
7.27 J.L. Adams, L.C. Botten, R.C. McPhedran: J. Opt. (Paris) *9*, 91 (1978)
7.28 J.L. Adams, L.C. Botten: "The Crossed Lamellar Transmission Grating"; Report DGRG 77/3, University of Tasmania (1977)

7.29 H. Blok, G. Mur: Appl. Sci. Res. *26*, 389 (1972)
7.30 J.L. Adams, L.C. Botten: "Double Gratings and Their Application as Fabry-Perot Interferometers", J. Opt. (Paris) *10*, 109 (1979)
7.31 L.C. Botten, J.L. Adams, R.C. McPhedran, G.H. Derrick: "Symmetry Properties of Lossless Diffraction Gratings", J. Opt. (Paris) *11*, 43 (1980)
7.32 P. Grivet: "Time Reversibility in an Optical Proof of the Reciprocity Theorem of Electromagnetism", in *Proc. Symposium on Modern Optics*, ed. by J. Fox, Microwave Research Institute Symposia Series, Vol.17 (Polytechnic Press, New York 1967) pp.467-479
7.33 D.S. Jones: *The Theory of Electromagnetism*, 1st ed. (Pergamon, Oxford 1964) pp.56-57, 562-566
7.34 M.M. Pradhan, R.K. Garg: Infrared Phys. *16*, 449 (1976)
7.35 M.M. Pradhan, R.K. Garg: Infrared Phys. *17*, 253 (1977)
7.36 L.C. Botten: "A Generalized Treatment of Multielement Interference Filters for the Far Infrared". Infrared Phys. *19*, 659 (1979)
7.37 L.C. Botten: "On the Use of Fibonacci Recurrence Relations in the Design of Long Wavelength Filters and Interferometers", The Fibonacci Quarterly, to be published
7.38 D. Maystre, M. Nevière: J. Opt. *9*, 301 (1978)
7.39 P. Vincent: Opt. Commun. *26*, 293 (1978)
7.40 M. Nevière, M. Cadilhac, R. Petit: IEEE Trans. AP-*21*, 37 (1973)
7.41 J.M. Elson: Appl. Opt. *16*, 2872 (1977)
7.42 J. Chandezon, D. Maystre, G. Raoult: J. Opt., to be published
7.43 G.A. Korn, T.M. Korn: *Mathematical Handbook for Scientists and Engineers*, 2nd ed. (McGraw-Hill, New York 1968) pp.535-560
7.44 G.H. Derrick, R.C. McPhedran, D. Maystre, M. Nevière: "A Theory for Crossed Gratings of Finite Conductivity"; Report TP 78/1, University of Sydney (1979)
7.45 M. Nevière, D. Maystre, R. McPhedran, G. Derrick, M. Hutley: 11th Congress of the Intern. Commission for Optics, Madrid (1978)
7.46 C.G. Bernhard: Endeavour *26*, 79 (1967)
7.47 P.B. Clapham, M.C. Hutley: Nature *244*, 281 (1973)
7.48 B.S. Thornton: J. Opt. Soc. Am. *65*, 267 (1975)
7.49 L.C. Botten, I.T. Ritchie: Opt. Commun. *23*, 421 (1977)
7.50 G.H. Derrick, R.C. McPhedran: Submitted to J. Opt.

Additional References with Titles

Some recent papers, are mentioned here which have held the authors' attention without always being thoroughly analyzed. Most of them are concerned with applications (7, 33-55, 57). The theoretical studies generally describe approximate, but efficient techniques of solving some particular problems (22, 29, 32, 56) or corroborating the efficiency of methods which have been described in this book (1, 15, 24, 31). Only a few papers (4, 5, 7, 10, 21) suggest a new approach.

These additional references are numbered without any particular order; nevertheless we have tried to connect each of them with a chapter.

Chapter 1

1 A.Wirgin: Sur la thèorie de Raleigh de la diffraction d'une onde par une surface sinusoidale. C.R. Acad. Sci. B. *179*, 288 (1979)

2 A. Wirgin: Sur trois variantes de la theorie de Rayleigh de la diffraction d'une onde par une surface sinusoïdale. C.R. Acad. Sci. A *259*, 289 (1979)

3 A. Wirgin: Aspects numériques du problème de la diffraction d'une onde par une surface sinusoidale. C.R. Acad. Sci. B *273*, 289 (1979)

4 A.K. Jordan, R.H. Lang: Electromagnetic scattering patterns from sinusoidal surfaces. Radio Sci. *14*(6), 1077-1088 (1979)

5 J.A. DeSanto: Sttering from a sinusoid: derivation of linear equations for the field amplitudes. J. Acoust. Soc. Am. *57*,5 (1975)

6 J.P. Hugonin, R. Petit, M. Cadilhac: On the use of plane wave expansions to describe the field diffracted by a grating. Submitted for publication in J. Opt. Soc. Am.

7 D. Maystre, M. Cadilhac: A phenomenological theory for gratings, perfect blazing for polarized light in nonzero deviation mounting. Proceedings of the International U.R.S.I. Symposium, Munich (August 1980)

8 D. Maystre: A new analitical property of ruled diffraction gratings: perfect blazing at Rayleigh wavelength. Proceedings of the International U.R.S.I. Symposium, Munich (August 1980)

9 T. Namioka, T. Harada, K. Yasuura: Diffraction gratings in Japan. Opt. Acta *26*, 1021 (1979)

10 J. Chandezon, D. Maystre, G. Raoult: A new theoretical method for diffraction gratings and its numerical application. J. Opt. *11*, 235-241 (1980)

11 H.A. Kalhor: EM scattering by a array of perfectly conducting strips by a physical optics approximation. IEEE Trans. AP-*28*, 277 (1980)

12 G.M. Whitman and F. Schwering: Reciprocity identity for periodic surface scattering. IEEE Trans. AP-*27*, 252-254 (1979)

13 A. Gavrielides, P. Peterson: Power losses in lamellar gratings. Appl. Opt. *18*, 4168 (1979)

14 A. Basu, J.M. Ballantyne: Random fluctuations in first-order waveguide grating filters. Appl. Opt. *18*, 2575 (1979)

15 K. Yasuura, K. Shimohara, T. Miyamoto: Numerical analysis of a thin film waveguide by mode-matching method. J. Opt. Soc. Am. *70*, 183 (1980)

16 M.G. Moharam, T.K. Gaylord, R. Magnusson: Bragg diffraction of finite beams by thick gratings. J. Opt. Soc. Am. *70*, 300 (1980)

17 R.S. Chu, J.A. Kong: Diffraction of optical beams with arbitrary profiles by a periodically modulated layer. J. Opt. Soc. Am. *70*,1 (1980)

18 J.R. Fox: General modal theory of scalar wave scattering by periodic structures. Opt. Acta *27*, 289-305 (1980)

Chapter 3

19 K. Utagawa: Theory of diffraction efficiency and anomalies of shallow metal gratings of finite conductivity. J. Opt. Am. *69*, 333 (1979)
20 L.C. Botten: A study of bimetallic gratings. J. Opt. *11*, 161-166 (1980)
21 A. Wirgin: A new theoretical approach to scattering from a periodic surface. Opt. Commun. *27*, 189 (1978)

Chapter 4

22 R. Magnusson, T.K. Gaylord: Solutions of the thin-phase grating diffraction equation. Opt. Commun. *25*, 129 (1978)
23 Wai-Hon Lee, W. Streifer: Radiation loss calculations for corrugated dielectric waveguides II. TM polarization. J. Opt. Soc. Am. *69*, 1671 (1971)
24 P. Vincent: New improvement of the differential formalism for high modulated gratings. SPIE Proc. *240* (1980, in print)

Chapter 5

25 J.R. Andrewartha, J.R. Fox, I.J. Wilson: Further properties of lamellar grating resonance anomalies. Opt. Acta *26*, 197-209 (1979)
26 Ricardo Depine, Juan M. Simon and M.C. Simon: Diffraction grating anomalies: an experimental study of phase shifts and resonances. Opt. Acta *25*, 895-904 (1978)
27 R. Ulrich, R. Zengerle: Optical bloch waves in periodic planar waveguides. Digest of papers presented at the Topical meeting on Integrated and Guided-wave Optics, Incline Village (January 1980)
28 Y. Handa, T. Suhara, H. Nishihara, J. Koyama: Scanning electron microscope--written gratings in chalcogenide films for optical integrated curcuits. Appl. Opt. *18*, 248 (1979)
29 K.C. Chang, T. Tamir: Simplified approach to surface-wave scattering by blazed dielectric gratings. Appl. Opt. *19*, 282 (1980)
30 K. Utagawa: Theory of diffraction efficiency and anomalies of shallow metal gratings of finite conductivity. J. Opt. Soc. Am. *69*, 333 (1979)
31 K.C. Chang, U. Shah, T. Tamir: Scattering and guiding of waves by dielectric grating with arbitrary profiles. J. Opt. Soc. Am. *70*, 804 (1980)
32 K.C. Chang, V. Shah, T. Tamir: Directional scattering by blazed dielectric gratings. Proceedings of the International U.R.S.I. Symposium, Munich (August 1980)

Chapter 6

33 A. Roger, M. Breidne: Grating profile reconstruction by an inverse scattering method. Opt. Commun., to be published.
34 L.F. Johnson: Evolution of grating profiles under ion-beam erosion. Appl. Opt. *18*, 2559 (1979)
35 H. Bräuninger, H. Kraus, H. Dangschat, K.P. Beuermann, P. Predehl, J. Trümper: Fabrication of transmission gratings for use in cosmic X-ray and XUV astronomy. Appl. Opt. *18*, 3502 (1979)

36 I.J. Wilson, B. Brown, E.G. Loewen: Grazing incidence grating efficiencies. Appl. Opt. *18*, 426 (1979)
37 Wai-Hon Lee: High-efficiency multiple beam gratings. Appl. Opt. *18*, 2512 (1979)
38 H. Bräuninger, P. Predehl, K.P. Benermann: Transmission grating efficiencies for wavelengths between 5.4 Å and 44.8 Å. Appl. Opt. *18*, 368 (1979)
39 P. Predehl, R.P. Haelbich, H. Bräuninger: Transmission grating efficiencies in the 50-250 Å range. Appl. Opt. *18*, 2906 (1979)
40 M. Josse, Don L. Kendall: Rectangular profile diffraction grating from single-crystal silicon. Appl. Opt. *19*, 72 (1980)
41 A. Gavrielides, P.R. Peterson: Wood's anomaly angular alignment gratings. Appl. Opt. *19*, 1229 (1980)
42 A.C. Brinkman, J.H. Dijkstra, W.F.P.A.L. Geerlings, F.A. Van Rooijen, C. Timmermann, P.A.J. de Korte: Efficiency and resolution measurements of X-ray transmission gratings between 71 and 304 Å. Appl. Opt. *19*, 1601 (1980)
43 P. Lindblom, B. Sandberg: New Eagle-type monochromator mounting with ruled diffraction grating at 45° off-plane. Appl. Opt. *19*, 1941 (1979)
44 A.K. Jordan, R.H. Lung: An inverse scattering method for sinusoidal surfaces. Proceedings of the International U.R.S.I. Symposium, Munich (August 1980)
45 M. Golomb: Diffraction gratings and solar selective thin film absorbers: an experimental study. Opt. Commun. *27*, 177 (1978)
46 T. Tamir: Analysis and design of blazed dielectric gratings. Digest of papers presented at the Topical meeting on Integrated and Guided-Wave Optics, Incline Village (January 1980)
47 D. Maystre, M. Breidne: On an equivalence property of symmetrical diffraction gratings. Proceedings of the International U.R.S.I. Symposium, Munich (August 1980)
48 I.J. Wilson: Spectrograph mount grating efficiency. J. Opt. *9*, 45-50 (1978)
49 A. Eberhagen: A very rapidly scanning monochromator with low internal losses but high spectral resolution and repetition rate for millimetre and submillimetre wavelengths and below. Infrared Phys. *19*, 389-394 (1977)
50 J.M. Elson: Low-efficiency diffraction grating theory. SPIE Proc. *240* (1980)
51 M. Breidne, D. Maystre: 100% efficiency of gratings in non-Littrow configuration. SPIE Proc. *240* (1980)
52 P. Bliek, R. Deleuil: Experimental investigations in microwave range of diffraction by classical and crossed gratings. SPIE Proc. *240* (1980)
53 J. Seligson, P. Baumeister: Buried diffraction gratings: a phenomenological theory. Appl. Opt. *18*, 742 (1979)
54 K.R. German: Diffraction gratings tuners for CW lasers. Appl. Opt. *18*, 2348 (1979)
55 W.R. Hunter, D.W. Anger: Effect of diffusion pump oil contamination on diffraction grating efficiency in the VUV spectral region. Appl. Opt. *18*, 3506 (1979)

Chapter 7

56 A. Wirgin: Simplified theory of the diffraction of an electromagnetic wave by a perfectly conducting biaxial periodic surface. Opt. Commun. *28*, 275 (1979)
57 R. Petit: Sur l'étude électromagnétique de l'absorption par les surfaces rugueuses. J. Phys. *42*, Colloque C1 (1981, in preparation)
58 A. Wirgin: Sur la réponse spectrale de certaines surfaces texturées. S. J. Phys. *42*, Colloque C1 (1981, in preparation)

Subject Index

Nonlinear Infrared Generation

Editor: Y.-R. Shen
1977. 134 figures. XI, 279 pages
(Topics in Applied Physics, Volume 16)
ISBN 3-540-07945-9

Contents: *Y.-R. Shen:* Introduction. – *R. L. Aggarwal, B. Lax:* Optical Mixing of CO_2 Lasers in the Far-Infrared. – *R. L. Byer, R. L. Herbst:* Parametric Oscillation and Mixing. – *V. T. Nguyen, T. J. Brigdes:* Difference Frequency Mixing via Spin Nonlinearities in the Far-Infrared. – *J. J. Wynne, P. P. Sorokin:* Optical Mixing in Atomic Vapors. – *T. Y. Chang:* Optical Pumping in Gases.

Picosecond Phenomena

Proceedings of the First International Conference on Picosecond Phenomena Hilton Head, South Carolina, USA, May 24–26, 1978
Editors: C. V. Shank, E. P. Ippen, S. L. Shapiro
1978. 222 figures, 10 tables. XII, 359 pages
(Springer Series in Chemical Physics, Volume 4)
ISBN 3-540-09054-1

Contents: Interactions in Liquids and Molecules. – Poster Session. – Sources and Techniques. – Biological Processes. – Poster Session. – Coherent Techniques and Molecules. – Solids. – High-Power Lasers and Plasmas. – Postdeadline Papers.

Raman Spectroscopy

of Gases and Liquids
Editor A. Weber
1979. 103 figures, 25 tables. XI, 318 pages
(Topics in Current Physics, Volume 11)
ISBN 3-540-09036-3

Contents: *A. Weber:* Introduction. – *S. Brodersen:* High-Resolution Rotation-Vibrational Raman Spectroscopy. – *A. Weber:* High-Resolution Rotational Raman Spectra of Gases. – *H. W. Schrötter, H. W. Klöckner:* Raman Scattering Cross Sections in Gases and Liquids. – *R. P. Srivastava, H. R. Zaidi:* Intermolecular Forces Revealed by Raman Scattering. – *D. L. Rousseau, J. M. Friedman, P. F. Williams:* The Resonance Raman Effect. – *J. W. Nibler, G. V. Knighten:* Coherent Anti-Stokes Raman Spectroscopy.

Ultrashort Light Pulses

Picosecond Techniques and Applications
Editor: S. L. Shapiro
1977. 173 figures. XI, 389 pages
(Topics in Applied Physics, Volume 18)
ISBN 3-540-08103-8

Contents: *S. L. Shapiro:* Introduction – A Historical Overview. – *D. J. Bradley:* Methods of Generation. – *E. P. Ippen, C. V. Shank:* Techniques for Measurement. – *D. H. Auston:* Picosecond Nonlinear Optics. – *D. v. d. Linde:* Picosecond Interactions in Liquids and Solids. – *K. B. Eisenthal:* Picosecond Relaxation Processes in Chemistry. – *A. J. Campillo, S. L. Shapiro:* Picosecond Relaxation Measurements in Biology.

Springer-Verlag
Berlin
Heidelberg
New York

M/